Homogeneous Transition-metal Catalysis

Homogeneous Transition-metal Catalysis

A GENTLE ART

CHRISTOPHER MASTERS
Formerly of Shell Chemicals UK Ltd
(presently with Christian Salvesen Ltd, Edinburgh)

LONDON NEW YORK
CHAPMAN AND HALL

First published 1981 by
Chapman and Hall Ltd
11 New Fetter Lane, London EC4P 4EE

Published in the USA by
Chapman and Hall
in association with Methuen, Inc.
733 Third Avenue, New York NY 10017

© *1981 Christopher Masters*

Printed in Great Britain at the
University Press, Cambridge

ISBN 0 412 22110 1 (hardback)
ISBN 0 412 22120 9 (paperback)

This title is available in both hardbound and paperback editions. The paperback edition is sold subject to the condition that it shall not, by way of trade or otherwise, be lent, re-sold, hired out, or otherwise circulated without the publisher's prior consent in any form of binding or cover other than that in which it is published and without a similar condition including this condition being imposed on the subsequent purchaser.

All rights reserved. No part of this book may be reprinted, or reproduced or utilized in any form or by any electronic, mechanical or other means, now known or hereafter invented, including photocopying and recording, or in any information storage and retrieval system, without permission in writing from the Publisher.

British Library Cataloguing in Publication Data

Masters, Christopher
 Homogeneous transition-metal catalysis.
 1. Catalysis
 2. Transition metal compounds
 I. Title
 541'. 395   QD505   80-49952

 ISBN 0-412-22110-1
 ISBN 0-412-22120-9 Pbk

To Gill and 'Jemima'

Contents

Preface *page* ix

1 General considerations 1
1.1 The art of catalysis, the basic process 1
1.2 Why transition metals? 5
1.3 Why homogeneous? 20
1.4 Molecule activation 22
1.5 Proximity interaction 28
1.6 The catalytic cycle 33
1.7 Summary 35

2 Homogeneous catalyst systems in operation 38
2.1 Hydrogenation 40
2.2 Isomerization 70
2.3 Carbonylation 89
2.4 Hydroformylation 102
2.5 Oligomerization 135
2.6 Polymerization 159
2.7 Oxidation 172
2.8 Metathesis 196

3 Homogeneous catalyst systems in development 219
3.1 Nitrogen fixation 219
3.2 Reductive oligomerization/polymerization of carbon monoxide 227
3.3 Alkane activation 239

4 Where to now? 250

References	256
Index	272

Preface

Soluble catalysts are used extensively in many branches of chemistry and are indeed a vital constituent of many natural processes. They find wide application throughout the chemical industry where they assist in the production of several million tonnes of chemicals each year. Since homogeneous systems, especially those incorporating transition metals, often function effectively under milder conditions than their heterogeneous counterparts, they are becoming increasingly important at a time when the chemical industry in particular, and society in general, is seeking ways of conserving energy and of making the best possible use of available resources.

My principal objective in writing this book is to engender sufficient enthusiasm for, and knowledge of, the subject in the reader that he or she will be encouraged to begin, or continue, to make their own contribution to advancing our knowledge of homogeneous catalysis.

After attempting to acquaint the reader with some of the ground rules I have tried to describe the present scope, and the future potential, of this fascinating field of chemistry by drawing both on academic and on industrial data sources. This approach stems from a personal conviction that future progress could be considerably hastened by a more meaningful dialogue between chemists working both in industrial and in academic research institutions. Wherever possible, examples of the commercial application of homogeneous catalyst systems have been included and no attempt has been made in any way to disguise the many unresolved questions and exciting challenges which still pervade this rapidly developing area.

Some may find the treatment of the subject matter somewhat

anthropomorphic; my only excuse is personal enthusiasm coupled with an abhorrence of totally reducing science to the impersonal treatment of physical data. The *Shorter Oxford English Dictionary* gives the general definition of art as 'skill as a result of knowledge and practice', and then goes on to talk about art as 'an occupation in which skill is employed to gratify taste or produce what is beautiful'. It was with this in mind that the title *Homogeneous Transition-metal Catalysis – a gentle art* was chosen.

I wish to take this opportunity to express my sincere gratitude to Professor Bernard Shaw for introducing me to the subject of homogeneous catalysis and for giving me so much encouragement and help in the early years. My thanks are also due to Shell Research B.V. for providing such excellent research facilities during my six years in Amsterdam and for granting me permission to undertake this book. Finally I would particularly like to thank Miss Hilda Harvey for her diligent, careful and conscientious typing of an often difficult manuscript.

Christopher Masters
Edinburgh
April 1980.

1 General Considerations

1.1 The art of catalysis, the basic process

The *Shorter Oxford English Dictionary* gives the general definition of art as 'skill as the result of knowledge and practice' and then goes on to talk about an art as 'an occupation in which skill is employed to gratify taste or produce what is beautiful'. Although few people would consider propanal 'beautiful' there is perhaps a certain degree of gratification in designing a substance which can persuade carbon monoxide, hydrogen and ethene to react together to form propanal with 85% selectivity, especially when the concentration of this substance is only some hundredth of that of the olefin [1]. Hydridocarbonyltris(triphenylphosphine)rhodium(1), $RhH(CO)(PPh_3)_3$, is one such substance, or catalyst, and the process via which it works is referred to as catalysis. However, before going on to discuss catalysis we must first pause to consider the basic thermodynamics of a chemical reaction.

A chemical reaction is a process in which one or more chemical substances are transformed into one or more different chemical substances. If we consider a general system in which two substances, A and B, react together to form two new substances, C and D then, representing the reaction as shown in Equation (1.1), thermodynamics tells us that the position of equilibrium depends on the difference in free energy, ΔG, between the reactants, A and B, and the products, C and D.

$$A + B \rightleftharpoons C + D \qquad (1.1)$$

$$\Delta G = G(C,D) - G(A,B)$$

For convenience it is usual to work in terms of standard free energy which for a compound is defined as the free energy by which that compound is formed from its elements when all reactants and products are in standard states (25 °C or 298.15 K at 1 atm).

At equilibrium

$$\Delta G = -RT \ln K_p \qquad (1.2)$$

where K_p is the equilibrium constant expressed in terms of the partial pressures of the reactants and products. For the water formation reaction

$$H_2(g) + \tfrac{1}{2}O_2(g) \longrightarrow H_2O(g)$$

$\Delta G° = -228.6$ KJ mol^{-1} and $K_p = 1.19 \times 10^{40}$ atm$^{-1/2}$, indicating that at equilibrium this system lies essentially completely to the right. However, if hydrogen and oxygen are carefully mixed in a pure state nothing happens – this is where a catalyst comes in. Add some finely divided platinum to the mixture and the reaction takes place so rapidly that the platinum is heated to incandescence and the gases generally explode. Similarly although $\Delta G°_{298}$ for the hydrogenation of ethene is -101 KJ mol^{-1} and $K_p = 5.16 \times 10^{17}$ atm^{-1} at 298 K, in the absence of a suitable catalyst mixtures of pure hydrogen and pure ethene are almost indefinitely stable.

Thermodynamics can tell us a lot about the equilibrium state of systems; it can tell us nothing about the speed at which that state is achieved. Catalysts can have no effect on the position of the final equilibrium. They can, however, have a drastic effect on the rate at which that equilibrium is attained, or, to quote W. Ostwald's definition [2] 'a catalyst is an agent that speeds up a chemical reaction without affecting the chemical equilibrium'. Ostwald's definition is only applicable to reversible reactions and does not incorporate any form of autocatalysis as found in oxidation reactions (see Section 2.7). A more general definition of a catalyst, along the lines of that suggested by P. Sabatier [3], is a substance or system which alters the rate of a reaction by becoming intimately involved in the reaction sequence without becoming a product. This definition is somewhat misleading since, to do its job effectively, a catalyst must be continuously

GENERAL CONSIDERATIONS

reformed during the course of the reaction. This property can be used to characterize a catalyst as being a substance which is both a reactant and a product of a chemical reaction.

No matter how we define a catalyst it is important to emphasize the limitations of catalysis. A catalyst can only speed up a reaction – it *cannot* alter the final position of the equilibrium. Before starting out on any catalyst design, it is imperative to check the thermodynamic feasibility of the desired reaction. In other words, if we wish to find a catalyst for the reaction

$$A + B \longrightarrow C + D \qquad (1.3)$$

we should first make sure than under reasonable conditions of temperature and pressure the thermodynamically predicted equilibrium concentrations of the products C and D are more than minimal. There is little point searching for a catalyst for the reaction if at equilibrium the products are present at concentrations of say 10^{-4} mole l^{-1}. No amount of catalyst design is going to increase that yield. From Equation (1.2) it follows that if $\Delta G° < 0$ the equilibrium will lie to the right and if $\Delta G° > 0$ it will lie to the left. Thus a reaction such as Equation (1.3) will be thermodynamically feasible if accompanied by an overall decrease in free energy. It is tempting to oversimplify the $\Delta G°$ criterion and regard reactions with positive $\Delta G°$s as impractical. To do so would be mistaken since reactions with small positive $\Delta G°$s can give appreciable yields of desired product under feasible experimental conditions and, conversely reactions with small negative $\Delta G°$s may give disappointingly low yields. In practical terms the $\Delta G°$ criterion may be summarized as follows [4]:

$\Delta G° < 0$ the reaction has promise
$0 < \Delta G° < 40$ KJ the reaction is questionable but may be worth investigating
$\Delta G° > 40$ KJ do not bother to look for a catalyst.

The importance of checking the thermodynamic feasibility cannot be overemphasized and undoubtedly one of the most important tools in the catalyst designers kit is a comprehensive set of thermodynamic data (see for example [4]).

Given that a reaction is feasible, how does a catalyst set about speeding it up? Essentially a catalyst offers reactants an alternative, lower-energy, pathway to products; it makes it easier for a system to go from some initial to some final state.

The effectiveness of a catalyst is determined by the difference in the ease with which the system can follow the catalysed and uncatalysed pathways. Each pathway usually has some particular step which controls the overall rate, i.e. the rate determining step. The rate of this step is determined by its free energy of activation which corresponds to the highest free energy barrier along the route from reactants to products. A catalyst sets out to make this free energy of activation as small as possible or at least less than that found in the uncatalysed mechanism. Pictorially the situation can be represented as shown in Fig. 1.1.

Up to now we have discussed catalysis in general terms making only passing reference to the actual physical nature of the catalyst. We have adopted what could be described as a managerial approach to catalysis whereby given a reaction

$$A + B \longrightarrow C$$

all we need to do to catalyse it is write the magic words 'catalyst' or 'cat' over the arrow. What catalyst or cat actually is – well we can leave that to the scientists. After all they are paid to understand such things! In fact, a catalyst can be something as simple as a proton or something as complex as an enzyme. It can be purely organic in nature, purely inorganic or a mixture of the

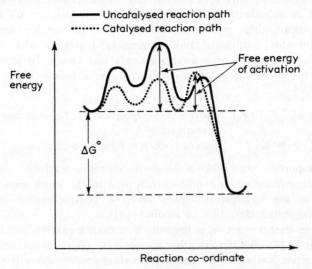

Fig. 1.1 *Free energy plot for a hypothetical uncatalysed and catalysed reaction*

GENERAL CONSIDERATIONS

two. It can be present in the same phase as the reactants, homogeneous, or in a different phase, heterogeneous. If present in the same phase we can talk about gaseous reactions or reactions occurring in solution. Clearly, to restrict our discussion to a reasonable length we need to draw some boundary lines around the type of systems we are going to consider. The key words in our boundary conditions are 'transition metal' and 'homogeneous'.

1.2 Why transition metals?

The principal reasons why transition metals contribute the essential ingredient in such a wide range of catalyst systems can be summarized under five main headings:

(a) Bonding ability
(b) Catholic choice of ligands
(c) Ligand effects
(d) Variability of oxidation state
(e) Variability of co-ordination number.

Transition elements are distinguished from main group elements in having partly filled d or f shells. The main transition group, or d-block elements, are those elements having partially filled d shells and it is with this group that we shall be principally concerned and to which we shall be referring when we use the term transition metal.

1.2.1 *Bonding ability*

A d-block metal ion has nine valence shell orbitals – s, p_x, p_y, p_z, d_{z^2}, $d_{x^2-y^2}$, d_{xz}, d_{yz}, d_{xy} – in which to accommodate its valence electrons and with which to form hybrid molecular orbitals in bonding with other groups. The availability of these valence orbitals to the transition metal put it in the position of being able to form both sigma, σ-, and pi, π-, bonds with other moieties or ligands. It is this ability to form both σ- and π-bonds which is one of the key factors in imparting catalytic properties to the transition metals and their complexes.

Consider the case of an olefin bonding to a transition metal centre and take, as an example, one of the first recognized organometallic compounds, i.e. Zeise's salt, $K[C_2H_4PtCl_3]$ [5,6]. The structure of the $[C_2H_4PtCl_3]^-$ ion is as shown below, *(1.1)*, i.e. basically a square planar arrangement with the ethene double bond normal to the co-ordination plane.

(1.1)

In such a square planar complex if we define the x axis as that being coincident with the Cl–Pt–Cl bond, then there are four metal ion valence shell orbitals, including the spherically symmetric $6s$ orbital, which have lobes lying along the metal–ligand bond axes, i.e. $6s$, $5d_{x^2-y^2}$, $5p_x$ and $5p_y$. These orbitals, or hybrid combinations thereof, are suitable for forming σ bonds by overlapping/interacting with ligand orbitals having suitable symmetry and energy. For the chlorine ligands a σ-bond is formed by overlap with the $3s$ chlorine orbital. The filled π-orbital of the ethene molecule has the correct symmetry/energy to form a σ-bond, with one of the empty metal orbitals lying along the y axis. This metal–olefin σ overlap is shown diagrammatically in Fig. 1.2a. (The metal orbital is represented as a metal hydride orbital of the type dsp^2.) Of the remaining metal orbitals, d_{yz} lies in the co-ordination plane defined by the $Pt(C_2H_4)$

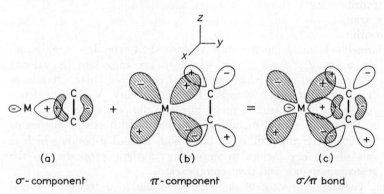

Fig. 1.2 *Molecular orbital representation of ethene bonded to a transition metal*

GENERAL CONSIDERATIONS

units and has the correct symmetry to form π-bonds with the ligand should suitable ligand orbitals be available. The unoccupied antibonding π-orbitals on the olefin fit the bill and thus a π-bond is formed by interaction of the filled d_{yz} orbital on the metal with the unoccupied antibonding orbital of the ethene. This metal–olefin π-overlap is illustrated in Fig. 1.2b and the combined bonding picture is shown in Fig. 1.2c [7,8]. In principle π-bonding is also possible between the metal and the chlorine via the empty chlorine 3d orbitals. However, the energy of these latter orbitals is too high, relative to the metal orbitals, for any significant overlap to occur and the chlorine–metal bond is basically σ in character.

The bonding components illustrated above are synergistic, i.e. they mutually reinforce or complement each other. In the σ-component, electron density flows from an olefin bonding orbital to the metal while, in the π-component, electron density is transferred from the metal to olefin antibonding orbitals. Both of these transferences result in a weakening, or reduction in bond order, of the olefin C–C bond [9,10].

Co-ordination of an alkene to a metal centre alters the electron density in the C–C bond and in many cases makes it more susceptible to nucleophilic attack by species such as OH^- and H^-. However, perhaps just as important, co-ordination to the metal keeps the olefin in one place so that the attacking groups know where to find it. We will come back to this latter point when we consider proximity interaction in Section 1.5.

A similar type of bonding to that occurring in metal–olefin complexes, i.e. involving both σ and π components, is found in transition metal carbonyl complexes as illustrated in Fig. 1.3. The σ component is formed by interaction between an empty metal σ-orbital and a filled sp carbon orbital. The π-component is formed by interaction between a filled metal $d\pi$-, or hybrid

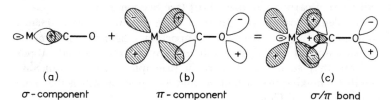

(a)   (b)   (c)
σ - component π - component σ/π bond

Fig. 1.3 *Molecular orbital representation of carbon monoxide bonded to a transition metal*

$d p\pi$-orbital, and one of the sets of empty antibonding $p\pi$-orbitals of the carbon monoxide. As was the case for the olefin, the σ-component results in a net transfer of electron density from the ligand to the metal and the π-component in a net transfer in the opposite direction. Again the bonding is synergistic and results in a decrease in the C–O bond order [11].

Bonding of a CO unit to a transition metal is not limited to unidentate or linear M–CO species. Complexes are also known in which CO bridges two, as in $Co_2(CO)_8$, or three, as in $Rh_6(CO)_{16}$, metal centres. In the complex $[Rh_7(CO)_{16}]^{3-}$ all three types of bonding interactions are present in the same molecule [12]. In all the above cases metal–carbonyl bonding exclusively occurs via the carbonyl carbon; the three modes of bonding being diagrammatically represented as shown in (1.2) to (1.4)

$$\underset{\text{linear}}{\underset{M}{\overset{O}{\underset{|}{\overset{\|}{C}}}}} \qquad \underset{\text{bidentate}}{\underset{M \quad M}{\overset{O}{\overset{\|}{C}}}} \qquad \underset{\text{tridentate}}{\underset{M \quad M}{\underset{M}{\overset{O}{\overset{\|}{C}}}}}$$

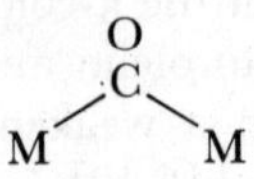

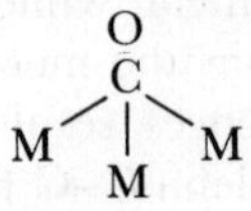

(1.2)　　　(1.3)　　　(1.4)

Recently it has been found that CO can also bridge two metal centres by making use of the CO π bonding orbitals, i.e. (1.5) [13–15].

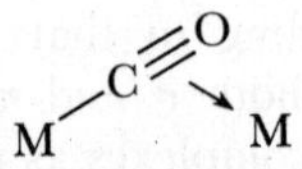

$$M \overset{O}{\underset{}{\overset{\|}{C}}} M$$

(1.5)

Although this latter type of bonding is relatively rare it could well be significant in the design of systems capable of catalysing the reduction of carbon monoxide [16] (see Section 3.2).

Trivalent phosphorus, and to a lesser extent trivalent arsenic, compounds are extremely important ligands in many transition-metal catalyst systems. Symmetry considerations indicate that such ligands can, in principle, bond to transition metals by making use of both σ- and π-orbitals in much the same way as C_2H_4 and CO.

The bonding picture is essentially the same as that described for carbon monoxide, the basic difference being that, in the case

GENERAL CONSIDERATIONS

of phosphorus, the π-component of the bonding is formed by overlap of a filled metal d, or dp hybrid, orbital with an empty $3d$, or hybrid $3d3p$, phosphorus orbital rather than with an empty $p\pi$-antibonding orbital as is the case with CO. There is some debate as to the relative importance of the π-component in metal–tertiaryphosphine bonds. Mason and Meek [17] have suggested that the importance of the π-bonding in determining molecular parameters in such complexes has, in the past, been overemphasized. They take the view that the electronic versatility of the tertiary phosphine ligand is largely a σ-effect since they could find little evidence for π-bonding effects in metal–tertiaryphosphine bond lengths.

From the above, it should be clear that d-block elements can readily form strong bonds with compounds containing π-electron systems or having orbitals of suitable symmetry/energy to form $d\pi$-bonds. Equally important, from the standpoint of catalysis, is the ability of transition metals to form strong, essentially σ-, bonds with several highly reactive species [18]. By so doing transition-metal complexes facilitate the preparation of these species under relatively mild conditions (often ambient temperature and pressure); they moderate their subsequent behaviour and often persuade them to act in a highly specific manner. In this context the two reactive species of greatest utility in catalytic reactions are hydride (H^-) and alkyl (R^-).

Thus chlorotris(triphenylphosphine)rhodium, $RhCl(PPh_3)_3$, in benzene readily reacts with molecular hydrogen to break the H–H bond and form a dihydrido complex, $RhH_2Cl(PPh_3)_3$, in which the two H atoms are directly bonded to the metal [19]. The metal–hydride bond is stable in the sense that it is detectable using normal spectroscopic techniques, e.g. ^1H n.m.r. and infrared. However, in the presence of an alkene, $RhH_2Cl(PPh_3)_3$ readily gives up its hydride ligands to the olefinic double bond. This sequence of reactions, which is discussed more fully in Section 2.1.1, is summarized below:

$$RhCl(PPh_3)_3 + H_2 \longrightarrow RhH_2Cl(PPh_3)_3 \rightleftharpoons RhH_2Cl(PPh_3)_2 + PPh_3$$

$$RhH_2Cl(PPh_3)_2 + RCH=CH_2 \longrightarrow \text{'}RhCl(PPh_3)_2\text{'} + RCH_2CH_3. \quad (1.4)$$

$RhCl(PPh_3)_3$ will also react with molecules such as CH_3I to give a species in which the methyl group is directly bonded to the metal centre

$$RhCl(PPh_3)_3 + MeI \longrightarrow RhMeClI(PPh_3)_3,$$

or with acetyl chloride to give a metal acetyl complex of the type $Rh(COMe)Cl_2(PPh_3)_3$ [20]. These are just a few examples of the ability of transition metals to 'stabilize' reactive species but, as we shall see in subsequent chapters, such species play key roles in a wide range of catalytic reactions involving many different substrates and catalyst systems.

1.2.2 Catholic choice of ligands

In the context of transition-metal co-ordination chemistry a ligand may be defined as any element or combination of elements which form(s) chemical bond(s) with a transition element. In many cases, much the preferred ligand of a transition element is itself as evidenced by the fact that all the d-block elements are metals. Transition elements are, however, very social entities in that they will readily form linkages with just about every other element in the periodic table and with almost any organic molecule we care to think of. It is this property which results in the transition elements having such a rich co-ordination chemistry and which is especially relevant to their role as catalysts.

Basically two types of ligands can be distinguished; those formally viewed as ionic, e.g. Cl^-, H^-, OH^-, CN^-, alkyl$^-$, aryl$^-$, $COCH_3^-$, and those formally viewed as neutral, e.g. CO, alkene, tertiary- secondary- primary-phosphine, -arsine or -phosphite, H_2O, amine. This distinction between ionic and neutral, although useful in determining oxidation states and describing reaction sequences, is somewhat formal since, in most transition-metal complexes, ionic ligands form covalent, rather than ionic, bonds with the metal centre and, in many cases, the charge separation in the metal ligand bond is small and indeed in some cases, less for 'ionic' than for 'neutral' ligands. Dipole moment measurements and X-ray spectroscopic techniques have indicated that in square planar platinum complexes the 'ionic' ligands H^- and CH_3^- are nearly neutral in terms of metal–ligand charge separation whereas 'neutral' ligands such as pyridines and tertiary phosphines can carry a fairly large positive charge ($\sim 0.3e$) [21].

These results highlight one of the problems encountered in discussing the reactions of transition-metal complexes; a lot of the

GENERAL CONSIDERATIONS

concepts and descriptions we use have evolved from a set of formal definitions. In many cases, as would be expected, these formal concepts work very well, thus most of the reactions of transition-metal hydride complexes can be rationalized by assuming that the H reacts as H^- and similarly that in metal alkyl species the alkyl reacts as R^-. It is, however, important to remember that such rationales are conceptual rather than actual. They are extremely useful and can often provide the basis of a working hypothesis. They must not be allowed to limit our imagination. Methyl bonded to a metal usually reacts as though it were Me^- but there is no reason why we should not design a system in which it reacts as $Me^{\cdot}$ or Me^+; indeed, as we shall see, such systems are known.

Up to now we have concentrated most of our attention on ligands, such as alkenes, alkyl, carbonyl, hydride, which do, or are potentially able to, take an active part in a catalytic cycle in the sense that they end up in the product(s) of the cycle, e.g. hydride and alkene in a catalytic hydrogenation sequence, or carbon monoxide in a catalytic carbonylation reaction. We will call such ligands, 'participative' ligands. As far as catalysis is concerned another, equally important group of ligands are those which do not *directly* participate in the catalytic cycle in the sense that they remain associated with the transition metal and do not appear as part of the product(s) of the reaction. Examples of such 'non-participative' ligands would be the chlorine and the triphenylphosphine groups in Equation (1.4). Although this latter group of ligands do not physically contribute to the direct products of the catalysed reaction they play a vital part in determining the activity and selectivity of the catalyst system [22,23].

1.2.3 *Ligand effects*

The ability of transition-metal catalysts to accommodate both participative and non-participative ligands within their co-ordination sphere offers us the possibility of being able to direct the course of a reaction, between participative ligands by modifying the structural/electronic properties of the non-participative ligands.

Formally, a ligand can influence the behaviour of a transition-metal catalyst by modifying the steric or electronic environment at the active site, i.e. the site at which the participative ligands combine. In practice, with all but the simplest ligands the resulting effect is a combination of both electronic and steric parameters. The weight given to either parameter is in many cases a question of personal preference or bias since there are, as yet, no well-developed theories encompassing both effects. There are, however, a number of concepts which can prove useful in interpreting, and in some cases predicting, the effect of non-participative ligands. Three such concepts are 'trans-effect', 'ligand electron donor–acceptor properties' and 'cone angle'. The first is concerned, in principle, with all ligands capable of bonding to a transition metal. The second two are essentially limited to phosphorus and, to a much lesser extent, arsenic ligands.

(a) Trans-effect

Because of the directional nature of most of the orbitals used by transition metals in forming metal–ligand bonds and because orbital–orbital interaction is at a maximum when the angle between the orbitals is zero, electronic effects, which occur principally via the bonding orbitals, are maximal between two ligands which are mutually *trans*. Thus, maximum electronic effect can be obtained by placing the non-participative ligands in a position *trans* to the participative ligand. The most commonly cited example of the *trans*-effect is in terms of the rate of ligand substitution in square planar platinum(II) complexes [24]. Thus the rate of substitution of chloride by pyridine in complexes of type (*1.7*) increase in the ratio 1:30:200:10 000 as we go from

(*1.7*)

$X = Cl$ to $X = C_6H_5$ to $X = CH_3$ to $X = H$ indicating that, in terms of *trans*-effect the order of these four ligands is $Cl < C_6H_5 < CH_3 < H$. The mechanism via which two mutually *trans* ligands interact will clearly depend on the nature of the ligands in question. Simple σ-bonding ligands, such as H^-, will interact with a *trans* ligand through the σ-bonding orbital system while π-acceptor ligands, such as CO and tertiary phosphine, can interact via both the σ- and π-orbital system. Although there is some debate as to the relative importance of the σ- and π-contributions to the *trans*-effect [17], there is no doubt that ligands such as H^- and $SnCl_3^-$, which come high in the *trans*-effect series, are effective in labilizing mutually *trans* groups in metal complexes. Introduction of a high *trans*-effect ligand into a catalyst system can often have a beneficial effect on the activity of the system when ligand, or substrate, dissociation is judged to be the rate determining step [19,25].

In the above discussion we have implied that the *trans*-effect of a ligand is essentially a function of its electronic characteristics. While, in the examples cited, electronic factors almost certainly dominate, for bulky ligands, steric effects can also make a significant contribution to the *trans*-effect.

In an attempt to separate electronic and steric effects in trivalent phosphorus ligands, Tolman has proposed two parameters [26]; the first based on the electron donor–acceptor properties of ligands and the second based on the size of the ligand.

(b) *Electron donor–acceptor properties – the electronic parameter, v*
[26,27]

The CO infrared stretching frequency of a carbonyl group attached to a transition metal varies with the nature and with the number of the other ligands present in the complex. In 1970, Tolman proposed [27] that this fact could be used to define an electronic factor, v, related to the electron donor–acceptor properties of any given trivalent phosphorus ligand.

Since complexes of the type $Ni(CO)_3L$ form very rapidly on mixing $Ni(CO)_4$ and L in a 1:1 ratio at room temperature, and since such complexes give rise to a very sharp carbonyl band, the A_1 band, in their infrared spectra, Tolman proposed choosing the frequency of this A_1 band as a measure of the ligand's donor–acceptor properties. He found that this frequency is essentially an additive function related to the nature of the three groups bound to the phosphorus atom and that the electronic parameter, v, synonymous with the A_1 band frequency, is given by the equation

$$v = 2056.1 + \sum_{i=1}^{3} \chi_i \text{ cm}^{-1}$$

where 2056.1 cm^{-1} is the A_1 frequency of the tri-t-butylphosphine (the most basic tertiary phosphine in the original series) and χ is a substituent contribution appropriate to each group attached to the phosphorus in the $PX_1X_2X_3$ ligand. Table 1.1 gives some

Table 1.1 *Some selected substituent contribution, χ, values* [26]

Group, X, in ligand of type $PX_1X_2X_3$	χ (cm^{-1})
t-Bu	0.0
Cyclohexyl	0.1
i-Pr	1.0
Bu	1.4
Et	1.8
Me	2.6
o-tolyl	3.5
m-tolyl	3.7
p-tolyl	3.5
Ph	4.3
OMe	7.7
H	7.3
OPh	9.7
C_6F_5	11.2
CF_3	19.6

Table 1.2 *Some selected electronic parameters, ν, values for ligands of the type $PX_1X_2X_3$ together with the corresponding A_1 band frequencies for complexes of type $Ni(CO)_3L$* [26] *(where $L = PX_1X_2X_3$)*

$PX_1X_2X_3$	ν (cm^{-1})	A_1 (cm^{-1})
PMe$_3$	2063.9	2064.1
PEt$_3$	2061.5	2061.7
P(i-Pr)$_3$	2059.1	2059.2
P(t-Bu)$_3$	2056.1	2056.1
P(p-Tol)$_3$	2066.6	2066.7
P(o-Tol)$_3$	2066.6	2066.6
PMePh$_2$	2067.3	2065.3
PEt$_2$Ph	2064.0	2063.7
P(OMe)Ph$_2$	2072.4	2072.0
P(OPh)$_3$	2085.2	2085.3

GENERAL CONSIDERATIONS

representative values of χ and Table 1.2 shows some calculated ν values compared to experimentally determined A_1 frequencies.

(c) Cone angle – the steric parameter, θ

To complement his electronic parameter Tolman [26,29] has developed a steric parameter, θ, applicable to trivalent phosphorus ligands. This arose from the observation that the ability of phosphorus ligands to compete for co-ordination positions on Ni(0), as determined by the position of equilibrium shown in Equation (1.5),

$$\mathrm{NiL_4} + n\mathrm{L^1} \rightleftharpoons \mathrm{NiL_{4-n}L_n^1} + n\mathrm{L} \qquad (1.5)$$

could not be adequately explained in terms of their electronic character [29]. The position of the equilibrium appeared to be determined by the physical bulk of the phosphorus ligand L. In order to get some feel for the size of the ligands, CPK molecular models* of ligands were constructed. With all the substituents arranged to occupy as little effective space as possible the 'cone angle' of the ligand was measured. The 'cone angle' being defined as the apex angle of a cylindrical cone, centred 2.28 Å (2.85 cm using molecular models with a scale of 1.25 cm/Å) from the centre of the phosphorus atom, which just touches the van der Waals radii of the outermost atoms of the model (see Fig. 1.4).

The ability of a ligand to compete for co-ordination sites in Ni(0) complexes was found to correlate with its cone angle – the larger the cone angle the poorer the ligands competitive ability. Similarly, the degree of substitution of CO for L in nickel

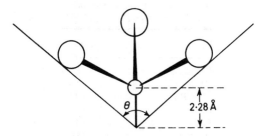

Fig. 1.4 *Tolman 'cone angle' for a symmetrical phosphorus ligand of the type PR_3*

*Space filled models where the boundary of each atom is related to the actual volume it occupies in space, i.e. its van der Waals radius.

carbonyl, when $Ni(CO)_4$ was treated with an eightfold excess of L and allowed to equilibrate, was found to correlate extremely well with the physically measured cone angle. These early studies of Tolman, later complemented by measurements of the equilibrium constant for the equilibrium [30]:

$$NiL_4 \rightleftharpoons [NiL_3] + L$$

have led to the tabulation of θ values for a wide range of trivalent phosphorus ligands [26]. Some selected values are given in Table 1.3. Since a good linear correlation exists between the degree of substitute (DS) of carbonyl groups from $Ni(CO)_4$ and the cone angle, θ, of the substituting ligands, θ values of new ligands, or ligands in which θ is difficult to measure from models, can be estimated by measuring DS. The experimental procedure is relatively simple; $Ni(CO)_4$ is treated with an eight-fold excess of ligand, L, in a sealed tube until equilibrium is reached. The degree of substitution is determined from the relative intensities of the carbonyl bands in the infrared spectrum of the resulting mixture. From the plot of DS values against θ, constructed for known ligands, the θ value of the new ligand can be read off.

Although cone angles are extremely useful for 'getting a feel' for the relative size of various ligands some caution needs to be exercised in using the absolute values since, not only are these based on the assumption that the metal–phosphorus bond length remains constant at 2.28 Å with the phosphorus retaining tetrahedral symmetry, but also they neglect the possibility of

Table 1.3 *Some selected cone angles, θ, for trivalent phosphorus ligands of the type $PX_1X_2X_3$* [26]

Ligand	θ (°)
PH_3	87
PMe_3	118
PHe_2Ph	122
PEt_3	132
PPh_3	145
$P(i\text{-}Pr)_3$	160
$P(cyclohexyl)_3$	170
$P(t\text{-}Bu)_3$	182
$P(o\text{-}tol)_3$	194

phosphorus ligand groups 'meshing' together in the metal complex molecule. For example on the basis of a cone angle of 170° for P(cyclohexyl)$_3$ we would not expect Pt{P(cyclohexyl)$_3$}$_3$ to be particularly stable or indeed capable of isolation. Yet, when Pt{P(cyclohexyl)$_3$}$_2$ is crystallized from heptane at -15 °C in the presence of excess P(cyclohexyl)$_3$ the solvate Pt{P(cyclohexyl)$_3$}$_3$ 1.5 heptane is readily isolated [28]. The X-ray structure of this molecule shows that the tri-cyclohexyl phosphine ligands adopt an orientation such that the cyclohexyl groups can mesh into each other thus effectively reducing the steric crowding and stabilizing the molecule.

Clearly, using Tolman's cone angles to predict the absolute stability of a given species can lead to error. Nevertheless, there is little doubt that judicious application of Tolman's electronic and steric parameters to ligand effect data, obtained from transition-metal catalysed reactions, can aid understanding of the relative significance of the two factors and can help in designing new ligand systems and metal–ligand combinations.

In the above discussion of ligand effects we have tended to concentrate on the influence a ligand can exert on the electronic and steric properties of a complex. A ligand can also influence the physical properties of a metal complex and in some cases this can be just as important in the design of a catalyst system. In complexes containing alkyl phosphine ligands the solubility of the complex in organic solvents can be increased by increasing the length of the alkyl chain. Thus for a given tertiary phosphine metal complex solubility in, say benzene, increases in going from PMe$_3$ to PEt$_3$ to PPr$_3$ to PBu$_3$. Alkyl phosphine complexes are generally more soluble than aryl phosphine complexes although the solubility of the latter can be increased by introducing alkyl groups into the aryl rings. The solubility of a complex in polar solvents such as water can be increased by introducing polar groups, e.g. carboxyl into the ligand system. The volatility of a given complex can be modified by correct choice of ligand, thus the volatility of cobalt–carbonyl–phosphine complexes of the type Co(CO)$_4$P{(CH$_2$)$_n$CH$_3$}$_3$ can be decreased by increasing the value of n without seriously modifying the electronic properties of the tertiary phosphine ligand.

1.2.4 Variability of oxidation state

In theory a transition metal can have access to as many formal positive oxidation states as it has valence d and s electrons. Thus Cr, $3d^5 4s^1$, can, in principle, exist in the positive oxidation states I to VI and in reality complexes of all 6 oxidation states, together with Cr(0) and indeed Cr(−II) are known.

The ability to form stable complexes with the metal in a variety of oxidation states is common among the transition elements, although not all of the elements form stable complexes for all of their available oxidation states. π-bonding ligands such as CO, tertiary phosphine and amines tend to prefer metals in low oxidation states while small essential σ-bonding ionic ligands such as H⁻ and F⁻ tend to prefer metals in high oxidation states. Having access to a wide range of oxidation states helps transition metals form complexes with a wide range of other elements and compounds. However, perhaps more important than this access is the ability to readily interchange between oxidation states during the course of a chemical/catalytic reaction. Consider the rhodium catalysed hydrogenation reaction mentioned earlier (Equation (1.4.)). For every catalytic cycle the rhodium undergoes a I → III → I oxidation/reduction cycle:

$$\text{`Rh}^I\text{ClL}_2\text{'} \xrightarrow{H_2} \text{Rh}^{III}H_2\text{ClL}_2 \xrightarrow{RCH=CH_2} \text{Rh}^{III}H(RCH_2CH_2)\text{ClL}_2 \xrightarrow{-RCH_2CH_3} $$

In a typical hydrogenation under ambient conditions the rhodium would have to be capable of going through this sequence about once a minute. Similarly in hydrogenation reactions catalysed by $RuClH(PPh_3)_3$ the mechanism involves the ruthenium undergoing a II → IV → II oxidation/reduction cycle. This ability to readily enter into redox cycles is especially marked for the Group VIII metals and is a major factor contributing to their wide ranging catalytic activity.

1.2.5 Variability of co-ordination number

Transition-metal complexes containing as many as nine ligands in the co-ordination sphere are well established, e.g. ReH_9^{2-} $WH_6(PR)_3$. More commonly, co-ordination numbers between four and six are encountered. The ability of a transition metal to accommodate several different ligands in its co-ordination sphere is clearly important if it is to catalyse reaction between one or more substrates. Thus in a hydroformylation reaction, in which an aldehyde is formed from an olefin, carbon monoxide and hydrogen, the transition metal must be able to accommodate olefin, carbonyl and hydride within its co-ordination sphere during the course of the reaction together with any other non-participative ligands which may be present. However, as in the case of oxidation state, of equal importance to being able to adopt different co-ordination numbers, and consequently different stereochemistries, is the ability to rapidly change between them. For example, during the hydrogenation reaction catalysed by $RhCl(PPh_3)_3$ the rhodium can be envisaged as going from 4 co-ordinate (square planar) to 6 co-ordinate (octahedral) to 5 co-ordinate (square pyramidal) to 6 co-ordinate (octahedral) to 5 co-ordinate (square pyramidal) to 4 co-ordinate (trigonal) and back to 4 co-ordinate (square planar or square pyramidal) during one cycle of the hydrogenation reaction. These changes are

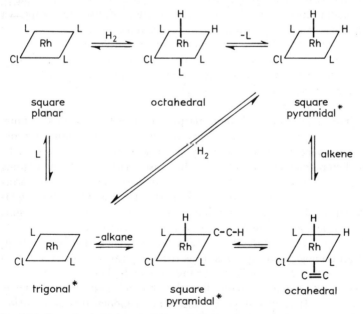

Fig. 1.5 *Stereochemical changes during the hydrogenation catalysed by $RhClL_3$ (L = PPh_3)*

illustrated in Fig. 1.5. The foregoing rationale for restricting our discussion of catalysis to complexes containing d-block transition metals may be summed up in one word – versatility. Complexes of the transition metals are some of the most versatile compounds known in chemistry.

1.3 Why homogeneous?

Before being able to answer this question we must first look at the two types of catalyst systems, i.e. homogeneous and heterogeneous, in a little more detail.

A homogeneous reaction is, by definition, one in which all the constituents of the reaction are present in the same phase. When discussing homogeneous catalysis, what is generally implied is that all the constituents of the reaction, including, of course, the catalyst, are present in a common liquid phase, i.e. in solution. What is further implied is that the catalyst is a discrete entity, i.e. in the case of a homogeneous transition-metal catalyst a single transition-metal complex or a combination of discrete complexes. A heterogeneous reaction, on the other hand, is one in which one or more of the constituents are in different phases. In heterogeneous catalysis the catalyst is usually present as a solid and the reactants as liquids or, more frequently, as gases. The catalysed reaction then takes place at the phase interface, i.e. on the catalyst surface.

The solid state or bulk nature of a heterogeneous catalyst system gives it certain advantages over its homogeneous counterpart. From a practical point of view the main advantage of a heterogeneous catalyst is the ease with which it can be separated from the reaction products, assuming of course these are either liquids or gases. This particular advantage has, up to recently, been of over-riding importance in the large scale industrial application of catalyst systems. Being bulk solids, heterogeneous catalysts generally have higher thermal stability than homogeneous systems, thus reactions can be run under more forcing thermal conditions giving rise to higher reaction rates – assuming equal activity for the homo- and hetero-geneous catalysts. If energy is cheap and time is money this is a distinct advantage. The thermal stability of heterogeneous catalysts can be a further advantage in catalyst regeneration. Unfortunately no catalyst lives for ever. In homogeneous catalyst systems side reactions decrease the concentration of the active species by converting it to some catalytically inactive complex such that at some stage in the reaction it is necessary to recover the metal, or inactive complex, and resynthesize the active species. Generally this cannot be done *in situ* and is frequently an expensive process. In heterogeneous catalyst systems deactivation is often the result of deposition of reaction by-products on the catalyst surface and regeneration can often be accomplished by simply burning off such by-products. Many heterogeneous catalyst systems can be simply and rapidly regenerated *in situ* which is a clear advantage in an industrial setting.

Thus, being a bulk solid has advantages if you are a catalyst; it also has disadvantages. A given reaction is catalysed by a particular site or combination of sites in a catalyst system, be it homogeneous or heterogeneous. In a solid state catalyst, especially one which is made up of a cocktail of elements deposited on a carrier, there are a multitude of different sites. Even in the pure metallic state the individual metal atoms can find themselves in a number of different environments. Of all these different sites, possibly only one type catalyses the reaction of interest. At best the other types of sites do not get involved: at worst they catalyse undesired side reactions. With a homogeneous transition-metal catalyst there is generally only one type of active site. Thus, in terms of activity per metal centre, homogeneous catalysts are frequently more active. They are also capable of being much more selective. These attributes are especially welcome at a time when the chemical industry, and society in general, is seeking ways to conserve energy and make the best use of available resources.

A major problem in understanding, and thus being in a position to effectively modify, the behaviour of solid state catalysts is the lack of physical techniques which facilitate accurate monitoring of the reactions taking place on the catalyst surface or which allow the nature of the active site(s) to be delineated. Spectroscopic techniques such as infrared and, more recently, LEED, ESCA and AUGER, have been used to study surface reactions and active sites [31]. However, owing to the complexity of such surfaces, considerable difficulty has been encountered in interpreting the results unambiguously. Most homogeneous reactions of interest occur in solution and involve discrete metal complexes. They are therefore amenable to study using infrared spectroscopy and, especially in the case of diamagnetic complexes, n.m.r. Both of these techniques are widely available and can be used to study catalysed reactions as they occur. Thus ^{1}H and ^{31}P n.m.r. spectroscopy has been used to characterize most of the metal species present in RhCl(PPh$_3$)$_3$ catalysed hydrogenation of olefins [19] and recently infrared spectroscopy has been used to demonstrate the presence of metal carbonyl cluster compounds during the rhodium-catalysed formation of ethene glycol from carbon monoxide and hydrogen [16]. The ability to study homogeneous catalyst systems in such detail has allowed the effect of changes in the ligands and/or the reaction conditions to be accurately followed and has resulted in a greater detailed understanding of homogeneous, compared to heterogeneous, catalyst systems. Clearly the two types of systems are closely related in that they both involve reactions occurring at metal centres and frequently insight obtained from studying homogeneous catalysts has helped in explaining results obtained with heterogeneous catalysts.

To return to our original question the reasons for choosing homogeneous systems can thus be summarized under four headings:

(a) selectivity
(b) activity
(c) ease of modification
(d) ease of study

and in a world where, in terms of utilization of resources, there is an increasing need to do things as efficiently as possible, homogeneous catalysts will have an increasingly important role to play.

1.4 Molecule activation

Having justified our choice of subject matter we return to the basic concept of catalysis. As pointed out in Section 1.1 a catalyst catalyses a reaction by offering the reactants an alternative, lower energy, pathway to products. It does this by 'activating' one or more of the reactants and then persuading them to react together. If we consider catalysis in terms of molecule activation then we

GENERAL CONSIDERATIONS

can distinguish two basic activation processes:

activation by coordination,
activation by addition.

1.4.1 *Activation by co-ordination*

Activation by co-ordination is the process in which a substrate XY interacts with the catalyst centre in such a way that the integrity of XY is maintained. In such a process, although the distribution of the electrons over the XY bond(s) may be radically altered, X and Y remain formally bonded to each other, and in any exchange process XY in its entirety is exchanged rather than either X or Y individually. Examples in homogeneous systems would be the co-ordination of carbon monoxide, an alkene or an allyl moiety at a single metal centre. The corresponding process in heterogeneous catalysis is known as 'non-dissociative adsorption'.

In the palladium catalysed oxidation of ethene, in acetic acid, co-ordination of ethene to the Pd(II) centre decreases the electron density on the olefin and makes it more susceptible to nucleophilic attack by acetate, i.e.

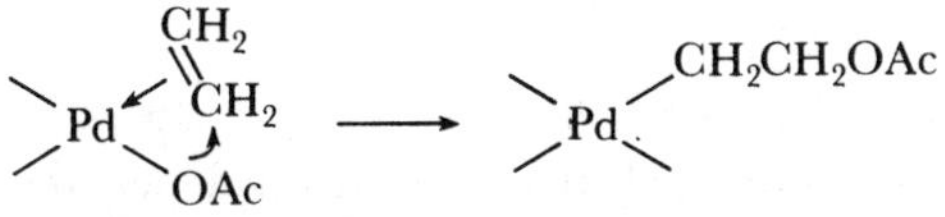

$$\text{Pd}\begin{pmatrix}CH_2\\ \| \\ CH_2 \\ OAc\end{pmatrix} \longrightarrow \text{Pd}\begin{pmatrix}CH_2CH_2OAc\\ \end{pmatrix}$$

Similarly co-ordination of an olefin to cyclopentadienedicarbonyliron, abbreviated Fp^+, renders the olefin susceptible to attack by a wide range of nucleophiles under mild conditions:

$$Fp^+ \!-\! \| \xleftarrow{Nu^-} \longrightarrow Fp\!\!-\!\!\diagup\!\!\diagdown\!\!Nu$$

Typical nucleophiles are $\bar{C}H(CO_2Et)_2$, R^-, X^- etc. Although not catalytic this latter reaction serves to illustrate the principle of activation by co-ordination and has, in its own right, proved very useful in organic synthesis [32].

In both of the above cases activation is principally electronic in character. Steric considerations may be equally, and are in some cases over-ridingly, important in activation by co-ordination processes. This is especially true in asymmetric synthesis catalysed, or moderated, by transition-metal complexes (see Section 2.1.3 and Section 2.4.5) and in stereospecific oligomerization (see Section 2.5.4).

1.4.2 *Activation by addition*

Activation by addition is the process in which a substrate XY interacts with a catalyst centre in such a way that the integrity of XY is destroyed, i.e. the bond(s) holding XY together is (are) formally broken. In the activation process either X or Y or both become bonded to the metal centre and in any exchange process either X or Y separately or together are exchanged. Examples in homogeneous systems would be the addition of hydrogen to a metal centre to give a di- or mono-hydride species or addition of HCN to give a hydrido-cyano complex. The corresponding process in heterogeneous catalysis is known as 'dissociative adsorption'.

In homogeneous systems three types of activation by addition process can be distinguished [18] namely:

(*a*) oxidative addition
(*b*) homolytic addition
(*c*) heterolytic addition.

(*a*) *Oxidative addition*

Oxidative addition [20,33] is defined as the addition of a substrate XY to a metal complex in such a way that both the formal oxidation state and co-ordination number is increased by two in the resulting complex:*

$$M^x L_y + XY \longrightarrow M^{x-2}(X)(Y)L_y$$

The reverse reaction is known as *reductive elimination*. Examples of oxidative addition of particular relevance to homogeneous catalysis are the addition of hydrogen to rhodium(I) or

---

*Frequently an oxidative addition reaction promotes, or occurs with concurrent, dissociation of one or more ligands such that the co-ordination number of the product finally detected is less than two greater than that of the starting complex. To accommodate this problem an increase in formal oxidation state of two together with the retention of X and Y within the co-ordination sphere of the metal is generally taken as sufficient in characterizing a reaction as oxidative addition.

GENERAL CONSIDERATIONS

platinum(II) complexes and the addition of methyl iodide at a rhodium(I) or iridium(I) centre:

$$Rh^{I}Cl(PPh_3)_x + H_2 \longrightarrow Rh^{III}ClH_2(PPh_3)_x \qquad x = 2 \text{ or } 3$$

$$Pt^{II}Cl(SnCl_3)(PPh_3)_2 + H_2 \longrightarrow Pt^{IV}ClH_2(SnCl_3)(PPh_3)_2$$

$$[M^{I}I_2(CO)_2]^- + MeI \longrightarrow [M^{III}I_3Me(CO)_2]^-. \qquad M = Rh \text{ or } Ir$$

(In assigning oxidation states to the metals in these and other hydrido- or alkyl-metal complexes the convention is to treat alkyl and hydrido groups as uni-negative ligands.)

A particularly apposite example of reductive elimination is the formation of an alkane from an alkyl hydrido complex, e.g.

$$Rh^{III}ClH(R)(PPh_3)_x \longrightarrow Rh^{I}Cl(PPh_3)_x + RH \qquad x = 2 \text{ or } 3$$

All of these examples involve the metals in a d^8 ($Rh^{I}4d^8$, $Pt^{II}5d^8$, $Ir^{I}5d^8$) to d^6 ($Rh^{III}4d^6$, $Pt^{IV}5d^6$, $Ir^{III}5d^5$) transformation. This is quite a common feature of oxidative addition/reduction elimination reactions and it is generally the later d-block elements, especially the group VIII metals, which are the most prone to undergo such reactions. Typical redox transitions are $d^{10} \rightleftharpoons d^8$ (e.g. $Pd^0 \rightleftharpoons Pd^{II}$), $d^8 \rightleftharpoons d^6$ (e.g. $Rh^{I} \rightleftharpoons Rh^{III}$) and $d^6 \rightleftharpoons d^4$ (e.g. $Ru^{II} \rightleftharpoons Ru^{IV}$). Having said this, it is important to point out that these reactions are not *limited* to the Group VIII elements or indeed to the transition metals. A well known reaction which constitutes an example of oxidative addition involving a main group metal is the addition of an alkyl halide to magnesium to form a Grignard reagent:

$$Mg^\circ + RX \longrightarrow Mg^{II}(R)X.$$

Since oxidative addition involves a two-fold increase in the formal oxidation state of the metal, ligands which increase the electron density at the metal centre, i.e. basic ligands, generally increase the rate of oxidation. Thus complexes with alkyl phosphine ligands, e.g. PEt_3, generally undergo oxidative addition more readily than the analogous complexes with aryl phosphine ligands such as PPh_3 [34]. Similarly, since the reaction involves a formal increase in co-ordination number, ligands which increase the steric crowding at the metal centre, e.g. tri-*t*-butylphosphine, will generally decrease the rates of catalytic reactions involving oxidative addition [35]. The results of such ligand modifications on the rates of catalytic reactions involving oxidative addition is discussed in Section 2 especially under hydrogenation.

(b) *Homolytic addition*
Homolytic addition [18,19] is defined as the addition of a substrate XY to two metal centres in such a way that the formal oxidation state of each metal centre increases by one:

$$2M^xL_y + XY \longrightarrow 2M^{x+1}(X)(Y)L_y.$$

Two examples of this type of addition which occur during transition-metal catalysed reactions involve cobalt. Addition of hydrogen to an aqueous solution of cobalt cyanide leads to the formation of a hydridopentacyanocobalt anion, $[CoH(CN)_5]^{3-}$, which is active in the reduction of both inorganic and organic substrates [19].

$$2Co^{II}(CN)_5^{3-} + H_2 \rightleftharpoons 2Co^{III}H(CN)_5^{3-}.$$

It has been suggested that this reaction may occur via addition of hydrogen to a dicobalt species formed by dimerization of $Co(CN)_5^{3-}$ [36].

$$2Co^{II}(CN)_5^{3-} \rightleftharpoons Co_2^{II}(CN)_{10}^{6-}$$

$$Co_2^{II}(CN)_{10}^{6-} + H_2 \rightleftharpoons 2Co^{III}H(CN)_5^{3-}.$$

The presently available data do not allow the two routes to be clearly distinguished although there seems little doubt that $Co^{III}H(CN)_5^{3-}$ is the active species in the subsequent reduction reactions. A case where a dimeric species is almost certainly involved is in the formation of $CoH(CO)_4$ from dicobaltoctacarbonyl by homolytic addition of hydrogen [37]

$$Co_2^0(CO)_8 + H_2 \rightleftharpoons 2Co^IH(CO)_4$$

As discussed in Section 2.4 this reaction constitutes one of the most important activation by addition reactions, in terms of the industrial application of homogeneous catalyst systems, since $Co_2(CO)_8$ is the catalyst precursor used in the cobalt catalysed hydroformylation of olefins:

$$RCH=CH_2 + CO + H_2 \xrightarrow[\substack{100-160^\circ C \\ 100-300 \text{ atm}}]{Co_2(CO)_8} RCH_2CH_2CHO + RCH(CHO)CH_3.$$

GENERAL CONSIDERATIONS

In 1976 over 3.5 million tonnes of aldehydes were produced using this, or closely related, systems.

In the above examples, homolytic addition results in the formation of two monomeric species. This is not a necessary requisite of the process since addition across a metal–metal bond with retention of the polymetallic structure could also constitute homolytic addition. Although no examples of such addition in catalytically significant systems have yet been reported, stoichiometric addition of HX across the molybdenum–molybdenum linkage in $Mo_2X_8^{4-}$ (X = Cl or Br) is a well established reaction [38]

$$Mo_2^{II}X_8^{4-} + HX \longrightarrow Mo_2^{III}(H)X_8^{3-} + X^-.$$

This type of activation could well be of importance in catalyst systems involving metal cluster compounds.

As with oxidative addition, homolytic addition results in a formal oxidation of the metal and similarly tends to be aided by electron donating ligands. Thus homolytic addition of hydrogen to $Co_2(CO)_6(PBu_3)_2$ occurs more rapidly than to $Co_2(CO)_8$.

(c) Heterolytic addition

Heterolytic addition [18] is defined as the addition of a substrate XY to a metal centre in such a way that there is no overall change in formal oxidation state, or co-ordination number, of the metal and only either X or Y becomes formally bonded to the metal:

$$M^xL_y + XY \longrightarrow M^xL_{y-1}X + Y^+ + L^-.$$

Essentially this process can be viewed as the substitution of one anionic ligand for another at the metal centre. The most widely quoted example of this type of activation in homogeneous catalysis is the addition of hydrogen to the ruthenium(III) hexachloro anion, $RuCl_6^{3-}$:

$$Ru^{III}Cl_6^{3-} + H_2 \rightleftharpoons Ru^{III}Cl_5H^{3-} + H^+ + Cl^-.$$

The resulting hydridochloride species will readily reduce Fe^{III} to Fe^{II} or Ru^{IV} to Ru^{III}. Thus ruthenium(III) chloride in aqueous hydrochloric acid catalyses the hydrogen réduction of iron(III) and ruthenium(IV) substrates at *ca.* 1 atm/80 °C [39].

Ruthenium(II) complexes will also participate in heterolytic addition [40]

$$Ru^{II}Cl_2(PPh_3)_3 + H_2 \rightleftharpoons Ru^{II}Cl(H)(PPh_3)_3 + H^+ + Cl^-. \quad (1.6)$$

This type of reaction is favoured by the presence of a suitable base, thus the addition of ethanol to a benzene solution of $RuCl_2(PPh_3)_3$ considerably enhances the effectiveness of the system as a hydrogenation catalyst. Other bases such as triethylamine, or indeed potassium hydroxide, are also effective in promoting the addition of hydrogen to $RuCl_2(PPh_3)_3$ to give $RuCl(H)(PPh_3)_3$ plus base.HCl [19].

In practice, it is frequently extremely difficult to distinguish between heterolytic addition and oxidative addition followed by reductive elimination. Thus, as an alternative to Equation (1.6), we could equally well write

$$Ru^{II}Cl_2(PPh_3)_x + H_2 \rightleftharpoons Ru^{IV}Cl_2(H)_2(PPh_3)_x \rightleftharpoons$$

$$Ru^{II}Cl(H)(PPh_3)_x + HCl.$$

Indeed there is some evidence to support such a route [41].

Similarly the addition of hydrogen to cis-$PtCl_2(PEt_3)_2$ involves at first sight a heterolytic addition:

$$cis\text{-}Pt^{II}Cl_2(PEt_3)_2 + H_2 \rightleftharpoons trans\text{-}Pt^{II}Cl(H)(PEt_3)_2 + HCl,$$

while in fact the equilibrium involves the platinum(IV) dihydride intermediate $Pt^{IV}Cl_2H_2(PEt_3)_2$ [42]. Whether all examples of heterolytic addition prove to be special cases of oxidative addition/reductive elimination reactions remains to be proven. In the meantime suffice it to say that this activation by addition process is generally favoured by the presence of a readily ionizable ligand on the original metal complex and, in the case of dihydrogen, by the presence of a suitable base.

Having activated our substrate molecules the next stage in the catalytic cycle is to allow them to react together. This process we can consider under the heading of 'proximity interaction'.

1.5 Proximity interaction

Proximity interaction is the term used to cover the process during which the activated substrate(s) present at the catalyst site

GENERAL CONSIDERATIONS

interact either with themselves or with an external substrate to give either a further activated intermediate or the product(s) of the catalytic cycle. Two proximity interaction processes may be distinguished, namely,

insertion/inter-ligand migration
elimination

1.5.1 *Insertion/inter-ligand migration*

In this process, two substrates, X and Y, present in the catalyst system bonded to one or more metal centres combine to form an intermediate XY which remains bonded to the metal centre(s):

$$\begin{array}{c} X \\ | \\ M-Y \end{array} \longrightarrow M-XY \qquad (1.7)$$

or

$$\begin{array}{cc} X & Y \\ | & | \\ M_1-M_2 \end{array} \longrightarrow \begin{array}{c} XY \\ | \\ M_1-M_2. \end{array} \qquad (1.8)$$

We shall restrict our discussion to systems involving a single metal centre (Equation (1.7)) as these are well characterized. Systems involving more than one metal centre, e.g. Equation (1.8), are less well established, although, as suggested in Section 3.3 and Section 4, they have considerable potential.

Examples of single metal insertion, or inter-ligand migration, reactions are to be found in almost all catalytic systems and they are covered as appropriate in Section 2. Here we describe just one such process to illustrate the principle.

In benzene solution under an atmosphere of carbon monoxide methylpentacarbonylmanganese(I) exists in equilibrium with acetylpentacarbonylmanganese(I) [43]

$$Mn(Me)(CO)_5 \underset{-CO}{\overset{+CO}{\rightleftharpoons}} Mn(COMe)(CO)_5.$$

Increasing the carbon monoxide partial pressure drives the equilibrium to the right. On the face of it what we are observing is the insertion of carbon monoxide into the metal–methyl bond. Mechanistic studies using ^{14}CO have shown that it is a carbonyl group already present on the manganese rather than an incoming carbon monoxide which ends up in the metal–methyl bond, i.e.

$$Mn(Me)(CO)_5 + {}^{14}CO \longrightarrow Mn(COMe)(CO)_4({}^{14}CO).$$

The kinetic data available for the reaction can best be accommodated by the sequence shown in Equation (1.9)

$$R\text{–}Mn(CO)_5 \rightleftharpoons \underset{(1.8)}{RCO\text{–}Mn(CO)_4} \underset{-L}{\overset{+L}{\rightleftharpoons}}$$

$$RCO\text{–}Mn(CO)_4L \quad (\text{where } L = CO).$$

(1.9)

The reaction proceeds via a 5 co-ordinate intermediate (*1.8*) to which the incoming ligand, L, adds. Thus, this insertion reaction is more properly seen as an inter-ligand migration reaction in which the methyl group migrates to a metal bonded carbonyl carbon:

$$\begin{array}{c}\text{OC}\\\text{OC}\end{array}\!\!\!\!\overset{\overset{\displaystyle O}{\overset{\displaystyle \|}{C}}}{\underset{\underset{\displaystyle O}{\underset{\displaystyle \|}{C}}}{Mn}}\!\!\!\!\begin{array}{c}Me\\CO\end{array} \rightleftharpoons \begin{array}{c}\text{OC}\\\text{OC}\end{array}\!\!\!\!\overset{\overset{\displaystyle O\!\!\diagdown\;\;\diagup Me}{C}}{\underset{\underset{\displaystyle O}{\underset{\displaystyle \|}{C}}}{Mn}}\!\!\!\!\begin{array}{c}\\CO\end{array}$$

The vacant co-ordination site resulting from the migration can be filled by solvent or by incoming ligand and such migration reactions are generally favoured by the presence of potentially co-ordinating solvents, e.g. ethers, or suitable ligands, e.g. CO or tertiary phosphine. In catalytic systems where this particular process is important, e.g. carbonylation, hydroformylation, the incoming ligand is usually carbon monoxide. In stoichiometric reactions, ligands such as $P(OPh)_3$ or PPh_3 are equally effective in promoting alkyl migration. During a ligand migration reaction there is no change in the formal oxidation state of the metal and accordingly the electronic demands of the metal, *vis-à-vis* associated ligands, are less stringent than, for instance, in oxidative addition reactions. Most of the *d*-block metals participate in some type of inter-ligand migration reaction and, as previously implied, this particular process probably constitutes one of the most important single type of reaction in homogeneous catalysis. In addition to the alkyl to carbonyl migration just described other examples include hydride migration to co-ordinated olefin in catalytic hydrogenation, acetate migration to olefin in the palladium catalysed oxidation reaction and alkyl to olefin migration in many polymerization systems. In alkyl to carbonyl migration processes the reaction probably proceeds through a three-centre transition state:

$$\underset{M}{\overset{O}{\underset{|}{\overset{|||}{C}}}}\diagdown R \longrightarrow \left[\underset{M}{\overset{O}{\underset{\diagdown}{\overset{\diagup}{C}}}} \diagdown R \right]^{\ddagger} \longrightarrow \underset{M}{\overset{O}{\underset{|}{\overset{\diagup}{C}}}} \diagdown R$$

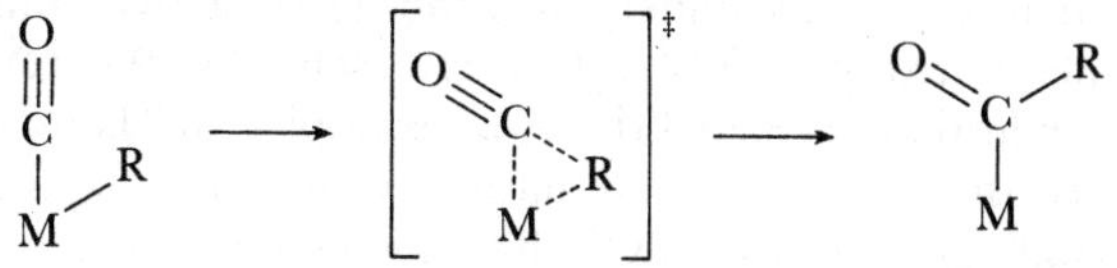

whereas in inter-ligand migrations involving co-ordinated alkene units four-centre transition states may be involved [44]

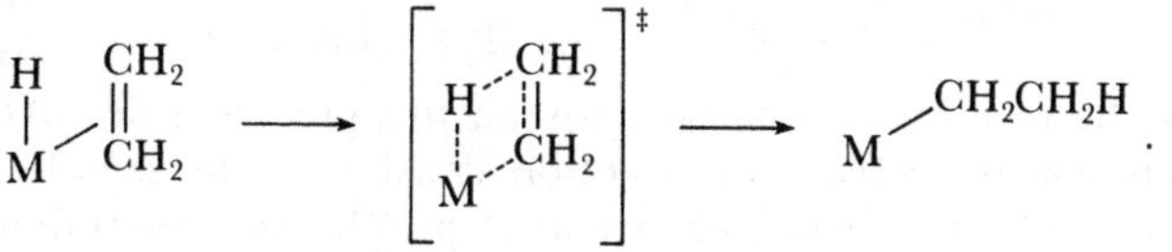

$$\underset{M}{\overset{H}{\underset{|}{}}} \diagdown \overset{CH_2}{\underset{CH_2}{||}} \longrightarrow \left[\underset{M}{\overset{H}{}} \diagdown \overset{CH_2}{\underset{CH_2}{||}} \right]^{\ddagger} \longrightarrow M \diagdown CH_2CH_2H$$

A further instance of proximity interaction, which may be included under the heading inter-ligand migration, is oxidative coupling as exemplified by reactions of the type;

$$M \overset{\diagup\!\!\!\diagup}{\underset{\diagdown\!\!\!\diagdown}{}} \longrightarrow M\!\!\left[\right.$$

and

$$M\!\!\left(\!\!\begin{array}{c}\diagup\!\!\!\diagup\\ \\ \diagdown\!\!\!\diagdown\end{array}\!\!\right) \longrightarrow M\!\!\left\langle\right..$$

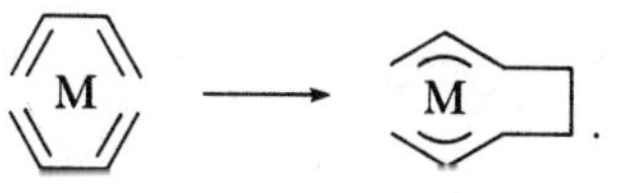

Such processes are particularly important in alkene oligomerization/polymerization systems and are discussed in detail in Sections 2.5 and 2.6.

In inter-ligand migration processes the product of the interaction remains bonded to the metal centre. In the second type of proximity interaction process, i.e. elimination, this is not necessarily the case and elimination frequently marks the end of a catalytic cycle.

1.5.2 *Elimination*

In this process either, two substrates, X and Y, present in the catalyst system bonded to one or more metal centres combine to form a product, XY, which leaves the co-ordination sphere of the metal or, a substrate, XY, bonded to one or more metal centres undergoes some change such that it gives rise to a stable product which can dissociate from the metal. Both processes constitute a termination reaction in the catalytic cycle. The first type of elimination process falls under the heading reductive elimination and as mentioned previously, this is essentially the reverse of oxidative addition [18,20,33]. An example in homogeneous catalysis is the reductive elimination which an alkyl hydrido complex undergoes in a catalytic hydrogenation sequence leading to product alkane:

$$\underset{|}{\overset{H}{M^x}} - R \longrightarrow M^{x-2} + RH.$$

Being the reverse of oxidative addition this process is favoured by ligands which decrease the electron density on the metal. It is, however, unusual for this particular step to be rate determining in catalytic processes. Other examples of reductive elimination occur in the rhodium catalysed carbonylation of methanol [45], with the elimination of acetyl iodide from *1.9*,

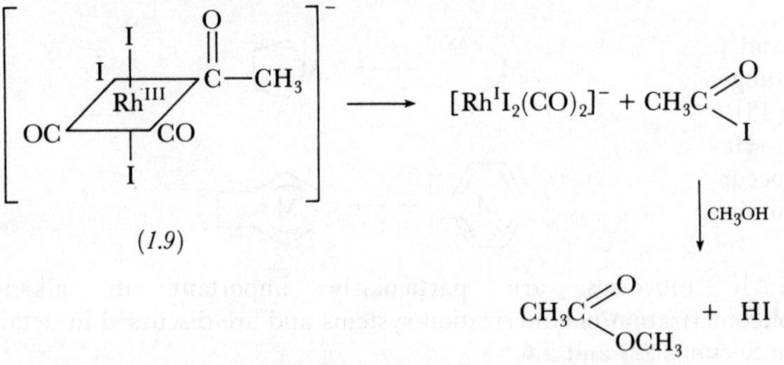

$$\left[\begin{array}{c} \text{Rh}^{III} \text{ complex} \\ (1.9) \end{array} \right]^{-} \longrightarrow [\text{Rh}^{I}\text{I}_2(\text{CO})_2]^{-} + \text{CH}_3\text{C}(\!=\!\text{O})\text{I}$$

$$\downarrow \text{CH}_3\text{OH}$$

$$\text{CH}_3\text{C}(\!=\!\text{O})\text{OCH}_3 + \text{HI}$$

and with the reductive elimination of the aldehyde in the cobalt catalysed hydroformylation reaction [46].

$$\text{Co}^{I}(\text{COR})(\text{CO})_2\text{L} + \text{H}_2 \longrightarrow [\text{Co}^{III}\text{H}_2(\text{COR})(\text{CO})_2\text{L}] \longrightarrow$$
$$\text{Co}^{I}\text{H}(\text{CO})_2\text{L} + \text{RC}(\!=\!\text{O})\text{H}.$$

The second type of elimination process, i.e. that in which the substrate undergoes some change such that it gives rise to a stable product, is generally known as β-elimination since it is usually a group in the β-position in the substrate which interacts [44]. Thus a metal alkyl complex usually converts to a metal–hydrido complex plus alkene by transfer of a hydrogen on the β-carbon of alkyl unit to the metal:

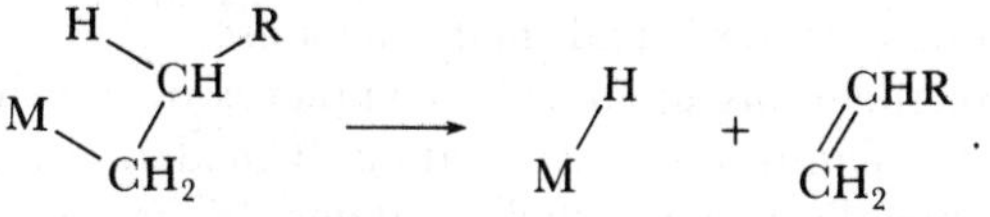

$$\mathrm{M} \underset{CH_2}{\overset{H\diagdown CH\diagup R}{\diagup}} \longrightarrow \mathrm{M} \diagup^H + \underset{CH_2}{\overset{CHR}{/\!/}}.$$

This process is particularly significant in the transition-metal catalysed isomerization of olefins and as a termination process in metal-catalysed alkene polymerization reactions. Essentially this latter type of elimination process is the reverse of the inter-ligand migration reaction previously discussed. Although less well documented, eliminations involving a group in the α-position of the substrate can also occur. Such a process, known as α-elimination, has been shown to occur in tungsten-methyl complexes [47]

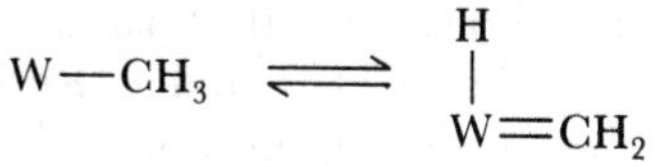

$$W-CH_3 \rightleftharpoons \underset{W=CH_2}{\overset{H}{|}}$$

and may well be important in generating active intermediates in tungsten and molybdenum catalysed disproportionation reactions [48] (see Section 2.8).

Having considered the various types of reactions which can occur during a catalytic process the next stage is to combine these individual elementary steps to form a catalytic cycle.

1.6 The catalytic cycle

A catalytic system can be seen as a series of reactions connected in a cyclic way such that during one journey round the circle substrates are converted to product(s). As far as the catalyst is concerned the journey results in no *net* change. We can construct a catalytic cycle either retrospectively in an attempt to rationalize a known catalytic process or prospectively as an aid to catalyst design/discovery. In either case a useful guide is the 16 and 18 electron rule first proposed by C. A. Tolman in 1972 [49].

1.6.1 *The 16 and 18 electron rule*

The 16 and 18 electron rule is based on the observation [50] that essentially all well characterized diamagnetic complexes of the later d-block elements have 16 or 18 metal valence electrons.* On the basis of this observation Tolman proposed two rules for organometallic complexes and their reactions:

(1) 'Diamagnetic organometallic complexes of transition metals may exist in a significant concentration at moderate temperatures only if the metal's valence shell contains 16 or 18 electrons. A significant concentration is one that may be detected spectroscopically or kinetically and may be in the gaseous, liquid or solid state.'

(2) 'Organometallic reactions, including catalytic ones, proceed by elementary steps involving only intermediates with 16 or 18 valence electrons.'

A corollary of the second rule is that in a series of reaction steps no steps are possible in which the number of valence electrons (NVE) changes by more than two, i.e. $\Delta NVE = 0, \pm 2$. Although some exceptions to the rules are known, e.g. $Pt^0\{P(cyclohexyl)_3\}_2$ $NVE = 14$ and $Rh^IH\{P(cyclohexyl)_3\}_2$ $NVE = 14$, the rules are widely applicable and extremely useful in predicting reaction mechanisms and constructing catalysis cycles. We have previously discussed ligand-assisted methyl migration in methylpentacarbonylmanganese. The overall reaction is

$$Mn(Me)(CO)_5 \xrightarrow{CO} Mn(COMe)(CO)_5$$

and in the absence of kinetic data we could write two possible elementary step reaction sequences

$$Mn(Me)(CO)_5 \xrightarrow{CO} Mn(Me)(CO)_6 \longrightarrow Mn(MeCO)(CO)_5$$

NVE 18   20   18

and

$$Mn(Me)(CO)_5 \longrightarrow Mn(MeCO)(CO)_4 \xrightarrow{CO} Mn(MeCO)(CO)_5$$

NVE 18   16   18

*In determining the number of valence electrons all ligands covalently bonded to the metal are considered to contribute two electrons. The metal is considered to contribute the number of electrons appropriate to its formal oxidation state. Thus $Rh^ICl(PPh_3)_3$ has $8 + (4 \times 2) = 16$ and $Mn^I(Me)(CO)_5$ has $6 + (6 \times 2) = 18$ valence electrons.

GENERAL CONSIDERATIONS

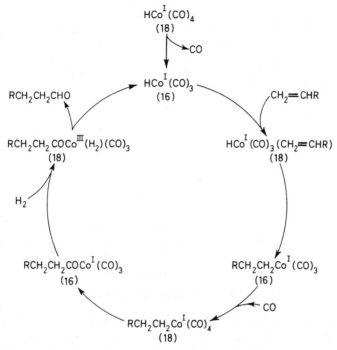

Fig. 1.6 *Catalysis cycle for the cobalt catalysed hydroformylation of a terminal alkene*

Consideration of the number of valence electrons associated with the individual complexes suggests that the latter sequence is the preferred route: a prediction supported by the experimental data. Similarly the rules can be used to assess the feasibility of a proposed catalysis cycle. In Fig. 1.6 the basic catalysis cycle for the cobalt catalysed hydroformylation of a terminal olefin is illustrated. The number of valence electrons attributable to each of the proposed species is shown in parenthesis.

1.7 Summary

In this first section we have attempted to present enough background information to allow us to look at some individual homogeneous catalyst systems presently in use (Chapter 2) or in active development (Chapter 3).

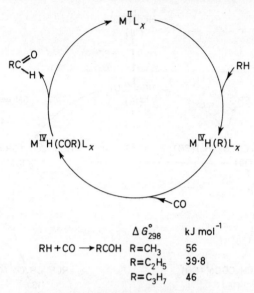

Fig. 1.7 *Hypothetical Catalysis Cycle*

As we have tried to point out, the potential of transition-metal complexes as homogeneous catalysts is enormous and varied. We should nevertheless always be aware of their limitations. A catalyst can only influence a process which in the first place is feasible – it cannot make water flow uphill. Thus the first stage in any search for a suitable catalyst must always be an assessment of the thermodynamical feasibility of the process. In Fig. 1.7 we show a hypothetical catalysis cycle, all the individual steps in the process have some degree of experimental precedent [46,51] below the cycle is the overall reaction proposed in the cycle with the $\Delta G°$ change. The chances of designing a system to catalyse the conversion of alkane/carbon monoxide mixtures to aldehydes would not appear good! This brief mention of thermodynamics brings us back to the theme with which we started this section and means it is time to move on to consider some actual homogeneous catalyst systems in operation.

Further reading

General texts and reviews

(a) *Thermodynamics*

Cox, J. D. and Pilcher, G. (1970), *Thermochemistry of Organic and Organometallic Compounds*, London: Academic Press.

Janz, G. J. (1967), *Thermodynamic Properties of Organic Compounds*, New York: Academic Press.

Stull, D. R., Westrum, E. F. and Sinke, G. C. (1967), *The Chemical Thermodynamics of Organic Compounds*, New York: John Wiley (including excellent compilation of thermodynamic data).

(b) *Transition metal and co-ordination chemistry*

Cotton, F. A. and Wilkinson, G. (1972), *Advanced Inorganic Chemistry*, 3rd Edn, London: John Wiley.

Barnard, A. K. (1965), *Theoretical Basis of Inorganic Chemistry*, London: McGraw-Hill.

Coates, G. E., Green, M. L. H. and Wade, K. (1968), *Organometallic Compounds Volume Two: The Transition Elements* by M. L. H. Green, 3rd Edn, London: Chapman and Hall.

Heck, R. F. (1974), *Organotransition Metal Chemistry, A Mechanistic Approach*, New York: Academic Press.

Pidcock, A. and McAuliffe, C. A. (1973), *Transition Metal Complexes of Phosphorus, Arsenic and Antimony Ligands*, New York: Wiley.

(c) *Homogeneous catalysis*

Schrauzer, G. N. (ed.) (1971), *Transition Metals in Homogeneous Catalysis*, New York: Marcel Dekker.

Ugo, R. (ed.) (1970, 1974, 1977), *Aspects of Homogeneous Catalysis*, Vols. 1, 2 and 3, Dordrecht: Reidel.

Tsuji, J. (1975), *Organic Synthesis by Means of Transition Metal Complexes*, Berlin: Springer-Verlag.

Alper, H. (ed.) (1976, 1978), *Transition Metal Organometallics in Organic Synthesis*, Vols. 1 and 2, New York: Academic Press.

Catalysis, Chem. Soc. Specialist Periodical Report, (1977, 1978), Vols. 1 and 2, London: Chemical Society.

Parshall, G. W. (1978), Industrial applications of homogeneous catalysis. *J. Mol. Catal.*, **4**, 243.

Kochi, J. K. (1978), *Organometallic Mechanisms and Catalysis*, New York: Academic Press.

2 Homogeneous catalyst systems in operation

The last thirty years have seen a steady growth in the use of homogeneous catalyst systems in the chemical industry. Although in terms of the total volumes of chemicals produced processes using heterogeneous catalysts still far outweigh those using homogeneous catalysts, the latter make an important contribution. It has recently been estimated [52] that there are presently more than 20 industrial processes which employ soluble metal compounds as catalyst and that in the United States alone some 8 million tonnes of chemical were manufactured using such processes in 1975. Certain chemicals, notably aldehydes, are almost exclusively produced using homogeneous catalysts. Of the 613 000 tonnes of acetaldehyde consumed in the USA in 1976, over 90% was produced by the palladium catalysed oxidation of ethene, i.e. the Wacker process [53], and in the non-Communist world between 3.5 and 4.5 million tonnes of aldehydes, or direct derivatives, are produced annually using soluble metal carbonyl catalysts [1]. Again in the United States, the three largest American processes for the manufacture of acetic acid, production of which is expected to exceed 1.5 million tonnes (in the USA) by 1980, are all based on homogeneous catalysts [54]. The most recent process developed for producing acetic acid from methanol via carbonylation employs a homogeneous rhodium catalyst. Monsanto, the originators of the process, report that the process has been licensed worldwide and suggest that when these plants come on stream, they will be capable of producing in excess of 1 million tonnes per year [45]. The two major thermoplastic materials, polyethylene and polypropylene, owe much of their success to the discovery by Ziegler in the early 1950s that mixtures of triethyl aluminium and titanium tetrachloride are active catalysts for the polymerization of ethene. In 1976 the world market for polypropylene exceeded 3 million tonnes, all of which was produced using modified Ziegler, or Ziegler–Natta type catalysts. Although there is some question as to the homogeneity of these polymerization systems, there is no doubt that studies on soluble complexes have been important in developing the present generation of commercial catalysts [53].

These examples serve to illustrate the scale of the operations in which homogeneous catalysts play a vital role. Equally important, although considerably smaller in terms of volume, is the use of homogeneous catalysts to effect highly specific organic transformations. Soluble nickel systems catalyse the oligomerization of olefins to give a wide variety of linear and cyclic products. Modification of the nickel complex by the addition of suitable ligands has led to some highly selective catalyst systems and the production of chemicals of potential interest to the pharmaceutical industry [55]. 1-Dihydroxyphenylalanine (1-DOPA) is one of the drugs used in the treatment of Parkinson's disease. A key step in its synthesis, the asymmetric hydrogenation of acetamidocinnamic acid, can be effectively accomplished using a soluble rhodium catalyst having optically active tertiary phosphine ligands [56,57]. In this section we shall discuss eight major areas where homogeneous catalyst systems have proved to be extremely effective. With the exception of the first two, hydrogenation and isomerization and the last, metathesis, major commercial plants in excess of 10 000 tonnes per annum capacity are in operation using homogeneous catalyst systems. Hydrogenation and isomerization are included since, not only are these two of the most thoroughly studied of the processes employing homogeneous catalysts, but they are also of considerable significance in most of the other areas discussed. Metathesis is included because it is one of the most intriguing of the catalytic systems discovered in recent years and, using heterogeneous catalysts, forms the base of an important new process for the manufacture of C_{12}–C_{18} α-olefins. Furthermore, while these particular homogeneous catalyst systems may not yet be involved in large-scale chemical manufacture they nevertheless have considerable, in some cases, e.g. 1-DOPA, demonstrated, potential in the smaller scale manufacture of specific molecules of interest to the pharmaceutical and agrochemical industries.

2.1 Hydrogenation

Hydrogenation, the addition of hydrogen (H_2) to an unsaturated moiety, e.g. an olefin, is one of the most extensively studied reactions involving homogeneous catalysis. Since the discovery by Iguchi in 1939 that certain rhodium(III) complexes will catalyse the hydrogen reduction of organic substrates [58], such as fumaric acid, homogeneous hydrogenation catalysts involving most, if not all, of the *d*-block metals have been reported [19]. Most attention has focussed on the Group VIII elements since it is these which generally give rise to the most active catalyst systems. It is with complexes of these metals that we shall be principally concerned. Homogeneous hydrogenation exhibits many of the features we have discussed in Chapter 1, the unsaturated substrate is generally activated by co-ordination and the hydrogen by addition; systems involving oxidative, homolytic and heterolytic addition of H_2 are known; by careful choice of catalyst system a remarkable degree of selectivity is attainable; many of the catalysts are active under ambient conditions. In the subsequent discussion an attempt has been made to choose catalyst systems which illustrate these various aspects.

Under the heading 'simple hydrogenation' homogeneous hydrogenation catalysts exemplifying the three types of hydrogen activation processes are described with especial reference to the hydrogenation of olefinic substrates. In the sub-section entitled 'selective hydrogenation', systems capable of selectively reducing one specific type of unsaturation or one site of unsaturation in a di- or poly-unsaturated molecule are discussed, and finally, under the heading 'asymmetric hydrogenation', we turn our attention to those catalysts capable of generating optically active molecules by stereospecifically adding hydrogen to a non-optically active substrate.

For a comprehensive account of the whole field of transition-metal catalysed homogeneous hydrogenation the reader is referred to the excellent works of James [19,59].

2.1.1 *Simple hydrogenation*

(a) Hydrogen activation by oxidative addition

In 1965 two groups [60,61] working independently reported that, in an organic solvent, tris(triphenylphosphine)chlororhodium(I),

HOMOGENEOUS CATALYST SYSTEMS IN OPERATION

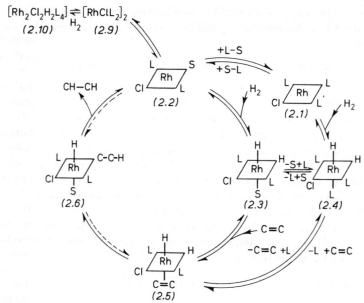

Fig. 2.1 *Catalysis cycle for RhClL$_3$ (L = PPh$_3$) catalysed alkene hydrogenation. S = solvent or possibly loosely bonded tertiary phosphine*

RhCl(PPh$_3$)$_3$, is an active catalyst for the hydrogenation of alkenes and alkynes at 25 °C and 1 atm of hydrogen gas. Since then much experimental effort has gone into characterizing this system. The currently accepted [59] catalysis cycle is shown in Fig. 2.1. The inner circle represents the major catalysis cycle. The active catalytic species appears to contain two tertiary phosphine ligands but how such a species is actually formed from the tri-tertiary phosphine complex (2.1) is still open to question. The original suggestion [62] that (2.1) rapidly dissociates in solution to form the solvated species (2.2) was somewhat discredited when it was shown, using molecular weight [63] and ^{31}P n.m.r. [64] measurements, that, in the absence of oxygen, (2.1) is essentially undissociated ($K \simeq 3 \times 10^{-3}$ M) in benzene solution. This fits in with Tolman's 16/18 electron rule (see Section 1.6) in that the equilibrium so suggested would involve a 16 to 14 electron species transformation;

$$\text{Rh}^{\text{I}}\text{ClL}_3 \rightleftharpoons \text{`Rh}^{\text{I}}\text{ClL}_2\text{'} + \text{L}.$$

NVE   16   14

In the presence of a suitable co-ordinating solvent, e.g. EtOH or indeed O_2, this objection could be overcome via a sequence of the type;

$$Rh^ICIL_3 + S \rightleftharpoons Rh^ICIL_3S \rightleftharpoons Rh^ICIL_2S + L.$$
(2.2)

NVE 16 18 16

In this context it is noteworthy that the rate of hydrogenation of cyclohexene with $RhCl(PPh_3)_3$ as catalyst in a benzene–ethanol (3:1) solvent is about twice that observed in pure benzene [65,66]. Furthermore, in practice it is difficult to rigorously exclude oxygen. Oxidative addition of hydrogen to the solvated species (2.2) gives the 18 electron rhodium(III) dihydrido complex (2.3). ^1H and ^{31}P n.m.r. data indicate that the two hydride ligands are mutually *cis* and also *cis* in relation to the tertiary phosphine ligands [67]. An alternative route to the solvated dihydrido species is via the sequence

$$Rh^ICIL_3 + H_2 \rightleftharpoons Rh^{III}ClH_2L_3 \underset{+L}{\overset{-L}{\rightleftharpoons}}$$

NVE 16 18

$$Rh^{III}ClH_2L_2 \underset{-S}{\overset{+S}{\rightleftharpoons}} Rh^{III}ClH_2L_2S$$

16 18

as shown in the top half of the outer catalysis cycle in Fig. 2.1. However formed, there is little doubt that a formally octahedral rhodium(III) *cis* dihydride species is actively involved in the catalysis cycle. The hydride ligand exerts a large *trans*-effect (see Section 2.1.3a) thus the *trans*-ligand, in either (2.3) or (2.4), is labilized and readily replaced by substrate olefin to give (2.5) in which all reactants necessary to complete the cycle have been brought together in an 'activated' form. Whether the outer or inner right-hand semi-circles in Fig. 2.1 are followed will depend to some extent on the nature of the catalyst precursor, i.e. the rhodium complex initially added to the solution.

Frequently a di-tertiary phosphine rhodium(I) complex is formed *in situ* by adding four mole equivalents of tertiary phosphine (two per rhodium) to a readily prepared rhodium(I) chloro-bridged complex containing labile olefin ligands, e.g. $[RhCl(C_2H_4)_2]_2$ or $[RhCl(C_8H_{14})_2]_2$. In such a case the inner cycle is preferred. When $RhClL_3$ is used initial activation probably occurs via the outer cycle. The next stage in the catalysis cycle is 'proximity interaction' involving hydride migration to the olefin in (2.5) to give a hydride-alkyl rhodium(III) species. Such a migration, visualized as proceeding via a four-centre transition state, can best occur when the hydride and alkene moieties are mutually *cis* [68] and coplanar [69]. Thus starting from (2.5) we have the sequence

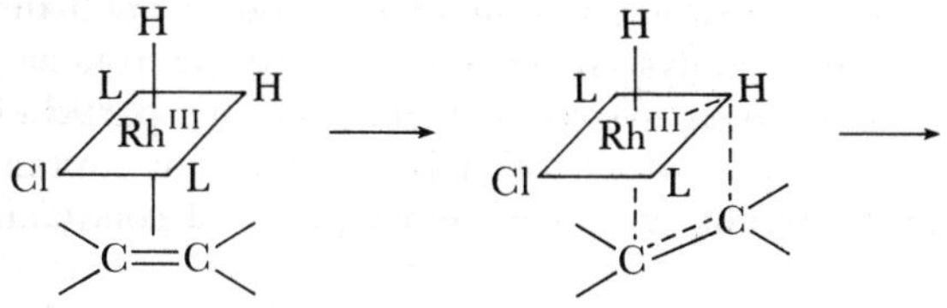

NVE   18   18

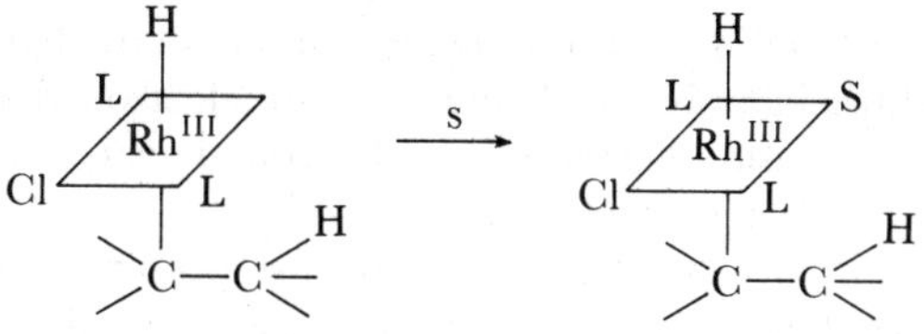

          16   18

In either (2.7) or the solvated species (2.8) the remaining hydride and alkyl units are mutually *trans* and before the final stage of the cycle can occur, they must move into mutually *cis* positions. Whether this isomerization occurs via (2.8) or directly from (2.7) is not clear since the resulting alkyl hydrido species (2.6) rapidly undergoes a 'reductive elimination' reaction giving product alkane and completing the cycle to the solvated rhodium(I) complex (2.2). The rate determining step, which occurs between (2.5) and (2.6), is thought to be the hydride to olefin migration reaction [70]. As indicated in the top left-hand corner of Fig. 2.1, the solvated di-tertiary phosphine species (2.2) can also dimerize to form a chloro-bridged dirhodium complex (2.9) which can rapidly add hydrogen to give (2.10). Formally, as indicated below, this dihydrido species contains rhodium in oxidation states (I) and

$$\begin{array}{c} \text{L} \\ \text{L} \diagdown \text{Cl} \diagup \text{H} \\ \text{Rh}^{\text{I}} \diagup \text{Rh}^{\text{III}} \\ \text{L} \diagup \text{Cl} \diagdown \text{H} \\ \text{L} \end{array}$$

(2.10)

(III). Solvent assisted bridge splitting would convert *(2.10)* into a mixture of *(2.2)* and *(2.3)* which accounts for the observation that *(2.9)* is also an active catalyst precursor. The concentration of dimeric species in a given system depends on the particular reaction conditions, e.g. nature of solvent, type of tertiary phosphine ligand. Generally, however, under practical conditions it is low.

As mentioned above, when L = PPh$_3$, the rate determining step is thought to involve ligand migration, i.e. hydride to olefin, in going from the dihydride olefin complex *(2.5)* to the hydrido alkyl complex *(2.6)*. Such migrations can be visualized as a nucleophilic attack of a hydrido ligand on an activated double bond [44]. Thus, ligands which increase the electron density on the hydrido ligand and decrease the electron density in the olefinic double bond should, neglecting steric effects, increase the rate of reaction. Since both hydrido and olefin are present in the same complex, a ligand which exerts opposite effects on the two is a tall, if not impractical, order. Results with substituted aryl phosphine ligands have shown that increasing the electron density on the metal increases the rate of hydrogenation. Thus RhClL$_3$, with L = P(p-methoxyphenyl)$_3$, gives rise to a catalyst which is nearly twice as efficient at hydrogenating cyclohexene than one with L = PPh$_3$. Replacing PPh$_3$ by P(p-fluorophenyl)$_3$ decreases the rate by a factor 5 [71]. Similarly, replacing the chlorine ligand by either bromine or iodine increases the overall rate of reaction. For the catalytic hydrogenation of cyclohexene in benzene using RhX(PPh$_3$)$_3$ the rate increases roughly in the ratio 1:2:3 for X = Cl, Br, I [62]. These results indicate that in this particular ligand migration reaction increasing the electron density at the metal centre, i.e. effectively increasing the hydridic character of the H ligand, is the dominant rate-enhancing factor. It is not, however, possible to go too far down this road in increasing the activity of the catalyst system since analogous complexes with alkylarylphosphines of the type PR$_n$Ar$_{3-n}$ or with trialkylphosphine ligands, both of which would be expected to increase the electron density at the metal centre, show very low catalytic activity [71,72]. This highlights one of the problems encountered in attempting to modify a catalyst which is involved in a complex catalysis cycle as illustrated in Fig. 2.1 – by tuning one part of the cycle you frequently snarl up some other part. In this case increasing the basicity of the tertiary phosphine ligand increases the stability of complexes of the type (*2.4*) to such an extent that they no longer take part in the catalysis cycle. Thus $RhClH_2(PEtPh_2)_3$, (*2.4*) with L = $PEtPh_2$, is a stable readily isolable complex [71]. Under conditions where (*2.4*) cannot be formed, i.e. when the Rh:L ratio = 1:2, the dimeric species (*2.9*) is formed and, with L = PR_3 or PR_nAr_{3-n} ($n \geq 1$), this species shows negligible catalytic activity. Thus in this case, as in most of the systems we shall be looking at, it is necessary to tread carefully when attempting to modify the catalytic activity often accepting being wise after, rather than before, the event.

Considering species (*2.5*) with L = triarylphosphine, we can see that the complex is fairly crowded (PPh_3 has a cone angle of 145°, see Section 1.2.3*b*). Thus we would expect the reaction to be very susceptible to steric parameters. This is indeed the case since increasing the steric hindrance about the C=C bond drastically decreases the rate of hydrogenation. Thus 1-methylcyclohexene is reduced some 50 times more slowly than cyclohexene, *trans*-olefins are reduced slower than *cis*-olefins and internal olefins are reduced much more slowly than terminal olefins [71,73]. Similarly, increasing the bulk of the tri-arylphosphine reduces the hydrogenation activity of the catalyst. The system with L = tri-orthotolylphosphine (cone angle = 194°) is some 700 times less active than that with L = tri-paratolylphosphine (cone angle = 145°) [19,26]. We have mentioned the importance of steric parameters in the olefinic substrate. Electronic parameters are also important; electron-withdrawing substituents on the double bond, provided they are not too numerous or too big, generally increase the rate of hydrogenation. Thus acrylonitrile and allyl acetate are reduced somewhat (*ca.* 30%) more rapidly than a non-substitute α-olefin such as hex-1-ene [74]. This electronic effect is readily rationalized in that decreasing the electron density in the double bond will make it more susceptible to nucleophilic attack in the rate-determining hydride to olefin migration reaction.

A particular advantage of the RhCl(PPh$_3$)$_3$ system is that under mild conditions, typically ambient temperature in benzene or benzene–ethanol solution with 1 atm pressure of hydrogen gas, it does not catalyse the reduction of other functional groups present in the olefinic substrate. Groups such as keto, cyano, nitro, aryl, carboxylic and indeed most commonly occurring, non-olefinic, organic functions are unaffected by the RhCl(PPh$_3$)$_3$/H$_2$ catalyst systems. This fact, coupled with the steric restrictions previously mentioned, makes RhCl(PPh$_3$)$_3$ an extremely selective hydrogenation catalyst having considerable potential in the preparation of complex organic molecules. However, before going on to expand on this selectivity potential in Section 2.1.2 we must first consider the second type of hydrogen activation process, namely homolytic addition.

(b) Hydrogen activation by homolytic addition

This type of activation process is exemplified by the pentacyanocobalt(II) catalyst system, first reported by Iguchi in 1942. An aqueous, or aqueous alcoholic (MeOH or EtOH), solution of cobalt cyanide rapidly absorbs hydrogen under ambient conditions to give a species capable of catalysing the hydrogenation of activated alkenes of the type

$$\mathrm{\underset{}{\overset{}{>}}C=C\underset{A}{\overset{}{<}}}$$

where A = CH=CH$_2$, —CO$_2$H, —CO$_2^-$, —CN, —C(O)NH$_2$, —C(O)R, aryl, α-C$_5$H$_4$N [17]. As far as organic substrates are concerned the system is highly selective in the sense that it is totally unreactive towards simple monoalkenes. Thus conjugated dienes can be selectively reduced to monenes, e.g. butadiene to butene. The principal catalytic species is the pentacyano-hydridocobalt anion formed via the homolytic cleavage of hydrogen:

$$2\text{Co(CN)}_5^{3-} + \text{H}_2 \longrightarrow 2\text{HCo(CN)}_5^{3-}. \qquad (2.1)$$

(2.II)

The 16/18 electron rule, applying as it does to diamagnetic complexes, is of limited use with this particular system since Co^{II}(CN)$_5^{3-}$ is a d^7, 17 electron, paramagnetic ion.

A number of groups [19] have studied the formation of the

hydrido complex, (*2.11*), from cobalt cyanide and hydrogen and two mechanisms involving dimeric species of the type, $[Co_2^{II}(CN)_{10}]^{6-}$ and $[(CN)_5CoH_2Co(CN)_5]^{6-}$, have been proposed [36,77]. However, most data [19] seem to favour a simple termolecular reaction as implied by Equation (2.1). Following the formal convention of treating co-ordinated H as H^- then the oxidation state of cobalt in (*2.11*) is three. It has, however, been suggested [78,79] on the basis of polarographic studies that it may be more accurate to look upon the cobalt in (*2.11*) as being in an oxidation state of two with the hydrogen being present as a stabilized atom. Thus it is proposed that (*2.11*) should be represented as $[Co^{II}(CN)_5(.H)]^{3-}$ rather than $[Co^{III}(CN)_5(:H)]^{3-}$. Certainly such a representation would fit in with the description of the activation process as 'homolytic' addition but more importantly, as will become clear when we consider the hydrogenation cycle, it conforms closely with the nature of the reactions involving $HCo(CN)_5^{3-}$. The question of 'actual' versus 'formal' oxidation state is frequently raised in organo-transition-metal chemistry. In most cases formal oxidation states are of sufficient conceptual use to justify their adoption. However, in this particular instance it is probably better to discard the formalism and think of (*2.11*) as $[Co^{II}(CN)_5(.H)]^{3-}$. To avoid any further confusion with the earlier formal definition of homolytic addition (Section 1.4.2*b*) we have refrained from designating the oxidation state of the cobalt in the following discussion.

The catalysis cycle for the $[Co(CN)_5H]^{3-}$ catalysed hydrogenation of activated olefins is shown in Fig. 2.2. The initial interaction of (*2.11*) with the activated olefin can be visualized as occurring via a four-centre transition state (*2.12*), in which the hydrogen atom is transferred to the carbon atom β to the activating group, A. Deuteration studies [77,80], together with the isolation of organo–cobalt complexes of the type (*2.13*), A = PhCO, α-C_5H_4N, have confirmed the direction of addition [80–82]. The cycle the system now follows depends on the nature of the activated olefin. Conjugated dienes and polyenes follow cycle (I) while most other activated substrates are hydrogenated via cycle (II) [83]. Below we consider two representative substrates, butadiene and cinnamic acid.

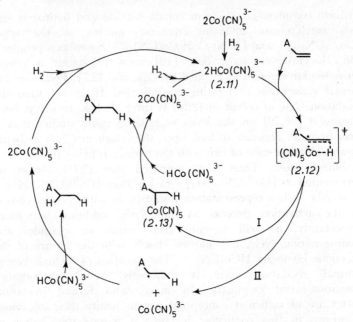

Fig. 2.2 *Catalysis cycle for the $Co(CN)_5^{3-}$ catalysed hydrogenation of activated alkenes ($A = CH=CH_2, CO_2^-, CO_2H, CN, Ph, \alpha\text{-}C_5H_4N$)*

(i) Hydrogenation via cycle (I) – butadiene

With a conjugate diene, such as buta-1,3-diene, an alkyl-cobalt species of the type (*2.13*) is formed via formal *cis* addition of the cobalt hydride moiety to the activated double bond. The nature of the monoene product formed via cycle (I) depends on the concentration of free cyanide present in the reaction medium. In the case of butadiene, when the hydrogenation is carried out in the presence of excess cyanide, i.e. when the cyanide to cobalt ratio is greater than 5:1, but-1-ene is the major (*ca.* 90%) product: when little free cyanide is present, i.e. the cyanide to cobalt ratio is less than 5, the major product (*ca.* 80%) is *trans*-but-2-ene. In both cases only a small amount (*ca.* 5%) of *cis*-but-2-ene is formed [84]. This dependence of product on cyanide concentration can be rationalized using the sequence shown in Fig. 2.3 [77,82,84]. Addition of $CoH(CN)_5^{3-}$ to butadiene initially gives, via (*2.12*), $A = -CH=CH_2$, the σ complex (*2.14*) in which the cobalt is co-ordinated to carbon 3, (*2.13*) $A = -CH=CH_2$. Loss of

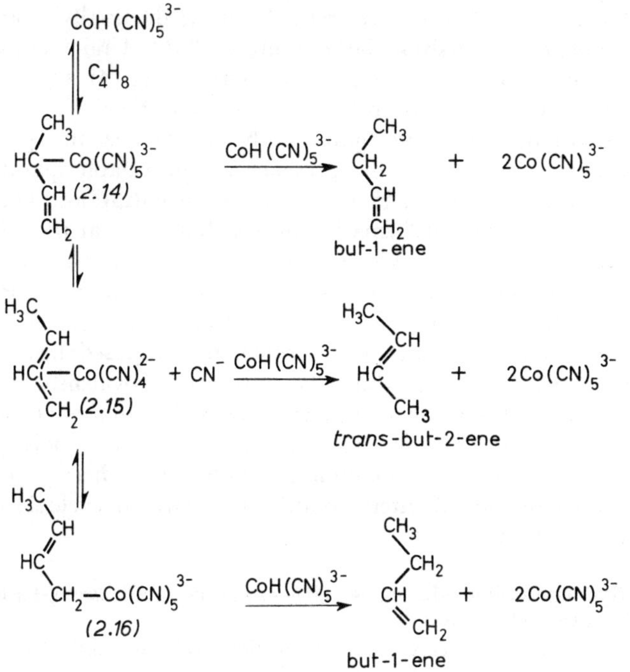

Fig. 2.3 *Sigma-pi-sigma equilibria during the $CoH(CN)_5^{3-}$ catalysed hydrogenation of buta-1,3-diene*

cyanide from (2.14) creates a vacant co-ordination site in the octahedral complex, the uncoordinated double bond starts to interact at this vacant co-ordination site and the C_4 unit transforms from being a σ-bonding ligand, occupying one co-ordination site, into being a π-bonding ligand, effectively occupying two co-ordination sites in the complex (2.15), with the methyl group pointing away, i.e. positioned *syn*, from the cobalt. This type of σ–π transformation which is extremely common in Group VIII metal complexes containing an allyl unit – we can regard the C_4 unit in (2.15) as a methyl substituted allyl – is further discussed in Section 2.2 when we consider isomerization. Addition of CN^- back to (2.15) can either reform (2.14) or form (2.16) in which the cobalt is now σ-bonded at C–4. Although the detailed mechanism by which $CoH(CN)_5^{3-}$ interacts with the σ and π species, (2.14) to (2.16), is not known, detailed spectroscopic [82] and kinetic studies [80,85] have shown that *trans*-but-2-ene is formed preferentially from the π-allyl complex (*2.15*) and but-1-ene from the σ complex (*2.16*). Under both sets of reaction conditions, only small amounts of (*2.14*) are present but that which is present probably gives rise to but-1-ene. Clearly the formation, and more importantly, as far as the product distribution is concerned, the presence in appreciable quantities, of (*2.15*) will be favoured by low cyanide concentrations. This fits in with the preferential formation of *trans*-but-2-ene at the CN:Co ratio of less than 5:1. The system is highly selective towards reduction of the butadiene and no further hydrogenation, or isomerization, of the product butenes occurs.

In the above case, all the intermediates proposed have been organo–cobalt species in which the substrate is covalently bonded to the cobalt. With the cobalt cyanide hydrogenation catalysts this is only true when the substrate is a conjugated polyene or diene. As previously mentioned, with all other activated substrates free radical intermediates, as shown in cycle II (Fig. 2.2) are involved.

(ii) Hydrogenation via cycle (II)–cinnamic acid (*trans*-1,phenyl 2, carboxyl-ethene)

In the cobalt-catalysed hydrogenation of cinnamic acid to β-phenylpropionic acid [77,86,87], initial hydrogen atom transfer, probably via a four-centre transition state of the type (*2.17*), (*2.12*) in Fig. 2.2 with A = carboxyl, gives rise to the phenylcarboxy radical, (*2.18*). Hydrogen abstraction/transfer from a second molecule of $CoH(CN)_5^{3-}$ gives the product (Equation (2.2))

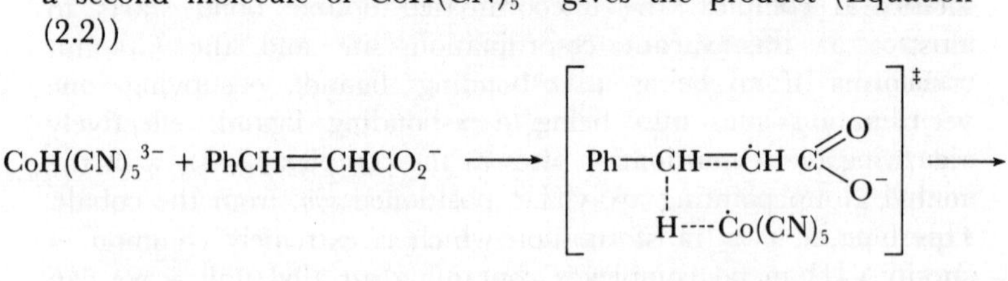

$$CoH(CN)_5^{3-} + PhCH{=}CHCO_2^- \longrightarrow \left[\begin{array}{c} Ph{-}CH{=}\dot{C}H\ C\overset{\displaystyle O}{\underset{\displaystyle O^-}{\diagup}} \\ | \\ H{\cdots}\dot{Co}(CN)_5 \end{array} \right]^{\ddagger} \longrightarrow$$

(*2.17*)

$$PhCH_2\dot{C}HCO_2^- + Co(CN)_5^{3-}$$

(*2.18*)

$$PhCH_2\dot{C}HCO_2^- + CoH(CN)_5^{3-} \longrightarrow PhCH_2CH_2CO_2^- + Co(CN)_5^{3-}.$$

(2.2)

Kinetic studies [80,87] together with radical trapping experiments [84], using this and related systems, have provided ample evidence for the occurrence of this radical pathway. The radical intermediate (*2.18*) may form an organo–cobalt complex of the type (*2.19*). Such species are not directly involved in the catalysis

$$\text{PhCH}_2\text{CHCO}_2^- \atop | \atop \text{Co(CN)}_5^{3-}$$

(*2.19*)

cycle; indeed, substrates which can form stable σ-complexes, e.g. acrylic, maleic and fumaric acid anions, effectively tie up the cobalt and inhibit the reduction [81].

Unlike the rhodium system, considered under hydrogen activation by oxidative addition, the cobalt cyanide hydrogenation catalyst is not readily amenable to tailoring via ligand modification. It does, however, amply make up for this possible shortcoming with its high selectivity towards activated olefins.

The third type of halogen activation process, i.e. activation by heterolytic addition, is exemplified by the behaviour of certain ruthenium(II) catalyst systems.

(*c*) *Hydrogen activation by heterolytic addition*
Aqueous hydrochloric acid solutions of ruthenium(II) chloride catalyse the hydrogenation of α,β unsaturated carboxylic acids or amides, e.g. maleic, fumaric, acrylic acids or acrylamide, but are ineffective when it comes to reducing simple olefins, or indeed most other organic substrates [88]. While the nature of the ruthenium(II) chloro complex active in the catalysis cycle is not yet clear, it is reasonably certain that hydrogen activation occurs via heterolytic addition – probably to a ruthenium(II) olefin complex. A catalysis cycle consistent with the presently available experimental data is shown in Fig. 2.4. No attempt is made to specify the nature of the ruthenium(II) species. The presence of a strongly electron-withdrawing substituent, e.g. —COOH or —CONH$_2$, is necessary to activate the alkene towards hydrogenation. Non-substituted olefins such as ethene or propene readily form 1:1 complexes (*2.20*) with ruthenium(II) in the aqueous hydrochloric acid media. Such complexes are, however, relatively stable and are not hydrogenated [19]. It is suggested that step (II) in the cycle is reversible and that the ligand

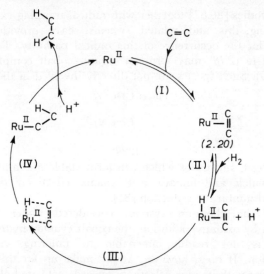

Fig. 2.4 *Catalysis cycle for the ruthenium(II) catalysed hydrogenation of activated olefins in aqueous hydrochloric acid solution*

migration (H to olefin) implied in steps (III) and (IV) can only occur when the olefin is highly activated towards nucleophilic attack. In this system, in contrast to the cobalt system previously discussed, the H ligand appears to react very much as H^-. Deuteration studies have shown the H_2 addition to the olefin occurs stereospecifically *cis* indicating that electrophilic attack by H^+ on the ruthenium–alkyl species occurs with retention of configuration at the metal bonded carbon atom.

Unfortunately, owing to the experimental difficulties encountered in working with this particular system, it has proved difficult to characterize many of the individual species involved and there is, therefore, considerable uncertainty surrounding the details of the catalysis cycle. Another ruthenium (II) system, where heterolytic addition of hydrogen is thought to be important, is that formed when ruthenium(II) complexes of the type $RuCl_2(PPh_3)_n$, $n = 3$ or 4, react with hydrogen at room temperature in benzene–ethanol solution [40]. The resulting hydrido complex, $RuCl(H)(PPh_3)_3$, is one of the most active homogeneous catalysts yet discovered for the hydrogenation of alk-1-enes. It will also catalyse the reduction of nitro compounds

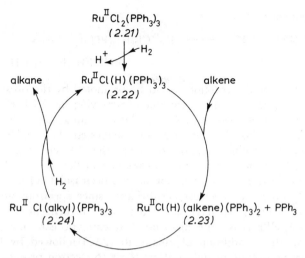

Fig. 2.5 Catalysis cycle for the $Ru^{II}Cl(H)(PPh_3)_3$ catalysed hydrogenation of alkenes

to amines [89–91], and aldehydes to alcohols [92]. In spite of being much more amenable to detailed study, e.g. by ^1H and ^{31}P n.m.r. techniques, than the $RuCl_2/HCl/H_2O$ system, there is still some degree of uncertainty surrounding the detailed catalysis cycle involving $RuCl(H)(PPh_3)_3$. The cycle illustrated in Fig. 2.5 is merely one which serves to rationalize most of the currently available data.

The active species is believed to be the hydrido-chlorotris(triphenylphosphine)ruthenium(II) complex, (2.22), formed by the heterolytic addition of hydrogen to $RuCl_2(PPh_3)_3$

$$Ru^{II}Cl_2(PPh_3)_3 + H_2 \longrightarrow Ru^{II}Cl(H)(PPh_3)_3 + H^+ + Cl^-.$$
(2.21) (2.22)

Formally this process involves no change in either the co-ordination number or oxidation state of the ruthenium with both (2.21) and (2.22) being penta-co-ordinate d^6 species having 16 valence electrons. However, as pointed out in Section 1.4.2c, experimentally it is often extremely difficult to distinguish between heterolytic addition and oxidative addition followed by reductive elimination. (2.22) could equally well be produced by the sequence

$$Ru^{II}Cl_2(PPh_3)_3 + H_2 \longrightarrow Ru^{IV}Cl_2(H)_2(PPh_3)_3 \longrightarrow$$
$$Ru^{II}Cl(H)(PPh_3)_3 + H^+ + Cl^-.$$

In both cases the reaction would be promoted by the presence of a base, e.g. triethylamine, capable of removing the HCl. However formed, it does seem that (2.22) plays an important role in the catalytic process. It was originally suggested that (2.22) lost a tertiary phosphine ligand, to give the 14 electron species '$Ru^{II}Cl(H)(PPh_3)_2$', prior to co-ordination of the alkene substrate. This 14 electron species has not, as yet, been isolated [93, 94] nor directly observed spectroscopically and, since spectrophotometric measurements [94] have indicated that in benzene at 25 °C $RuCl(H)(PPh_3)_3$ is little dissociated at catalytic concentration of 10^{-3} M, direct addition of alkane to (2.22) followed by loss of tertiary phosphine ligand, a $16 \rightarrow 18 \rightarrow 16$ electron process seems more plausible [59]. In strongly co-ordinated solvents displacement of PPh_3 by solvent prior to alkene co-ordination, a process analogous to that described for the $RhCl(PPh_3)_3$ catalyst system, may well occur. A hydrido to alkene ligand migration, probably via a four-centre transition state (2.25), converts the hydrido–alkene complex (2.23) into the ruthenium-alkyl species (2.24) with reassociation of a tertiary phosphine or solvent molecule.

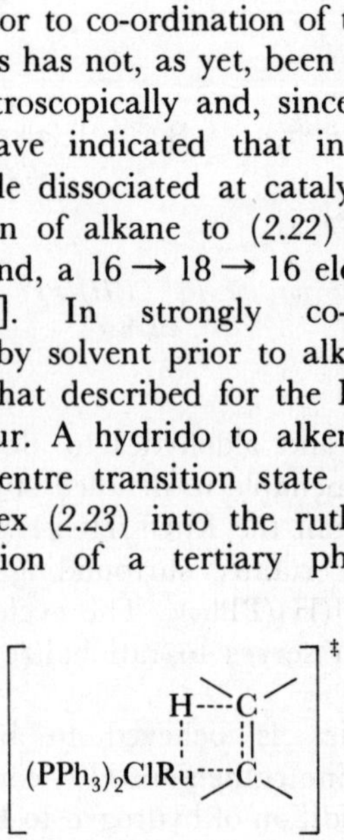

$$\left[\begin{array}{c} H\text{----}C \\ \vdots \quad \| \\ (PPh_3)_2ClRu\text{----}C \end{array} \right]^\ddagger$$

(2.25)

Hydrogenolysis of (2.24) gives the product alkane and regenerates the catalytic species $RuClH(PPh_3)_3$, (2.22). This last part of the catalysis cycle, which includes the rate-determining step, could occur via heterolytic cleavage of hydrogen

$$Ru^{II}Cl(alkyl)(PPh_3)_3 + H_2 \longrightarrow Ru^{II}(H)(alkyl)(PPh_3)_3 + H^+ + Cl^- \longrightarrow$$
$$Ru^{II}Cl(H)(PPh_3)_3 + \text{Halkyl}$$

but more probably occurs via the oxidative addition/reductive elimination sequence shown in Fig. 2.6 [19], with the

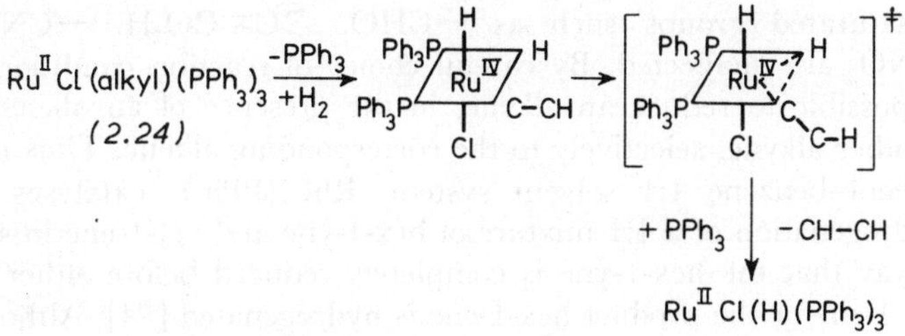

Fig. 2.6 *Oxidative addition/reductive elimination sequence leading to the elimination of alkane from chloro(alkyl)tris(triphenylphosphine)ruthenium(II)*

rate-determining step being the oxidative addition of hydrogen to the ruthenium(II) complex *(2.24)*.

The three systems discussed have been chosen to illustrate the three formal types of hydrogen activation found in homogeneous hydrogenation reactions catalysed by complexes of the *d*-block elements. While including some of the more important catalyst systems we have done no more than scratch the surface of the myriad of homogeneous systems presently known [19]. Up to now, we have tended to concentrate on mechanistic aspects: in the next section we concentrate more on practical applications.

2.1.2 *Selective hydrogenation*

Homogeneous systems capable of catalysing the reduction of most commonly occurring unsaturated organic groups under relatively mild conditions are now known [19,95]. It is, however, the selectivity of such systems, i.e. their ability to reduce one particular site of unsaturation in the presence of other sensitive or potentially reducible sites, rather than their universality, which makes them particularly useful in synthetic organic chemistry. We have already touched briefly on this aspect when we discussed the $RhCl(PPh_3)_3$ system. Here we look a little more closely at some of the types of selective transformations possible.

(a) Hydrogenation of simple alkenes and alkynes
Homogeneous catalysts capable of selectively reducing alkenes and alkynes in the presence of other function groups are $RhCl(PPh_3)_3$, $RuCl_2(PPh_3)_3$ and $RhH(CO)(PPh_3)_3$. Under mild conditions, 25 °C and 1 atm of hydrogen pressure, all three are highly selective towards the alkene or alkyne function. Other

unsaturated groups such as —CHO, $>$C=CO$_2$H, —CN or —NO$_2$ are unaffected. By careful choice of reaction conditions it is possible to reduce an alkyne, in the presence of an alkene or another alkyne, selectively to the corresponding alkene. Thus, in a phenol–benzene 1:1 solvent system, RhCl(PPh$_3$)$_3$ catalyses the hydrogenation of a 1:1 mixture of hex-1-yne and oct-1-ene in such a way that the hex-1-yne is completely reduced before either the oct-1-ene or the product hex-1-ene is hydrogenated [74]. Although the absolute rates of alkyne hydrogenation are generally lower than those of corresponding to alkene hydrogenation, in alkene–alkyne mixtures the alkyne function is generally preferentially reduced because of its greater bonding ability with the metal centre [19,85].

All three catalyst systems are extremely sensitive to steric changes in the substrate and have a distinct preference for terminal alkenes or alkynes. This is especially marked for RuCl$_2$(PPh$_3$)$_3$, the most active of the three, which catalyses the hydrogenation of terminal alkenes at a rate approximately 10 000 times that of the corresponding internal olefins [96]. RhCl(PPh$_3$)$_3$ is slightly less selective than RuCl$_2$(PPh$_3$)$_3$, although it still prefers alk-1-enes, and oct-2-ene is hydrogenated at approximately a quarter of the rate of oct-1-ene [74]. The general trend of reduction rate is alk-1-enes > *cis*-alk-2-enes ≫ *trans*-alk-2-enes > *trans*-alk-3-enes. As far as cyclic substrates are concerned, exocyclic double bonds are usually reduced more rapidly than endo-cyclic ones. RhH(CO)(PPh$_3$)$_3$ also shows high selectivity for the reduction of terminal olefins including non-conjugated di-olefins [97].

These selectivity effects, which can be rationalized in terms of the steric bulk of the tertiary phosphine ligands, have been utilized to great effect in selective hydrogenations, e.g. nootkatone (*2.26*) can be hydrogenated to 11,12-dihydronootkatone (*2.27*) with 95% selectivity using RhCl(PPh$_3$)$_3$ in benzene at 25 °C [98] and the steroid androsta-1,4-diene-3,17-dione (*2.28*) can be reduced to the corresponding 4-ene (*2.29*) with over 90% selectivity using RuCl$_2$(PPh$_3$)$_3$ in benzene at 40 °C [99].

(2.28) →[RuCl$_2$(PPh$_3$)$_3$][H$_2$/C$_6$H$_6$] (2.29)

(b) Hydrogenation of conjugated dienes, polymers and activated monenes
As previously mentioned, the catalyst system formed when cobalt cyanide is dissolved in aqueous hydrochloric acid is highly specific for the hydrogenation of conjugated dienes, polyenes and certain activated monoenes. This system does, however, have the disadvantage of being normally prepared in aqueous solution in which many organic substrates are only sparingly soluble. Other homogeneous catalysts, which selectively reduce dienes to monoenes, are generally less active but three which have the advantage of functioning in organic solvents are Fe(CO)$_5$ [100], (η^5-C$_5$H$_5$)M(CO)$_3$H (M = Cr, Mo or W) [101] and *trans*-Pt(SnCl$_3$)H(PPh$_3$)$_2$ [102]. Typical reaction conditions involve temperatures in excess of 100 °C and hydrogen pressures of the order of 30 atm. The iron and Group VI*a* metal carbonyl systems are reportedly activated by irradiation with ultraviolet light [103–108]. Cr(CO)$_3$(MeCN)$_3$, on u.v. irradiation, catalyses the hydrogenation of conjugated dienes to *cis*-monoenes under ambient conditions of temperature and pressure [107]. The platinum–tin system requires a hydrogen pressure in excess of 30 atm and a temperature in the range 90–110 °C. Under such conditions dienes are reduced to monoenes with reasonable selectivity, e.g. hepta-1,5-diene to heptenes 91%, hexa-1,5-diene to hexenes 70% [109]. Such selectivity is not, however, found with all diene systems [110]. Given the restriction of the aqueous, or aqueous alcoholic, solvent system cobalt cyanide remains the catalyst of choice as far as selectivity is concerned.

(c) Hydrogenation of aromatic and heterocyclic compounds
Most homogeneous catalyst systems are ineffectual when it comes to reducing aromatic hydrocarbons. For industrial applications some use has been made of Ziegler systems, i.e. combinations of transition-metal salts, generally halides, with alkyls and chloro-alkyls of aluminium. Several of these systems are highly active, e.g. Co(2-ethylhexanoate)$_2$/AlEt$_3$ will reduce phenol to cyclo-hexanol with greater than 90% selectivity at 25 °C and *ca.*

40 atm H_2 in heptane [111]. Unfortunately, as we shall discuss in Section 2.6 when we consider polymerization, these systems are difficult to characterize and it is not always clear whether we are dealing with a homogenous or heterogeneous catalyst system. Under forcing conditions, e.g. *ca.* 200 °C, 150 atm, in the presence of carbon monoxide, i.e. hydroformylation conditions, (see Section 2.4.1) $Co_2(CO)_8$ gives rise to a complex, probably $CoH(CO)_4$, capable of catalysing the hydrogen reduction of a wide range of aromatic and heterocyclic compounds [112]. As found with the $Co(CN)_5^{3-}$ catalysed hydrogenation of activated olefin, these aromatic hydrogenation reactions are thought to involve free radical intermediates rather than organo–cobalt complexes [113]. Allylcobalt(I) complexes of the type η-$C_3H_5CoL_3$ (L = tertiary phosphine or phosphite) have been reported [114] to catalyse the reduction of aromatics to cyclohexanes under ambient conditions. Such catalysts generally have an extremely limited life and low activity (*ca.* 15 catalyst turnovers in 24 h). A more promising system is that formed when $[RuCl_2(\eta^6\text{-}C_6Me_6)]_2$ is treated with sodium carbonate in propan-2-ol. The resulting hydrido-chloro-bridged dimer, $[(\eta^6\text{-}C_6Me_6)Ru(\mu\text{-}H)_2(\mu\text{-}Cl)Ru(\eta^6\text{-}C_6Me_6)]Cl$, catalysed the hydrogenation of benzene and substituted aromatics at 50 °C and 50 atm with catalyst turnover numbers in excess of 9000 [115].

(d)   Hydrogenation of aldehydes and ketones

Keto hydrogenation catalysts based on cobalt carbonyl are the most extensively studied because of their involvement in hydroformylation reactions (see Section 2.4). Under typical hydroformylation conditions, *ca.* 130 °C with a CO/H_2 pressure in the range 200–350 atm, $CoH(CO)_4$, formed *in situ* from $Co_2(CO)_8$, will catalyse the conversion of alkenes to aldehydes. At higher temperatures, in the range 160–350 °C, the aldehydes are hydrogenated to alcohols [116]. A plausible mechanism [117] for this hydrogenation involves nucleophilic attack of the hydrido ligand at carbonyl carbon via a four-centre transition state (*2.30*).

$$\left[\begin{array}{c} RCH \!=\! O \\ | \quad\quad | \\ H \text{------} Co(CO)_4 \end{array} \right]^{\ddagger}$$

(*2.30*)

Given such a transition state, increasing the hydridic nature of the cobalt hydrogen, by replacing the cobalt carbonyl ligands by more basic ligands, should increase the ease of hydrogenation. This proved to be the case, thus $CoH(CO)_3(PBu^n_3)$ will catalyse the hydrogenation of aldehydes to alcohols at $ca.$ 150 °C and 30 atm CO/H_2 pressure. The activity of the catalyst depends on electron donor ability of the tertiary-phosphine ligand, with $CoH(CO)_3(PPh_3)$ being markedly less active than $CoH(CO)_3(PBu^n_3)$ [118,119]. The analogous rhodium carbonyl systems show similar behaviour although they are generally more active than the cobalt catalysts. Both of these systems will be discussed in more detail when we consider hydroformylation in Section 2.4.

Cationic rhodium complexes of the type $[Rh(diene)L_2]^+$, where L = tertiary-phosphine, are considerably more active as regards keto hydrogenation. The active species is thought to be $[RhH_2L_2S_x]^+$, where S = solvent, formed by hydrogenation of the diene ligand and oxidative addition of hydrogen to the resulting rhodium(I) complex. With L = PMe_2Ph the complex will catalyse the reduction of acetone to propan-2-ol at 25 °C and 1 atm [120]. Similarly, rhodium complexes of the type $RhCl_2(BH_4)(DMF)py_3$ in DMF are reported to catalyse the hydrogenation of ketones under ambient conditions [121].

(e) *Hydrogenation of nitro groups*

As is the case with aromatic and hetrocyclic compounds, hydrogenation of nitro groups, using homogeneous catalysts, generally requires somewhat forcing conditions, i.e. temperatures in excess of 100 °C and hydrogen pressures in excess of 40 atm. Having said this the rhodium-borohydrido complex mentioned above will reduce nitrobenzene to aniline at room temperature and 1 atm H_2 [122]. An extremely selective catalyst for the hydrogenation of aliphatic and aromatic nitrocompounds to amines is $RuCl_2(PPh_3)_3$. In the presence of a base, this complex will catalyse the hydrogenation of aromatic nitro compounds without either hydrogenating the ring or affecting other substituents such as CN, CO_2R, OR or halide [89,90].

In the above brief discussion we have attempted to illustrate the scope, as well as the selectivity, of homogeneous hydrogenation catalysts. However, no discussion of selectivity could be complete without some mention of the most selective aspect of such catalyst systems – their ability to generate a chiral product from a prochiral substrate by asymmetric hydrogenation.

2.1.3 *Asymmetric hydrogenation*

Asymmetric synthesis [59,123,124] is the process by which an achiral unit, or collection of such units, is converted into a chiral entity in such a way that the resulting stereoisomeric products, or enantiomers, are obtained in unequal amounts. The efficiency of an asymmetric synthesis is measured in terms of the percentage excess of one enantiomer over the other, i.e. per cent enantiomeric excess (%ee) where %ee = %R − %S. Per cent enantiomeric excess is synonymous with the more commonly quoted term optical purity [123]. Asymmetric synthesis is particularly important in the pharmaceutical and agrochemical industries since it is frequently found that the two optical isomers of the same molecule show markedly different pharmacological and biological activity [125,126]. We have already mentioned the example of 1-DOPA [56,57] in the treatment of Parkinson's disease. A similar optical preference has recently been found with synthetic pyrethroid pesticides [127].

There are basically two ways of accomplishing an asymmetric synthesis: either a second chiral centre is created in a molecule under the influence of an existing chiral centre, e.g.

$$\text{Ph}-\underset{\underset{\text{O}}{\|}}{\text{C}}-\underset{\underset{\text{O}}{\|}}{\text{C}}-\text{O}-\text{C}^*\text{RR}^1\text{R}^2 \xrightarrow[\text{2. H}^+]{\text{1. MeMgI}} \text{Ph}-\underset{\text{Me}}{\overset{\text{OH}}{\underset{|}{\text{C}^*}}}-\underset{\underset{\text{O}}{\|}}{\text{C}}-\text{O}-\text{C}^*\text{RR}^1\text{R}^2$$

or a chiral reagent or catalyst acts on a prochiral substrate to create a new chiral centre. Asymmetric hydrogenation, as shown in Equation (2.3), is the best developed example of the latter process involving a catalytic system

$$\text{RCH}=\text{CR}^1\text{R}^2 + \text{H}_2 \xrightarrow[\text{catalyst}]{\text{chiral}} \text{RCH}_2\text{C}^*\text{HR}^1\text{R}^2. \qquad (2.3)$$

When the chiral catalyst is a transition-metal complex the chirality can be introduced in essentially two ways, i.e. the metal itself can be the centre of chirality, if all the associated ligands are different, or the chirality can be present in one or more of the ligands attached to the metal. As far as hydrogenation is concerned it is the latter type which is important. Almost all the homogeneous asymmetric hydrogenation catalysts presently known are based on rhodium complexes having tertiary phosphine ligands. With such systems two approaches have been adopted to the problem of introducing optical activity; viz. using tertiary-phosphine ligands of the type PRR^1R^2 which are asymmetric at phosphorus, or using ligands in which the asymmetry is contained in one of the groups attached to the phosphorus, e.g. PR_2R^* where $R = Ph$ and $R^* =$ menthyl or neomenthyl. In an attempt to make the most of both approaches, ligands of the type $P^*R^*R^1R^2$, e.g. (−)Menthylphenylmethylphosphine, have been prepared [128], but up to now most attention has focussed on the separate approaches and it is to these that we confine the rest of the discussion.

(a) Asymmetry at phosphorus

The first examples [129,130] of asymmetric hydrogenation using optically active tertiary phosphine ligands of the type $P^*RR^1R^2$ were reported in 1968 using rhodium complexes containing the ligand (S)(+)methylphenyl-n-propylphosphine, P^*MePh,Pr^n. Initial experiments were conducted with preformed complexes of the type $RhCl_3L_3$, which, on treatment with hydrogen in the presence of a base, were considered to give $RhClL_3$ (*c.f.* Section 2.1.1(*a*)). In subsequent experiments the catalyst was formed *in situ* by adding the tertiary-phosphine ligand to a labile rhodium(I) olefin complex, e.g. μ-dichlorobis(1,5-hexadiene)dirhodium(I), $[RhCl(1,5HD)]_2$ (this latter dodge is extremely useful in preparing a wide range of tertiary phosphine rhodium(I) halogeno complexes *in situ* since dimeric complexes of the type $[RhX(olefin)_2]_2$ are readily prepared and easily stored). With atropic acid as substrate enantiomeric excesses of the order of 15% were obtained [129]. Using the triethylamine salt of atropic acid, in place of the free acid, resulted in an increased optical yield of 28% ee [22].

$$\underset{HOOC}{\overset{Ph}{>}}C=CH_2 \quad \xrightarrow[\substack{HPh/EtOH/Et_3N \\ 60°C\ 20\ atm\ H_2}]{RhCl_3L_3^*} \quad \underset{HOOC}{\overset{Ph}{>}}\overset{*}{C}\underset{H}{\overset{CH_3}{<}}$$

ee = 15%

In any asymmetric synthesis it is clearly important that the chiral directing function should exert as much influence as possible on the prochiral substrate during the course of the reaction. In terms of catalytic asymmetric hydrogenation this means that the unsaturated molecule should be bound as intimately as possible to the catalyst system since, for an alkene, hydrogenation on one side of the double bond gives one enantiomer while hydrogenation on the other side gives the other enantiomer. The term intimately bound is not intended to be synonymous with strongly bound. What is important is that the substrate should have a marked preference for one particular orientation with respect to the catalyst system rather than that it should necessarily be strongly bound in that orientation, although frequently the two go hand in hand. In designing a suitable ligand this requisite can be approached either by spacial or electronic tailoring or, more likely, by a combination of the two.

In spacial tailoring the co-ordination site, or cavity, in which the substrate will be accommodated is shaped such that the substrate fits in one preferred orientation. In electronic tailoring the electronic properties of the co-ordination site, or cavity, are made such that they match in a preferred manner with those of the substrate. These two parameters, which are crudely illustrated in Fig. 2.7, imply that, for optimal effectiveness, it is necessary to carefully match the catalyst ligand systems with the individual needs of a particular substrate. Furthermore that, in so doing, we should seek to utilize as many substrate features as possible. The best examples of catalyst/substrate matching are found in enzyme systems which are renowned for their high stereoselectivity. However, the increase in optical purity obtained in going from atropic acid to the carboxylic acid salt is at least partially due to the carboxylate group co-ordinating with the metal site and thus

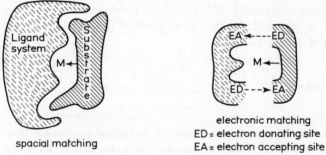

spacial matching

electronic matching
ED = electron donating site
EA = electron accepting site

Fig. 2.7 *Spacial and electronic matching between substrate and ligand system*

increasing the intimacy of the interaction. Although catalyst tailoring via synthetic ligand systems is in its infancy, the chiral ligand 2-methoxy-phenylcyclohexylmethylphosphine (ACMP), (2.31) was designed [56,57,131–134] for the reduction of α-acylaminoacrylic acid substrates (2.32) in the hope that, while the olefinic and carboxylate groups of the acid could bond to the metal centre, the acylamino group could hydrogen bond to the methoxy group(s) of the ACMP ligand(s). Whether such interaction actually occurs during the catalysis cycle is a moot point. There is, however, no disputing the optical efficiency of this particular ligand systems as witnessed by the results presented in Table 2.1 [131–134].

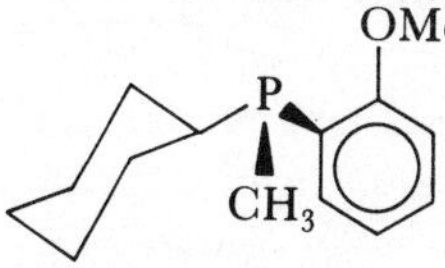

(2.31)

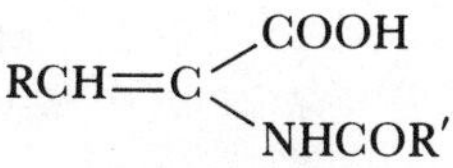

$RCH=C\begin{smallmatrix}COOH\\NHCOR'\end{smallmatrix}$

(2.32)

A major practical limitation in using chiral phosphine ligands in which the asymmetry is centred at the phosphorus is the complexity of the synthetic process required to produce the ligand which involves, of necessity, a resolution step. Starting from $PhPCl_2$ a seven stage reaction sequence is required to produce a chiral phosphine of the type P^*PhCH_3R and, even starting from the resolved phosphine oxide, $Ph(Me)(O-menthyl)PO$, three steps are required to produce ACMP with an overall yield of ca. 25% [123]. The second type of chiral phosphine

Table 2.1 *Asymmetric homogeneous hydrogenations of α-acylaminoacrylic acids, (2.32) with the rhodium-ACMP catalyst system*

R	R'	Optical purity (% ee) of product amino acid
m-OMe, p-OH-phenyl	Ph	90
m-OMe, p-OAc-phenyl	Me	88
Ph	Me	85
Ph	Ph	85
p-Cl-phenyl	Me	77

$[(ACMP)_2Rh(diene)]^+ X^-$, diene = 1,5-cyclooctadiene, $X^- = BF_4^-$, PF_6^- or BPh_4^- in methanol at 50 °C and ca. 3 atm H_2

ligands, i.e. those in which the asymmetry is present in a group attached to the phosphorus rather than at the phosphorous itself, have the advantage that they can be prepared from readily available, inexpensive, chiral precursors.

(b) Asymmetry attached to phosphorus

Some three years after the first reports of asymmetric hydrogenation using chiral phosphorus tertiary-phosphine ligands it was found that chiral phosporus was not a prerequisite of such ligands; introducing the asymmetry into one of the groups attached to the phosphorus was also effective. Thus a rhodium system incorporating neomenthyldiphenylphosphine, abbreviated as (+)-NMDPP, (2.33) catalyses the hydrogenation of β-methyl-

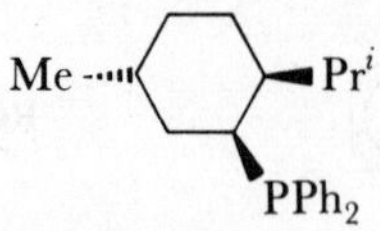

$$\text{Me} \cdots \text{cyclohexane} \text{—} Pr^i$$
$$\text{PPh}_2$$

(2.33)

cinnamic acid to give (S)-(+)-3-phenylbutanoic acid in 61% ee [135]

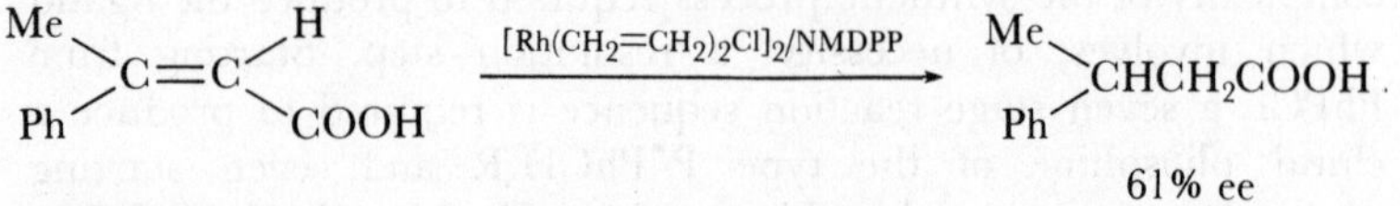

$$\underset{Ph}{\overset{Me}{\diagdown}}C=C\underset{COOH}{\overset{H}{\diagup}} \xrightarrow{[\text{Rh}(CH_2=CH_2)_2Cl]_2/\text{NMDPP}} \underset{Ph}{\overset{Me}{\diagdown}}CHCH_2COOH$$

61% ee

NMDPP has the considerable practical advantage that it is readily prepared from diphenylphosphine and the chiral precursor (−)-menthol, both of which are inexpensive and commercially available. Similarly, possibly the most widely cited chiral phosphine ligand, 2,3-*o*-isopropylidene-2,3-dihydroxy-1,4-bis(diphenylphosphino)butane (2.34), commonly known as (−)-DIOP and first reported by Kagan and co-workers [136], can be fairly simply prepared from commercially available (+)-tartaric acid

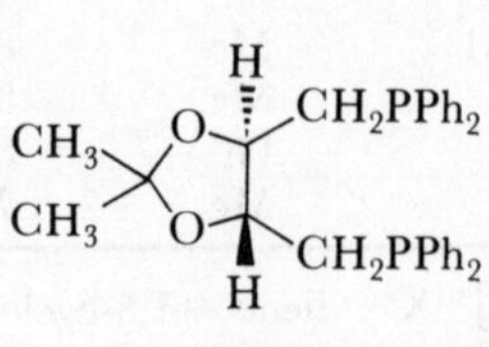

$$\begin{array}{c} \text{CH}_3 \\ \text{CH}_3 \end{array} \!\!\!\! \diagdown\!\!\!\! \begin{array}{c} \text{O} \\ \text{O} \end{array} \!\!\!\! \diagup\!\!\!\! \begin{array}{c} \overset{H}{\underset{\vdots}{}} \text{CH}_2\text{PPh}_2 \\ \overset{}{\underset{H}{}} \text{CH}_2\text{PPh}_2 \end{array}$$

(2.34)

HOMOGENEOUS CATALYST SYSTEMS IN OPERATION

A rhodium complex system, prepared by adding one mole per rhodium of this bidentate ligand to $[Rh(cyclooctene)_2Cl]_2$ in a benzene–ethanol solvent system containing triethylamine, catalysed the hydrogenation of (2.35) to give the saturated acid (2.36) with an optical purity of 72% [136]

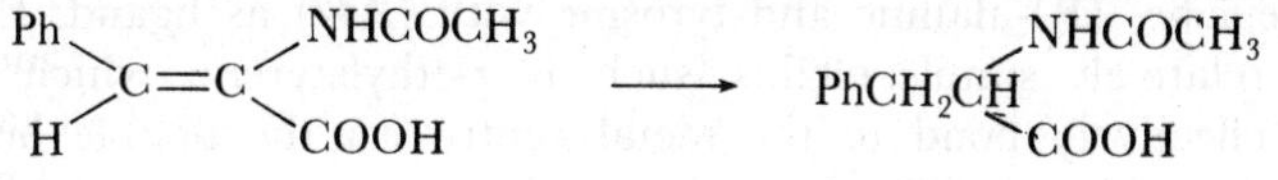

$$\underset{(2.35)}{\underset{H}{\overset{Ph}{>}}C=C\underset{COOH}{\overset{NHCOCH_3}{<}}} \longrightarrow \underset{(2.36)}{\underset{72\% \text{ ee}}{PhCH_2C\underset{COOH}{\overset{NHCOCH_3}{<}}H}}$$

Since these initial reports, a number of chiral ligand systems have been developed in which the chirality is not directly located on the phosphorus atom. Several of the more successful ones are shown below:

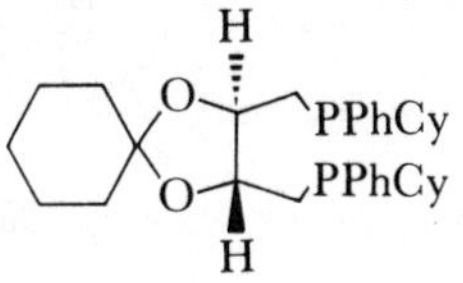

(2.37) [137]

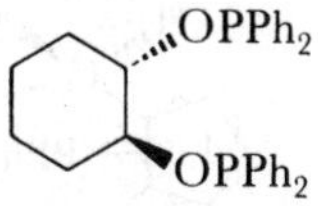

(2.38) [138]

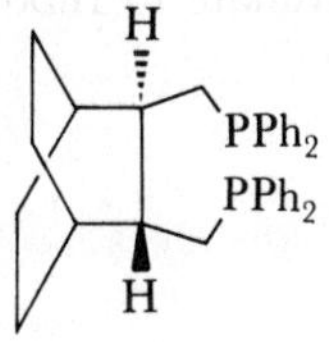

(2.39) [139]

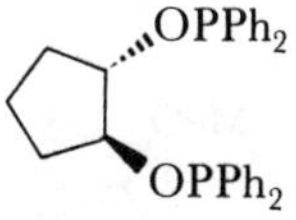

(2.40) [140]

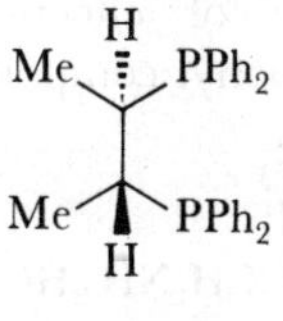

(2.42) [141]

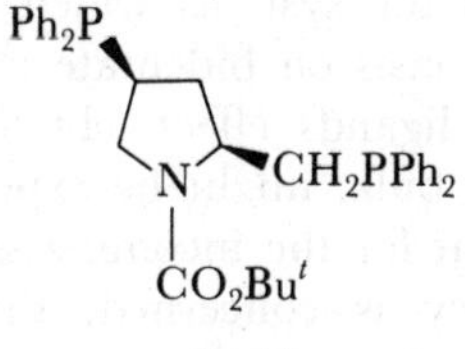

(2.43) [142]

Using rhodium complexes incorporating these and closely related ligand systems as catalysts, high optical purities have been obtained by hydrogenating a range of prochiral substrates. Percentage enantiomeric excesses of well into the nineties have been achieved in the preparation of (R)-phenylalanine, (R)-leucine, (R)-alanine and tyrosine with (2.42) as ligand. Even with relatively simple olefins, such as α-ethylstyrene, which can only effectively bond to the metal centre via the double bond, optical yields of 60% have been obtaining using a rhodium catalyst system incorporating (2.40) [124].

A recent development, which could have considerable potential in a wide spectrum of transition-metal catalysed reactions, is the use of organometallic complexes as ligands. Thus, the substituted ferrocene complex, (R)-α-[(S)-1′,2bis(diphenylphosphino)-ferrocenyl]ethyl alcohol, (2.44), known as (R)-(S)-BPPFOH, can co-ordinate to rhodium via the two phosphorus centres to give an

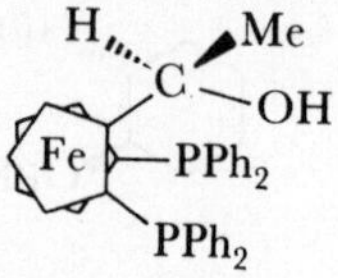

(2.44)

extremely effective asymmetric hydrogenation catalyst which is even capable of effecting the asymmetric hydrogenation of prochiral carbonyl compounds [143], e.g.

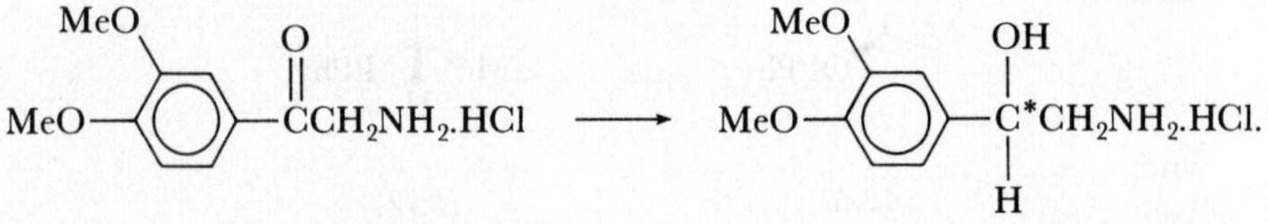

ca. 90% ee

In these, and many of the systems which have followed DIOP, there has been an emphasis on bidentate chelating ligands. Such large cis co-ordinating ligands effectively tie up one side of the complex, and, by their bulk, might be expected to create a tight asymmetric environment for the incoming substrate. However, as far as optical efficiency is concerned, they have shown little improvement over monodentate systems of the type (2.31). What they do show is that high optical efficiency can still be obtained when the chiral centre is well removed from the metal centre, e.g. (−)DIOP where, in a complex, the chiral carbon centres are 3 atoms removed from the metal.

There are still many unanswered questions surrounding asymmetric hydrogenation and associated ligand/substrate effects. What is clear, is that dramatic changes in optical yield can be produced by relatively minor changes in either the ligands or the substrate. Similarly, it would appear from the Monsanto work [131–134] that much can be gained by matching ligand to substrate. That subtle changes in the catalyst system can result in such large changes in optical yield is not too surprising when you consider that a 99.9% optical yield corresponds to only a *ca.* 13 KJ mol^{-1} difference in the free energy of the diastereomeric transition states [124]. Like most of homogeneous catalysis, catalytic asymmetric hydrogenation is a gentle art.

2.1.4 *Transfer hydrogenation*

All of the hydrogenation reactions we have discussed up to now have been of the type

$$A + H_2 \longrightarrow AH_2 \qquad (2.4)$$

with the hydrogen being supplied as molecular hydrogen. An alternative type of hydrogenation [144,145] is represented by the reaction

$$A + DH_2 \longrightarrow AH_2 + D \qquad (2.5)$$

where the hydrogen is supplied by a donor molecule, DH_2, which itself undergoes dehydrogenation during the course of the reaction. Generally the donor molecules are present as solvent since there is usually an advantage in their being present in excess over the unsaturated acceptor molecule, A. This is not, however, a prerequisite of the reaction. Suitable donor molecules are, in essence, any molecular capable of giving up hydrogen. In practice most success has been achieved using alcohols, e.g. iso-propanol, acids, e.g. formic acid, cyclic ethers, e.g. dioxane, and aromatic cyclic amines, e.g. pyrrolidine and piperidine. The hydrogenation of cyclopentene using dioxane as donor in the presence of $RhCl(PPh_3)_3$, as catalyst precursor, is one well-studied

[146–148] example of this class of reactions

$$\text{1,4-dioxane} + \text{cyclopentene} \longrightarrow \text{1,4-dioxene(dione)} + \text{cyclopentane}$$

Since there are many mechanistic similarities between transfer hydrogenation and hydrogenation involving molecular hydrogen, many of the catalyst systems previously discussed are active in transfer hydrogenation. We have already mentioned $RhCl(PPh_3)_3$; $RuCl_2(PPh_3)_3$ is also active for a range of transfer hydrogenation reactions [149.150].

The essential difference between hydrogenation and transfer hydrogenation, as represented by Equations (2.4) and (2.5), respectively, is that in transfer hydrogenation there is no change in the overall unsaturation in the system, in its simplest form

'alkane(1)' + alkene(1) $\longrightarrow$ alkene(2) + 'alkane(2)'.

This places certain limitations on transfer-hydrogenation reactions. Not only must the hydrogen donor, 'alkane(1)', be capable of interacting productively, i.e. giving up at least one atom of hydrogen, with the metal centre, but it must also be able to compete reasonably successfully with alkene(1) for vacant catalyst sites. In practical terms, this means that the donor molecule must incorporate a function capable of co-ordinating to the metal site, e.g. the oxygen or nitrogen in the donors cited. Furthermore, the donor dehydrogenation product, alkene(2), should, once formed, leave the catalyst sphere in favour of the chosen hydrogen acceptor, alkene(1). These restrictions tend to show up in the complex kinetics of such systems; product, alkene(2), inhibition of reaction is not unusual [148].

Transfer hydrogenation is, of essence, a two-product reaction and as such is generally less appealing from a practical and commercial standpoint. It can, however, be useful if, for example, safety considerations mitigate against the presence of gaseous hydrogen or if, in the reaction medium in question, suitable donors and acceptors are already present as in, for example, an oil fraction produced by refining/reforming processes [151]. Although, in the latter case, heterogeneous rather than homogeneous catalyst systems would be preferred.

In this section on hydrogenation we have tried to highlight the major aspects of the reaction. The coverage is by no means comprehensive but at least it gives the flavour of the topic. As mentioned at the beginning of the section, there are no large scale commercial hydrogenation processes, as yet, using homogeneous catalyst systems, although they are widely used in the speciality chemical, e.g. pharmaceutical, industry. Also, aspects of the hydrogenation reaction are found in many catalytic reactions, such as hydroformylation, where homogeneous systems have achieved commercial notoriety. Before going on to discuss such systems, we must first pause to consider another type of reaction which, although like hydrogenation, is not presently being practised on a large scale, using homogeneous catalyst systems, nevertheless plays a key role in many homogeneously catalysed processes. The reaction we have in mind is isomerization.

Further reading on homogeneous hydrogenation

General texts and reviews

James, B. R. *Homogeneous Hydrogenation*, New York: John Wiley.

James, B. R. (1979), Homogeneous hydrogenation. *Adv. Organometal Chem.*, **17**, 319.

McQuillin, F. J. (1976), *Homogeneous Hydrogenation in Organic Chemistry*, Dordrecht: Reidel.

Kwiatek, J. (1971), in *Transition Metals in Homogeneous Catalysis* (ed. G. N. Schrauzer), p. 13, New York: Marcel Dekker.

Heck, R. F. (1974), in *Organotransition Metal Chemistry, A Mechanistic Approach* (ed. R. F. Heck), p. 55, New York: Academic Press.

Webb, G. (1978), in *Catalysis*, Vol. 2, p. 145, London: Chemical Society.

Kieboom, A. P. G. and van Rantwijk, F. (1979), *Hydrogenation and Hydrogeneolysis in Synthetic Organic Chemistry*, Delft: Delft University Press.

Rylander, P. N. (1973), *Organic Synthesis with Noble Metal Catalysts*, Ch. 2, New York: Academic Press.

Harman, R. E., Gupta, S. K. and Brown, D. J. (1973), *Chem. Soc. Rev.*, **73**, 21.

Marko, L. and Heil, B. (1974), Asymmetric homogeneous hydrogenation and related reactions. *Catal. Rev.*, **8**, 269.

Morrison, J. D., Masler, W. F. and Newberg, M. K. (1979), Asymmetric homogeneous hydrogenation. *Adv. Catal.*, **25**, 81.

Pearce, R. (1978), in *Catalysis*, Vol. 2, p. 176, *Specialist Periodical Report* London: Chemical Society.

2.2 Isomerization

Isomerization is the process by which chemical bonds or groups are redistributed within a given molecule. At the end of the process nothing, in a sense of a physical entity, has been added to, or removed from, the substrate. So unlike hydrogenation, and indeed unlike all the other catalysed reactions we shall be considering, this process only affects the molecular structure and not the chemical formula of the substrate. The two best studied types of isomerization which are amenable to catalysis by homogeneous transition-metal complexes are

alkene isomerization which involves the redistribution of carbon–carbon double bonds within a molecule, and

skeletal isomerization which involves the rearrangement of carbon–carbon double or single bonds within a carbocyclic, generally a strained carbocyclic, molecule.

We shall discuss these two isomerization processes in turn.

2.2.1 *Alkene isomerization*

Just about all the *d*-block elements capable of co-ordinating an alkene catalyse alkene isomerization to some extent. However, as with many other homogeneously catalysed reactions, it is complexes of the Group VIII metals, i.e. Ru, Os; Co, Rh, Ir; Ni, Pd, Pt, which have attracted most attention. During its metal catalysed isomerization an alkene undergoes two activation processes, i.e. activation by co-ordination and activation by addition. It is the activation by addition process which is responsible for moving the double bond or, to put it another way, moving the hydrogen atoms around. Double bond migration essentially involves movement of a hydrogen, from the alkyl group adjacent to the double bond, to the α-carbon of the double bond:

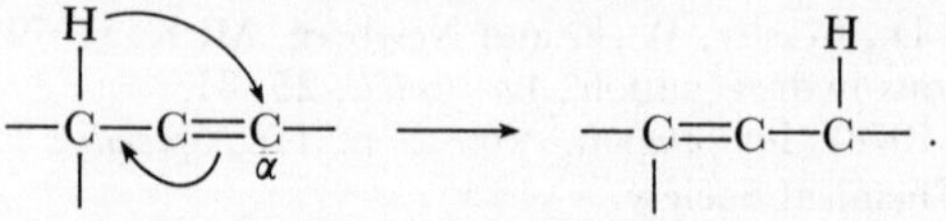

$$\begin{array}{c} \text{H} \\ | \\ -\text{C}-\text{C}=\text{C}- \\ | \quad \alpha \end{array} \longrightarrow \begin{array}{c} \text{H} \\ | \\ -\text{C}=\text{C}-\text{C}- \\ | \end{array}.$$

It is this process which the metal complex catalyses. Mechanistically two main types of isomerization reactions have emerged, viz. those involving metal–alkyl and those involving metal–allyl intermediates.

(a) Isomerization via metal-alkyl intermediates

The essential species in this sequence is a metal–hydride complex. Co-ordination of the alkene gives a hydrido–alkene–metal complex which rapidly undergoes a hydride ligand migration reaction to form a metal–alkyl species. The migration, which probably occurs via a four-centre transition state, is identical to that previously described as occurring during catalytic hydrogenation (Section 2.1.1) and, providing the alkene and hydrido groups are mutually

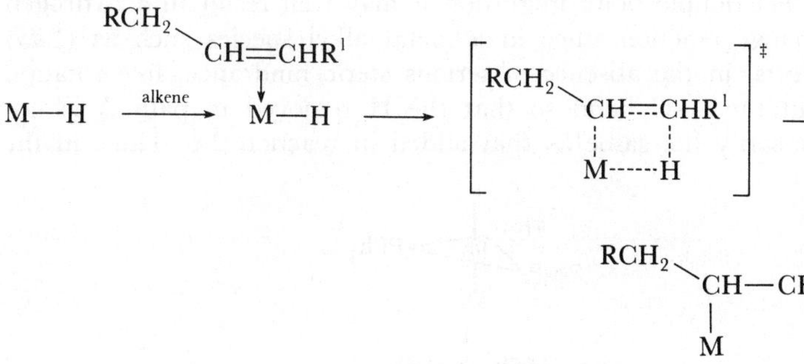

$$M-H \xrightarrow{\text{alkene}} \begin{array}{c} RCH_2\diagdown \\ CH=CHR^1 \\ \downarrow \\ M-H \end{array} \longrightarrow \left[\begin{array}{c} RCH_2\diagdown \\ CH\cdots CHR^1 \\ \vdots \quad \vdots \\ M\cdots H \end{array} \right]^{\ddagger} \longrightarrow \quad (2.6)$$

$$\begin{array}{c} RCH_2\diagdown \\ CH-CH_2R^1 \\ | \\ M \end{array}$$

(2.45)

cis, is extremely rapid for almost all transition-metal hydrido–alkene complexes. Once formed, the metal–alkyl species, (2.45), can undergo a β-elimination proximity interaction via either path A or path B.

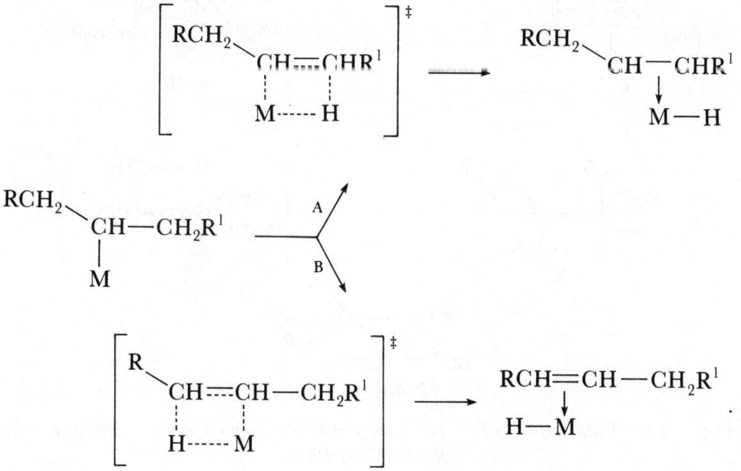

$$\left[\begin{array}{c} RCH_2\diagdown \\ CH\cdots CHR^1 \\ \vdots \quad \vdots \\ M\cdots H \end{array} \right]^{\ddagger} \longrightarrow \begin{array}{c} RCH_2\diagdown \\ CH-CHR^1 \\ \downarrow \\ M-H \end{array}$$

$$\begin{array}{c} RCH_2\diagdown \\ CH-CH_2R^1 \\ | \\ M \end{array} \quad \begin{array}{c} A \nearrow \\ \\ B \searrow \end{array}$$

$$\left[\begin{array}{c} R\diagdown \\ CH\cdots CH-CH_2R^1 \\ \vdots \quad \vdots \\ H\cdots M \end{array} \right]^{\ddagger} \longrightarrow \begin{array}{c} RCH=CH-CH_2R^1 \\ \downarrow \\ H-M \end{array}$$

Path A is merely the reverse of the initial process, reaction (2.6), and results in no net migration of the double bond. Path B, in which the H abstraction occurs at sites other than that at which the initial H addition occurred, results in a net migration of the double bond one position along the chain. Providing the alkene co-ordinated to the metal can exchange with alkene present in the reaction medium, the above sequence of reactions constitutes a catalytic isomerization cycle. Although Path A does not result in any net double bond migration it may well result in a hydrogen exchange reaction since in a metal–alkyl species such as (2.45) there is, in the absence of serious steric hindrance, free rotation about the C–C bond so that the H removed in path A is not necessarily the same as that added in reaction 2.6. Thus, in the

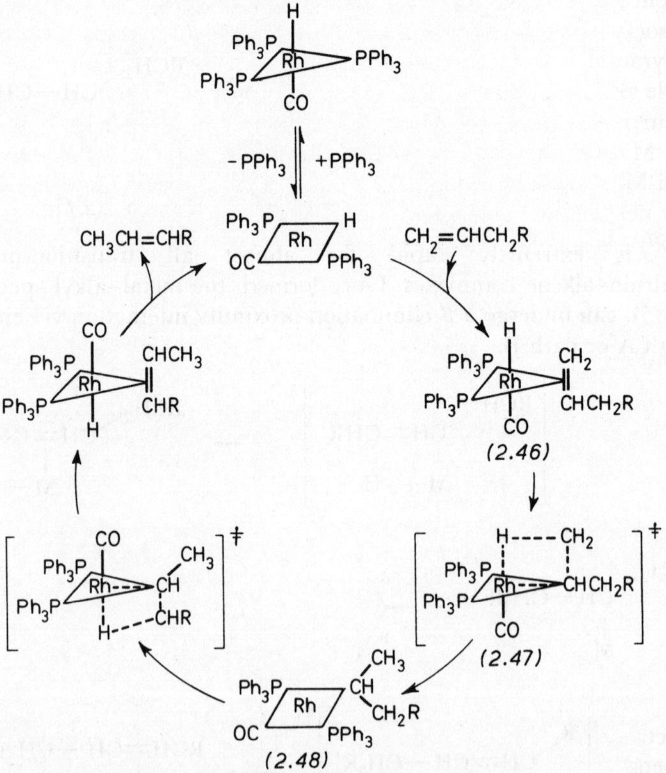

Fig. 2.8 *Catalysis cycle for the alkene isomerization catalysed by $RhH(CO)(PPh_3)_3$*

presence of a metal deuteride, deuterium incorporation into the non-isomerized olefins is frequently observed. In paths A and B either the *cis*- or the *trans*-alkene can be formed and, assuming the system is allowed to reach equilibrium, it will be the thermodynamically most stable mixture of isomers which is obtained.

The metal–hydrido complex can be either added as such to the reaction medium or be formed *in situ* from a metal complex and a suitable hydride source such as gaseous hydrogen or aqueous acid. $RhH(CO)(PPh_3)_3$ [152] is one example of the former and $Co_2(CO)_8/H_2$ [153] and $Ni[P(OEt)_3]_4/H_2SO_4$ [154] are examples of the *in situ* type. The catalysis cycle for $Rh(H)(CO)(PPh_3)_3$ is shown in Fig. 2.8. As indicated, the active species is the 16 valence electron species, *trans*-$RhH(CO)(PPh_3)_2$, formed by dissociation of a tertiary phosphine ligand from the 18 electron bipyramidal pentacoordinate rhodium(I) complex. In the catalysis cycle the hydride addition, via (*2.47*) to give (*2.48*), is shown as occurring at the least substituted carbon atom of the double bond, i.e. Markovnikov addition. The alternative mode of addition, i.e. anti-Markovnikov, via (*2.49*) to give (*2.50*), would not result in any net double bond migration since β-hydrogen elimination from (*2.50*) can only give the terminal alkene complex (*2.46*)

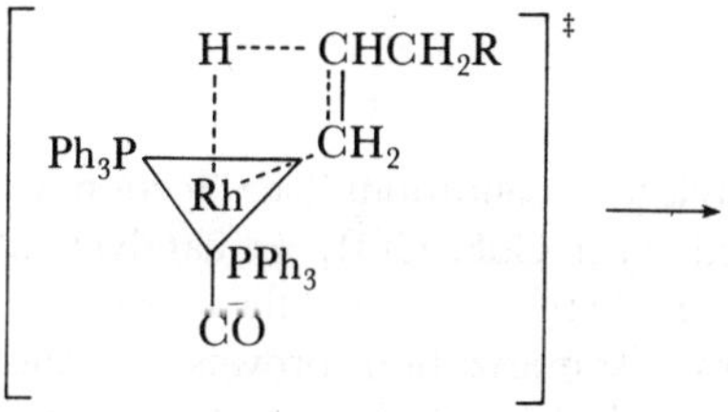

(*2.49*)

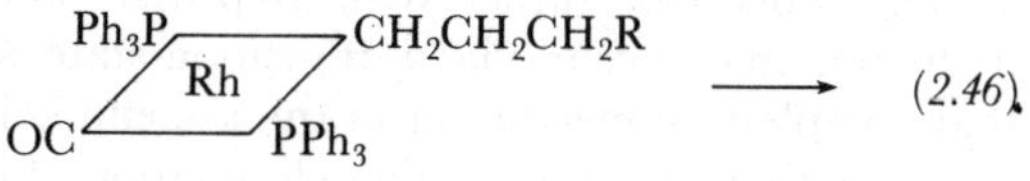

(*2.50*) ⟶ (*2.46*)

Electronically, the direction of addition would be expected to depend on the hydridic character of the hydrogen, with greater negative charge localization on hydrogen favouring anti-Markovnikov addition and vice versa. Steric considerations would favour the linear alkyl complex (*2.50*) over the branched species (*2.48*) since both complexes contain two bulky triphenylphosphine ligands and any move in the direction of reducing steric crowding would be welcomed. In the case of the rhodium complex, both factors mitigate against Markovnikov addition and as a result neither this, nor indeed most other rhodium(I) complexes, are particularly good isomerization catalysts. In fact, in the presence of hydrogen gas, the rate of isomerization is usually negligible compared to the rate of hydrogenation.

With the $Co_2(CO)_8/H_2$ system on the other hand, the active catalytic species is $CoH(CO)_4$ formed by homolytic addition of hydrogen:

$$Co_2(CO)_8 + H_2 \longrightarrow 2CoH(CO)_4.$$

In $CoH(CO)_4$ the hydrogen can be regarded as protic rather than hydridic; in aqueous solution $CoH(CO)_4$ behaves as a strong acid, ionizing to give the carbonylate ion $[Co(CO)_4]^-$. Thus $CoH(CO)_4$ has a much greater tendency than $RhH(CO)(PPh_3)_2$ to give the Markovnikov addition product, (*2.51*), with a terminal olefin:

$$\left[\begin{array}{c} RCH_2CH\text{---}CH_2 \\ | \quad\quad\quad\; | \\ (CO)_xCo\text{----}H \end{array} \right]^\ddagger \longrightarrow \begin{array}{c} RCH_2CH\text{---}CH_3 \\ | \\ (CO)_xCo \end{array} \quad x = 3 \text{ or } 4$$

(*2.51*)

In fact, under certain hydroformylation conditions (see Section 4) the mixture of aldehydes formed with $CoH(CO)_4$ as catalyst is essentially independent of isomer distribution in the substrate olefin [153] due, in part, to the isomerization prowess of the cobalt catalyst. On the basis of hydroformylation studies, using deuterated olefins, it has been suggested that with the cobalt system concerted, rather than step-wise, addition/elimination of hydrogen may occur via a transition-state such as (*2.52*), such that complete isomerization in the π-olefin cobalt carbonyl species occurs without the intervening elimination of free olefin

$$\begin{array}{c} RCH_2CH=CH_2 \\ | \\ (CO)_xCo\text{---}H \end{array} \rightleftharpoons \left[\begin{array}{c} RCH\text{---}CH\text{---}CH_2 \\ | \quad\quad\quad\; | \\ H\text{-----}Co\text{-----}H \\ (CO)_x \end{array} \right]^\ddagger \rightleftharpoons \begin{array}{c} RCH=CH\text{---}CH_3 \\ | \\ H\text{---}Co(CO)_x \end{array}$$

(*2.52*)

As far as the $Ni[P(OEt)_3]_4/H_2SO_4$ system is concerned, the active catalytic species appears to be the 16 valence electron, formally nickel(II), species, $NiH[P(OEt)_3]_3$, formed by the protonation/ligand dissociation sequence

$$Ni^0L_4 + H^+ \longrightarrow [Ni^{II}(H)L_4]^+ \longrightarrow [Ni^{II}(H)L_3]^+ + L$$

NVE 18 18 16

where $L = P(OEt)_3$. The ligand dissociation step is rate-limiting and the resulting four co-ordinate nickel species is an extremely efficient isomerization catalyst in acid solution. With the less basic phosphite ligand (on Tolman's electronic parameter scale, see Section 1.2.3b, $P(OEt)_3$ has $\nu = 2077\ cm^{-1}$ cf. $PPh_3 = 2068.9\ cm^{-1}$ and $PBu_3 = 2060.3\ cm^{-1}$) the nature of the hydrogen ligand is probably midway between that of the H in $CoH(CO)_4$ and that of the H in $RhH(CO)(PPh_3)_3$ and both Markovnikov and anti-Markovnikov addition can reasonably readily occur. Sterically, as with the rhodium system, anti-Markovnikov addition is preferred. Although the cone angle of the $P(OEt)_3$ ligand is considerably less than that of PPh_3 (109° versus 145°), there are three $P(OEt)_3$ ligands in the active nickel species compared to two PPh_3 ligands in the rhodium system. With the closely related $Ni[P(OEt)_3]_4/CF_3COOH$ system in benzene, strong support for the hydride addition/elimination mechanism has been obtained from studies using 1,2-d_2-1-pentene [155].

Owing to the general instability of alkene-hydrido metal complexes there are only a few reports of the direct observation of the equilibrium

$$\underset{|}{\overset{H}{M}}\text{—alkene} \rightleftharpoons M\text{—alkyl}.$$

One such report is that concerning $[(\eta^5\text{-}C_5H_5)Rh(H)(C_2H_4)(PMe_3)]^+$, which, in CD_3NO_2 solution, is an equilibrium with the corresponding ethyl complex $[(\eta^5\text{-}C_5H_5)Rh(C_2H_5)(PMe_3)]^+$ as evidenced by ^1H n.m.r. spectroscopy [156]. At $-20\ °C$ the exchange is slow on the n.m.r. time scale and signals appropriate to both species are detected. At room temperature the signals of the C_5H_5 and PMe_3 protons are sharp, while those of the hydridic H atom and of the C_2H_4 protons are broad, indicating that the equilibrium of Equation

(2.7) is being set up

$$[(\eta^5\text{-}C_5H_5)RhH(C_2H_4)(PMe_3)]^+ \rightleftharpoons [(\eta^5\text{-}C_5H_5)RhC_2H_5(PMe_3)]^+. \quad (2.7)$$

In all of the above isomerization systems, the active metal-hydride species was either added as such or generated *in situ* from a suitable hydrogen source. In the second mechanistic class of metal-catalysed isomerization reactions the metal-hydride species is generated from the substrate alkene.

(b) Isomerization via metal-allyl intermediates
An alternative pathway leading to alkene isomerization is via the formation of a metal-allyl complex. In this sequence, which is analogous to that previously described as occurring during the $CoH(CN)_5^{3-}$ catalysed hydrogenation of buta-1,3-diene (see Section 2.1.1*b*), initial alkene co-ordination is followed by hydrogen abstraction from an sp^3 hybridized carbon centre adjacent to the double bond to give a hydrido-metal-allyl species of the type (2.54). In such a species, generally referred to as an allyl complex, the C_3 unit is π-bonded to the metal, all three carbon atoms are within bonding distance of the metal centre and the unit effectively occupies two co-ordination sites in the complex.

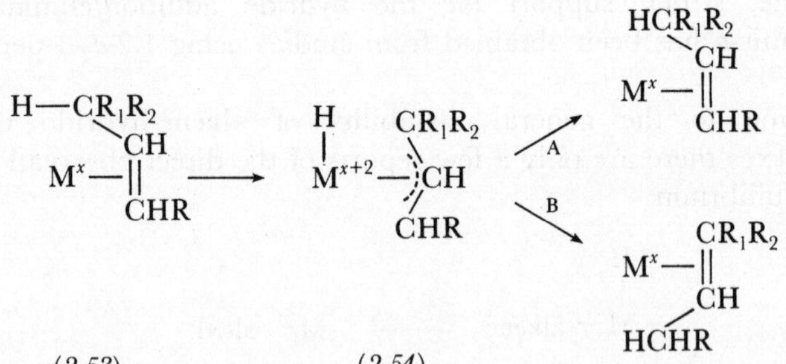

H—CR$_1$R$_2$
    \CH
M^x—||
    CHR
(2.53)

H CR$_1$R$_2$
| \
M^{x+2}—:CH
   /
  CHR
(2.54)
   A→
HCR$_1$R$_2$
  \CH
M^x—||
  CHR
 B→
CR$_1$R$_2$
M^x—||
  \CH
  HCHR

Whether or not double bond migration occurs depends on how the hydride is added back to the allyl unit. Re-addition at the site of abstraction, path A, results in no net movement of the double bond. Re-addition at the other end of the C-3 unit, path B, moves the double bond one step along the carbon skeleton. A five-centre transition state of the type (2.55) can be visualized for both the abstraction and addition processes:

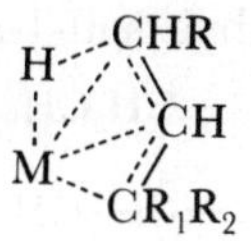

$$\begin{array}{c} \text{H}\cdots\text{CHR} \\ \vdots\diagup\diagdown\text{CH} \\ \text{M}\cdots\diagup \\ \text{CR}_1\text{R}_2 \end{array}$$

(2.55)

In going from the π-olefin complex (2.53) to the hydrido-π-allyl species (2.54) there is an increase of two units in both the formal oxidation state and co-ordination number of the metal centre thus this sequence constitutes a further example of the ubiquitous 'oxidative addition' reaction.

Unlike the metal-alkyl isomerization system previously discussed, the isomerization catalysis cycle involving π-allyl-metal species involves a redox sequence, i.e.

$$M^x \longrightarrow M^{x+2} \longrightarrow M^x$$

and thus this mechanism is only available to metal complexes in low oxidation states for which the $x + 2$ state is readily attainable. Indeed, the best characterized examples of alkene isomerizations involving π-allyl intermediates are those catalysed by Fe(0) carbonyl complexes. A further distinction between the metal-alkyl and the metal-allyl sequence is the distance within the substrate that a substrate hydrogen moves. In the π-allyl mechanism a substrate hydrogen moves between C-1 and C-3 of the C-3 units, i.e. the process involves a 1,3-hydrogen shift. This is clearly seen by reference to the above scheme. In the metal-alkyl mechanism a 1,2-hydrogen shift is involved. The 1,2 shift is best appreciated by referring to the two metal-alkyl species in the sequence shown below

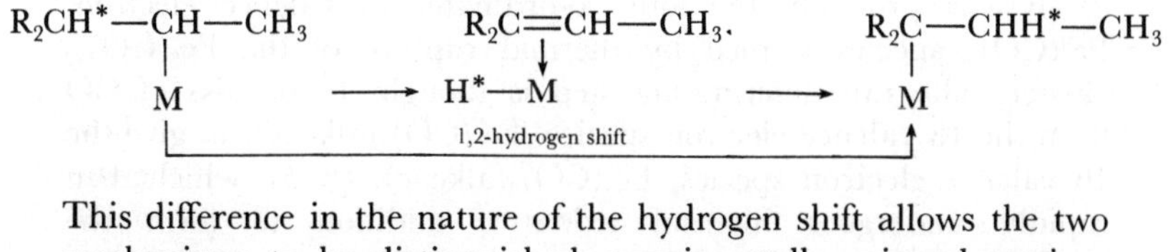

$$\underset{\text{M}}{\text{R}_2\text{CH}^*\text{—CH—CH}_3} \xrightarrow{\text{1,2-hydrogen shift}} \underset{\text{H}^*\text{—M}}{\text{R}_2\text{C}\text{=CH—CH}_3.} \longrightarrow \underset{\text{M}}{\text{R}_2\text{C—CHH}^*\text{—CH}_3}$$

This difference in the nature of the hydrogen shift allows the two mechanisms to be distinguished experimentally using deuterium labelled olefins. Thus the π-allyl metal-hydride mechanism has been firmly established for the $Fe_3(CO)_{12}$ catalysed isomerization of alkenes by using a alkene mixture consisting of

3-methyl-but-1-ene and 3-ethyl-pent-1-ene-3d_1, (2.56) [157]:

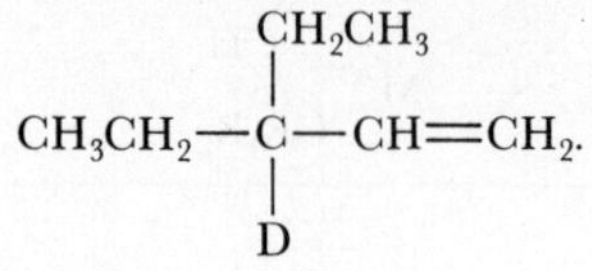

$$CH_3CH_2-\underset{\underset{D}{|}}{\overset{\overset{CH_2CH_3}{|}}{C}}-CH=CH_2.$$

(2.56)

At 80 °C 3-methyl-but-1-ene is isomerized to a mixture of 2-methyl-but-1-ene and 2-methyl-but-2-ene (the but-1-ene product is initially formed but as time progresses this is converted into the thermodynamically more stable but-2-ene product) and 3-ethyl-pent-1-ene into 3-ethyl-pent-2-ene (2.57)

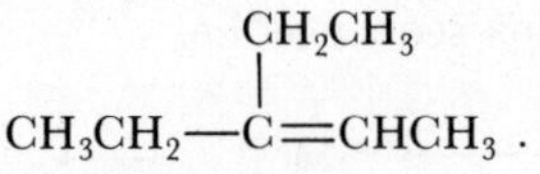

$$CH_3CH_2-\overset{\overset{CH_2CH_3}{|}}{C}=CHCH_3.$$

(2.57)

After isomerization the deuterium was only found located in the three methyl groups of the pent-2-ene product, indicating that the isomerization of (2.56) to (2.57) is accompanied exclusively by 1,3-hydrogen shift, i.e. occurs via a π-allyl metal-hydride intermediate. Furthermore, since no deuterium incorporation is found in the butene isomers, it can be concluded that the shift is intramolecular as required by the sequence shown on page 76. Further analysis of these results shows that multiple isomerization occurs; in other words, several double bond shifts can occur while the alkene is bonded to the metal, i.e. olefin isomerization of complexed alkene is faster than the rate of exchange between free and complexed alkene. A set of catalysis cycles illustrating the formation of the three isomers is shown in Fig. 2.9. The active catalyst is probably the four co-ordinate, 16 valence electron, $Fe^0(CO)_4$ species formed by thermal rupture of the $Fe_3(CO)_{12}$ cluster. The rate-determining step is thought to be loss of CO from the 18 valence electron species, $Fe^0(CO)_4$(alkene), to give the 16 valence electron species, $Fe^0(CO)_3$(alkene), (2.58), which then rapidly undergoes internal oxidative addition to form the π-allyl-hydrido complex (2.59).

In the closely related photocatalysed alkene isomerization using $Fe(CO)_5$, a co-ordinatively unsaturated carbonyl complex is produced by photoassisted dissociation of a carbonyl ligand. The

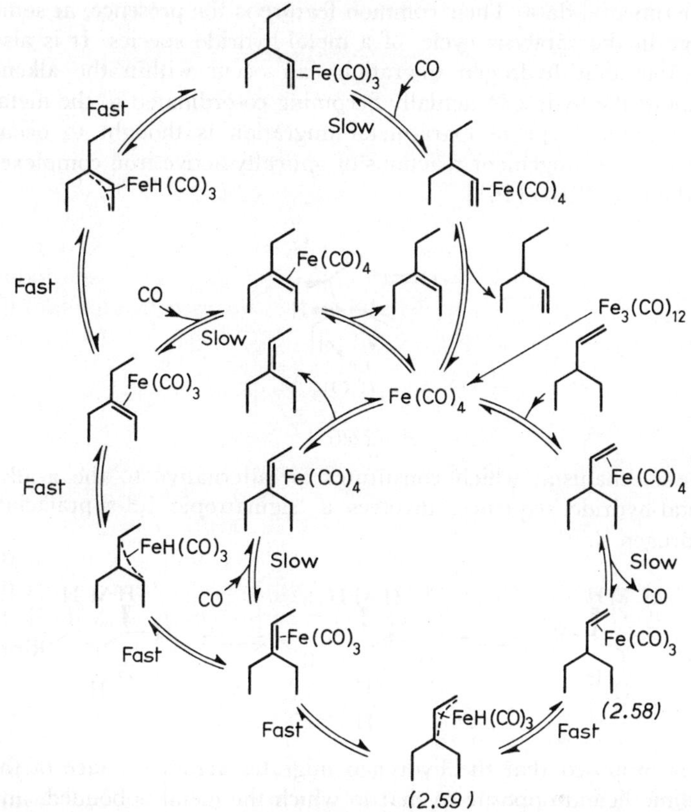

Fig. 2.9 *Catalysis cycle for the isomerization of 3-ethyl-pent-1-ene catalysed by* $Fe_3(CO)_{12}$

active catalyst is thought to be an π-allyl-hydrido complex of the type $Fe(\pi\text{-allyl})(H)(CO)_3$ [158].

The π-allyl mechanism is also thought to be important in the bis-(benzonitrile)dichloropalladium(II) catalysed isomerization of alkenes [159] but with the palladium(II) systems the experimental evidence is less clear cut [160] than with the $Fe_3(CO)_{12}$ system.

(c) *Isomerization via other intermediates*

The two alkene isomerizations mechanisms outlined above satisfactorily explain the bulk of the currently available experimental data. Their common feature is the presence, at some stage in the catalysis cycle, of a metal-hydrido species. It is also possible that hydrogen migration can occur within the alkene without the hydrogen actually becoming co-ordinated to the metal centre. This type of extra-metal migration is thought to occur during re-arrangement reactions of optically active iron complexes of the type (*2.60*) [161]

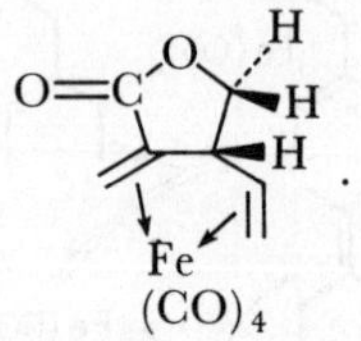

(*2.60*)

The mechanism, which constitutes an alternative to the π-allyl metal-hydride sequence, involves a 'sigmatropic 1,3-suprafacial-hydrogen shift'

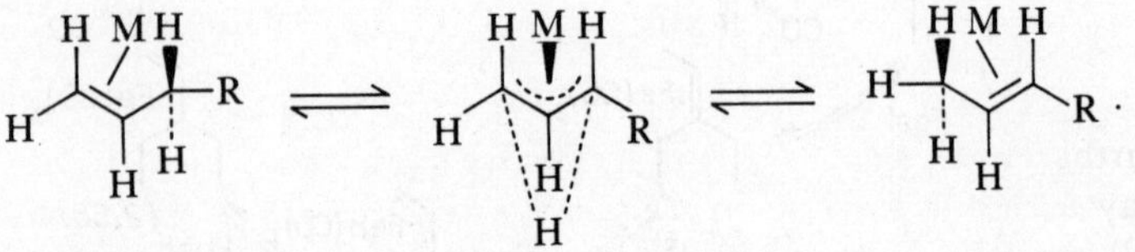

It is proposed that the hydrogen migrates across the face of the olefinic ligand opposite to that to which the metal is bonded, and does not, therefore, become directly involved with the metal bonding orbitals. Whether such a process plays a role in the iron carbonyl catalysed isomerization of simple olefins awaits experimental verification. There does, however, seem little doubt that it offers an energetically viable alternative to the π-allyl metal-hydride mechanism.

Another mechanistic sequence, not necessary involving a metal-hydride complex, is via a metal–carbene species of the type (*2.61*);

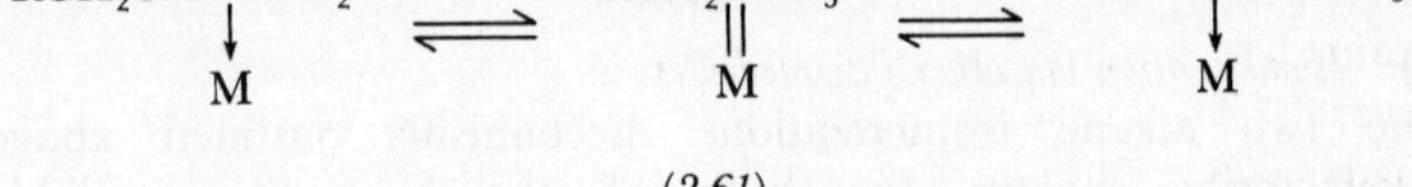

$$RCH_2CH=CH_2 \rightleftharpoons RCH_2CCH_3 \rightleftharpoons RCH=CHCH_3$$
$$\downarrow \qquad\qquad || \qquad\qquad \downarrow$$
$$M \qquad\qquad M \qquad\qquad M$$

(*2.61*)

The formation of (2.61) is envisaged as occurring via a transition state of the type (2.62), involving a 1,2-suprafacial-hydrogen shift [162]

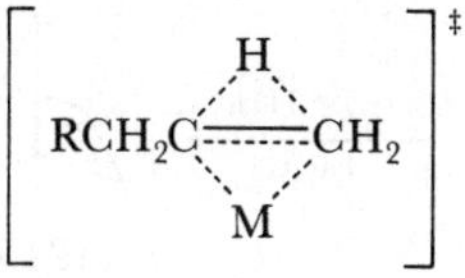

$$\left[RCH_2C \overset{H}{\underset{M}{\diamond}} CH_2 \right]^{\ddagger}$$

(2.62)

Again there is, as yet, no direct experimental evidence for such a process occurring during metal catalysed alkene isomerization although reasonable evidence has been obtained to show that the carbene ligand in (2.63) rapidly converts to the corresponding olefin ligand, PhCH=CHR [163]

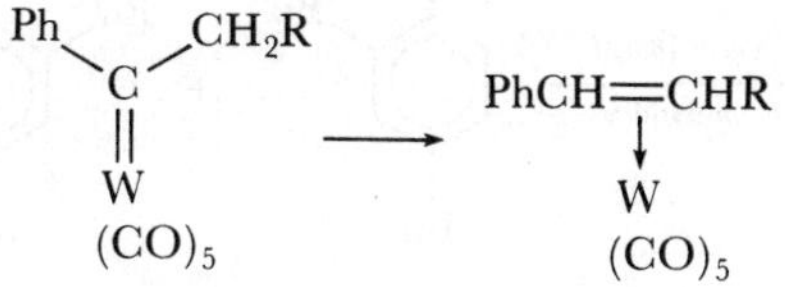

$$\underset{(CO)_5}{\underset{W}{\overset{Ph}{\underset{\|}{C}}}}CH_2R \longrightarrow \underset{(CO)_5}{\underset{W}{\overset{PhCH=CHR}{\downarrow}}} \qquad (2.8)$$

(2.63)

Furthermore, metal–carbene complexes are now considered to play a significant role in olefin metathesis reactions (see Section 2.8) and the reverse of the type of process outlined in Equation (2.8) could well be one route to the generation of such species in these reactions.

In all of the isomerization systems we have outlined above, isomerization occurred as the result of double-bond/hydrogen-migration. An equally important class of transition-metal catalysed isomerization reactions are those involving a fundamental skeletal re-arrangement of the substrate molecule such that the product molecule contains a different number, and distribution, of single and multiple bonds.

2.2.2 *Skeletal isomerization*

There are a number of organic molecules which, although highly strained in relation to one or more of their valence isomers, are nevertheless quite stable at room temperature [164,166]. Thus quadricyclene, (2.64), although highly strained in comparison to its valence isomer norbornadiene, (2.65), is quite stable at room temperature and even at 140 °C has a half life ($t_{1/2}$) of greater than 14 h [167],

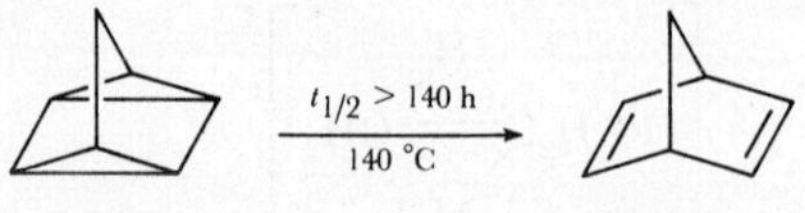

$$\text{(2.64)} \xrightarrow[140\,°\text{C}]{t_{1/2}\,>\,140\,\text{h}} \text{(2.65)} \tag{2.9}$$

Similarly, a highly strained molecule such as tri-*t*-butylprismane, (2.66), has a half life of 18 h in pyridine at 110 °C [166]. The high energy barrier to isomerization in these type of systems has been rationalized in terms of the need to conserve orbital

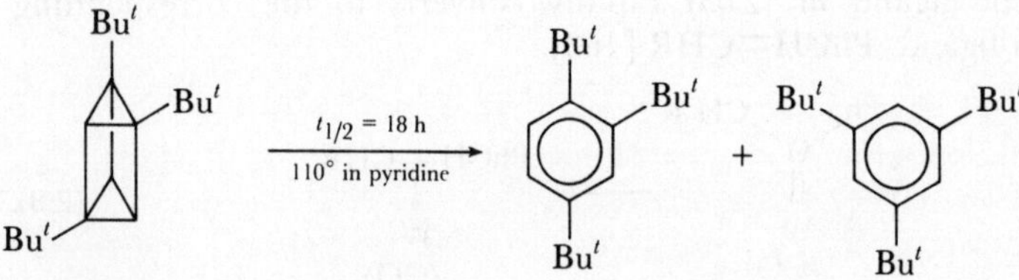

$$\text{(2.66)} \xrightarrow[110°\text{ in pyridine}]{t_{1/2}\,=\,18\,\text{h}} \quad + \quad \tag{2.10}$$

symmetry during a concerted process, since the type of transformations outlined above are, using the Woodward–Hoffmann approach [168], 'symmetry-forbidden.' In the late sixties it was discovered that several transition-metal complexes were extremely efficient at catalysing such symmetry forbidden skeletal isomerization. One of the first reports concerned the ring opening of quadricyclene, Equation (2.9). In the presence of certain homogeneous rhodium, palladium or platinum complexes, e.g. [Rh(norbornadiene)$_2$Cl]$_2$, quadricyclene has a half life of only 45 min. at −26 °C [169]. Similarly, in the presence of silver nitrate the substituted prismane compound (2.66) has, in methanol solution, a half life of less than a minute, being rapidly converted to the isomer mixture shown below [170]:

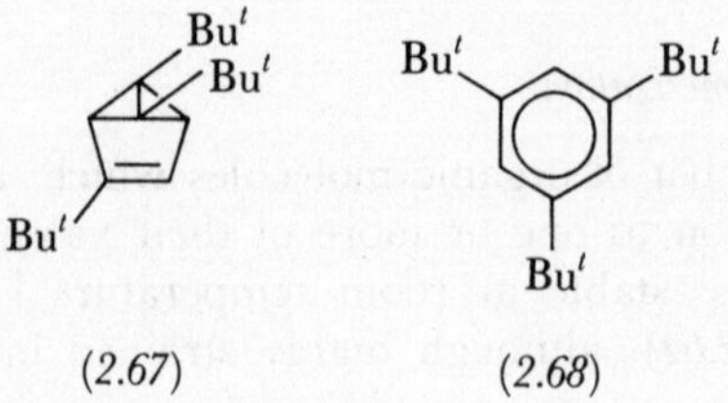

(2.67)   (2.68)

HOMOGENEOUS CATALYST SYSTEMS IN OPERATION 83

(2.69)   (2.70)

Further isomerization converts (2.67) into (2.68) and (2.69) into (2.70). Two basic mechanistic sequences have been proposed to account for these, and associated, metal catalysed skeletal isomerization reactions; one involving a concerted process and one involving a step-wise sequence.

(a) *Skeletal isomerization via a concerted process*
A metal can catalyse a symmetry forbidden transformation by offering the system a way round the symmetry constraint [171]. However, before being able to understand how a system can get round the constraint with the aid of a metal, we must first briefly examine the nature of the constraint. Considering the quadricyclene isomerization shown in Equation (2.9), we can reduce the problem to

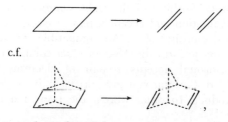

c.f.

i.e. the conversion of cyclobutane into two molecules of ethene. Symmetry considerations [171] allow us to label the σ-orbital bonding combinations of the cyclobutane as SS and SA and the σ-antibonding combinations AS and AA, where S and A refer to symmetric and antisymmetric relative to certain chosen symmetry elements in the molecular combination. Relative to the same symmetry elements the π-orbital bonding combinations of the two product ethene molecules can be designated SS and AS and the π-antibonding combinations SA and AA. We can now construct an orbital correlation diagram for the process as shown in Fig. 2.10.

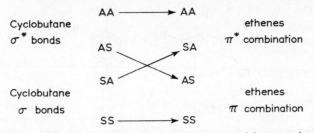

Fig. 2.10 *Correlation diagram for the conversion of a cyclobutane molecule into two ethene molecules*

Molecular orbital symmetry conservation requires that molecular orbitals maintain their symmetry relative to the chosen symmetry elements on progressing along the reaction co-ordinate. Thus, the cyclobutane SA-σ combination transforms into the olefinic SA π^*-combination and the AS-σ^* combination into the AS-π combination. The resulting orbital crossing corresponds to a symmetry forbidden reaction. To overcome the symmetry constraint, we effectively need to remove an electron pair from the filled SA combination and 'change' its symmetry to AS so that it can form the bonding π-AS combination in the product. This is where the transition metal, with its partially filled d-orbitals, can come in. An empty d-orbital of the right symmetry can interact with the SA combination and accept a pair of electrons. If at the same time a filled d-orbital of an AS compatible symmetry is in the right orientation to form the AS-π product combination, then, as the empty d-orbital accepts a pair of electrons from the substrate, the filled d-orbital can give up its pair of electrons to the incipient product. In other words, the transition metal can

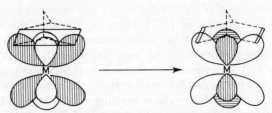

Fig. 2.11 *Schematic representation of the metal catalysed isomerization of quadricyclene;* ⌒ *corresponding to filled and* ⌒ *to empty metal d-orbitals*

offer the system a route round the symmetry impasse. For the quadricyclene transformation this intervention, following the treatment of Mango [171], can be schematically represented as shown in Fig. 2.11.

An alternative mechanism to the concerted sequence outlined above is a non-concerted pathway, or step-wise mechanism, involving the initial breaking of only one of the carbon–carbon bonds through either an oxidative addition reaction or a heterolytic addition process.

(b) Skeletal isomerization via a step-wise process

Sigma bonds in strained carbocyclic systems such as cyclopropane or cyclobutane are rich in p character and are, therefore, well placed to interact with the d-orbital system of a transition-metal complex. The bonding is essentially analogous to that described (Section 1.2.1) for an alkene–metal complex and involves a σ-component, resulting from electron donation from the strained carbon–carbon bond into an empty metal d-orbital of suitable symmetry, together with a weak π-component resulting from back donation from a filled metal orbital into an antibonding σ^*-orbital of the carbocycle [166].

This situation, which is clearly very similar to that prevailing in the transition state of the concerted process we have just considered, can lead to oxidative addition products which can then serve as intermediates in step-wise skeletal re-arrangement processes. Strong evidence in support of such an oxidative addition pathway is provided by the isolation of acyl rhodium(III) complexes from a number of rhodium(I) catalysed isomerization reactions. Thus, addition of the rhodium(I) chloro-bridged carbonyl dimer, $[RhCl(CO)_2]_2$, to quadricyclene results, in addition to isomerization of the quadricyclene to norbornadiene, in the formation of an acyl-rhodium(III) complex (2.71) isolated as a chlorobridged, polymeric material, with n probably equal to four [172]. It is thought that initial oxidative addition of one of the cyclopropyl σ-bonds of the quadricyclene gives (2.72) which can then undergo a ligand migration (alkyl to carbonyl carbon) to form the stable acyl derivative (2.72). Clearly, rhodium(III) complexes of the type (2.72) could play a significant role in a non-concerted rhodium catalysed isomerization of quadricyclene

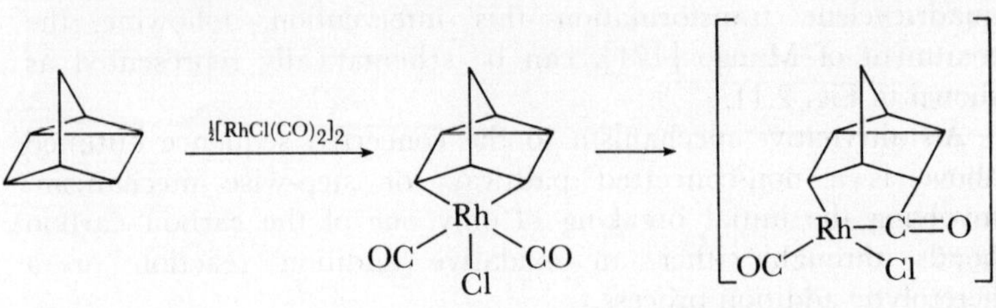

$$\text{quadricyclene} \xrightarrow{\frac{1}{2}[RhCl(CO)_2]_2} \begin{array}{c} \text{Rh} \\ OC \underset{Cl}{|} CO \end{array} \longrightarrow \left[\begin{array}{c} Rh-C=O \\ OC \quad Cl \end{array} \right]_n$$

c.f.   (2.72)   (2.71)

$$\xrightarrow{+Rh^I} \quad Rh^{III} \quad \xrightarrow{-Rh^I} \quad \text{norbornadiene}$$

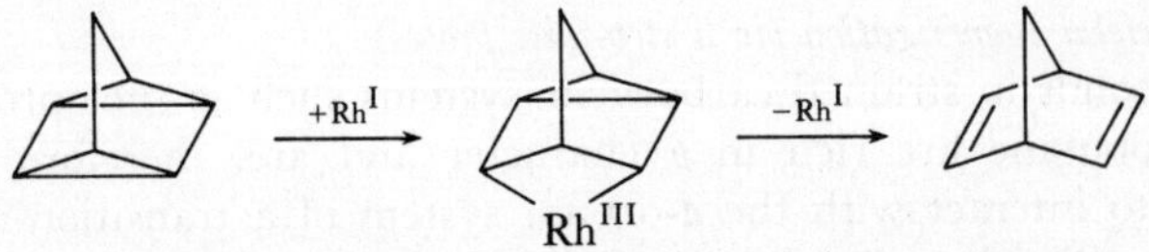

Similar oxidative addition pathways have been proposed for the rhodium(I) catalysed isomerization of cubane [173]

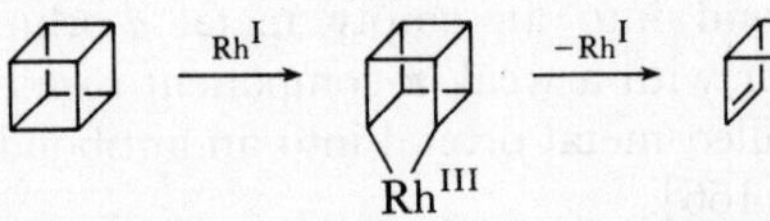

$$\text{cubane} \xrightarrow{Rh^I} \quad Rh^{III} \quad \xrightarrow{-Rh^I} \quad$$

and for the rhodium(I) or palladium(II) catalysed isomerization of homocubanes to dienes [174]

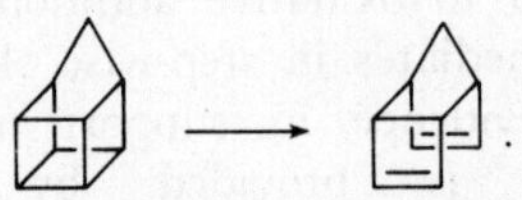

In both cases, the corresponding rhodium(III) acyl complexes, (2.73) and (2.74), have been isolated [173,175] from the stoichiometric reaction between the strained carbocycle and $[RhCl(CO)_2]_2$.

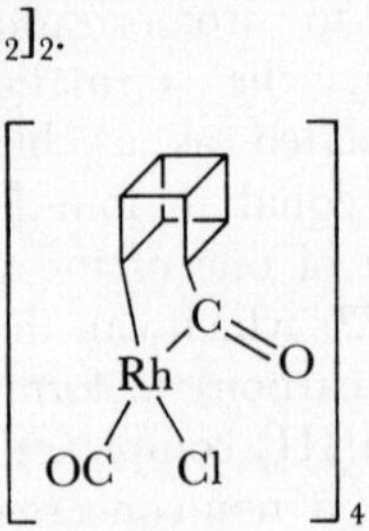

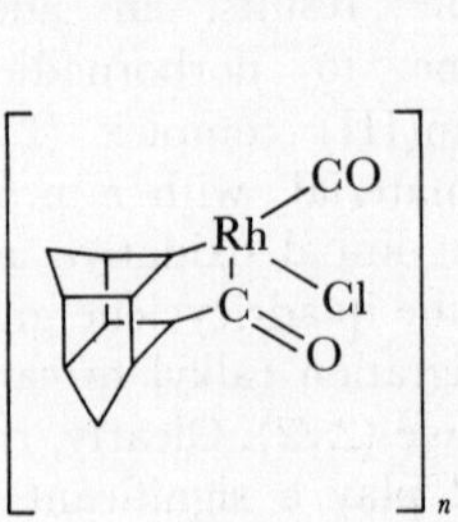

$$\left[\begin{array}{c} C=O \\ Rh \\ OC \quad Cl \end{array} \right]_4 \quad\quad \left[\begin{array}{c} Rh \underset{C}{\diagdown} \underset{\diagdown O}{\overset{CO}{\diagup}} Cl \end{array} \right]_n$$

         (2.73)   (2.74)

HOMOGENEOUS CATALYST SYSTEMS IN OPERATION

In addition to the concerted oxidative addition type of mechanism outlined above, transition metals can also catalyse skeletal rearrangement via a step-wise pathway involving metallocarbocations. In this type of mechanism the strained carbocycle is activated by internal heterolytic addition at the metal centre. Thus in the Ag(I) catalysed rearrangement of bicyclopropenyls to Dewer-benzenes, Equation (2.12), the silver-carbocation (2.75) formed, at least formally, by heterolytic addition of a cyclopropenyl σ-bond to the metal centre, is the most likely initial intermediate. Rearrangement, via bond migration, gives the carbocation (2.76) which then collapses to the product Dewer-benzene [176]

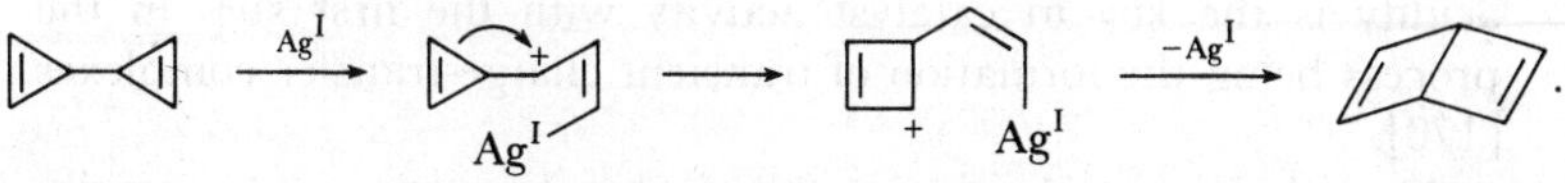

$$(2.12)$$

Evidence for both carbocations has been obtained from 'solvolytic interception', i.e. using a suitable nucleophile to trap the carbocation. Thus, from the silver perchlorate catalysed isomerization of (2.77) in methanol at −20 °C (2.78) has been isolated [177] and, under similar conditions, the methoxy carbocycle (2.80) has been recovered from the silver catalysed isomerization of (2.79) [178].

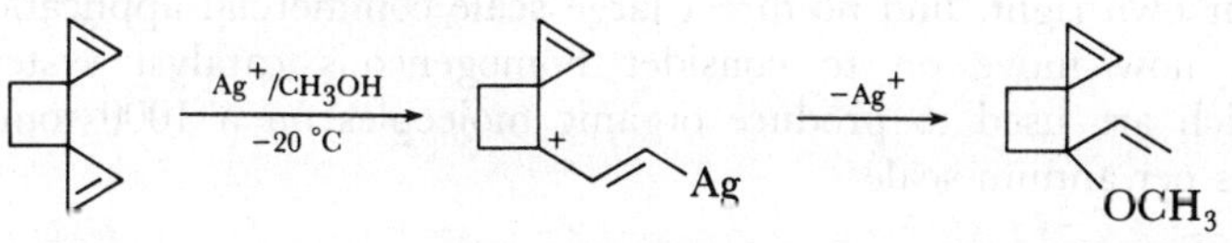

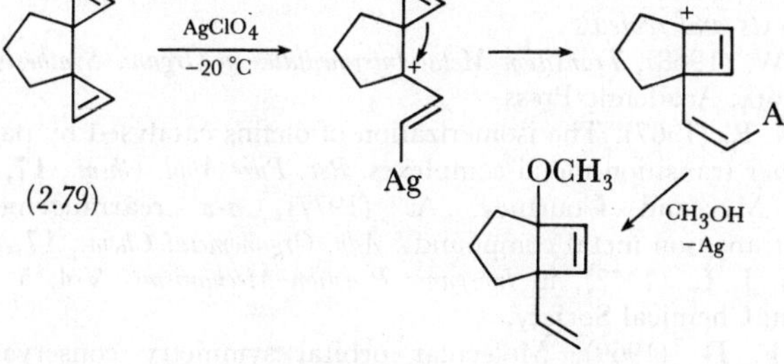

Catalysis of skeletal rearrangement processes are by no means the sole preserve of transition metals. In many cases simple protonation of the strained carbocycle, followed by bond rupture and rearrangement of the resulting carbononium ion, offers a facile route round the symmetry constraint. Thus, for the rearrangement of tri-t-butylprismane (2.66), Equation (2.10), 0.003 M hydrochloric acid is an effective catalyst ($t_{1/2}$ = 35 min), however, only the 1,2,4-substituted benzene product is obtained in contrast to the four products obtained with silver nitrate as catalyst (*see above*). With the prismane rearrangement a wide variety of catalysts, including Hg(II), Sn(II), Zn(II), Pb(II) and even 1,3,5-trinitrobenzene, are effective. It is suggested that Lewis acidity is the key to catalyst activity with the first step in the process being the formation of transient charge-transfer complexes [170].

As with most transition-metal catalysed reactions, the precise mechanistic steps, in so much as these are ever precisely defined, depend on the nature of the substrate and the metal catalyst, however, irrespective of the doubts which may surround the mechanism(s), there are no doubts attached to the ability of transition-metal complexes to catalyse a diverse range of isomerization/rearrangement processes.

In the last two sections we have concentrated on homogeneously catalysed reactions which, while important in their own right, find no direct large scale commercial application. We now move on to consider homogeneous catalyst systems which are used to produce organic molecules on a 1000 tonnes plus per annum scale.

Further reading on transition-metal catalysed isomerization

General texts and reviews

Bird, C. W. (1968), *Transition Metal Intermediates in Organic Synthesis*, p. 69, New York: Academic Press.

Davies, N. R. (1967), The isomerization of olefins catalysed by palladium and other transition-metal complexes. *Rev. Pure Appl. Chem.*, **17**, 83.

Tsutsui, M. and Courtney, A. (1977), σ-π rearrangements of organotransition metal compounds. *Adv. Organometal Chem.*, **17**, 241.

Davidson, J. L. (1977), in *Inorganic Reaction Mechanisms*, Vol. 5, p. 346, London: Chemical Society.

Mango, F. D. (1969), Molecular orbital symmetry conservation in transition metal catalysis. *Adv. Catal.*, **20**, 291.

Paquette, L. A. (1975), Preparative aspects of silver ion catalysed rearrangements of polycyclic systems. *Synthesis*, 347.

Bishop, K. C. (1976), Transition metal catalysed rearrangements of small ring organic molecules. *Chem. Rev.*, **76**, 461.

2.3 Carbonylation

The two basic technical breakthroughs in the field of homogeneous catalysed carbonylation reactions both occurred in Germany during the late 1930s and early 1940s. In the 1930s, Otto Röelen of Ruhrchemie was working on Fisher–Tropsch type chemistry, i.e. the synthesis of hydrocarbons from carbon monoxide and hydrogen, when he discovered [179,180] that alkenes could be converted to aldehydes by treatment with CO and H_2 in the presence of a cobalt catalyst at elevated temperatures and pressure

$$\mathrm{\searrow\!\!C\!=\!C\!\!\swarrow} + CO + H_2 \xrightarrow{Co} H-\underset{|}{\overset{|}{C}}-\underset{|}{\overset{|}{C}}-C\underset{H}{\overset{O}{\diagup}}$$

This process, named hydroformylation, since it formally involves the addition of formaldehyde (H—CHO) to the alkene, has been developed over the last 40 years to the stage where it now constitutes one of the major industrial applications of homogeneous catalyst systems.

The second major breakthrough occured at about the same time when W. Reppe [181] of I. G. Farben discovered that Group VIII metal–carbonyl complexes were capable of catalysing a whole series of carbonylation reactions involving alkynes, alkenes or alcohols as substrates and giving rise to, depending on the co-reactant, carboxylic acids, when the co-reactant is water, or carboxylic acid derivatives, when the co-reactant is either an alcohol or an amine. Some typical examples of 'Reppe reactions' are given below. Those involving carbon monoxide and water are usually referred to as hydroxycarboxylation reactions.

$$HC\equiv CH + CO + H_2O \longrightarrow H_2C\!=\!CHCOOH$$

$$HC\equiv CH + CO + ROH \longrightarrow H_2C\!=\!CHCOOR$$

$$R'C\equiv CH + CO + ROH \longrightarrow H_2C\!=\!C(R')COOR$$

$$RCH\!=\!CH_2 + CO + H_2O \longrightarrow RCH_2CH_2COOH$$

$$RCH=CH_2 + CO + R'OH \longrightarrow RCH_2CH_2COOR'$$
$$RCH=CH_2 + CO + R'NH_2 \longrightarrow RCH_2CH_2CONHR'$$
$$RCH_2OH + CO \longrightarrow RCH_2COOH.$$

In this section we shall consider each of these three types of 'Reppe reactions', i.e. those involving acetylenic substrates, those involving olefinic substrates and those involving alcoholic substrates, in turn, and in Section 2.4 we will go on to look at homogeneously catalysed hydroformylation reactions. Since Reppe chemistry is a subject in its own right, we can do no more than scrape the surface in the next ten or so pages and a list of more comprehensive texts is given at the conclusion of the section.

2.3.1 Alkyne carbonylation

When an aqueous organic solution of acetylene is treated with carbon monoxide at 150 °C and 30 atm in the presence of catalytic amounts of nickel tetracarbonyl, acrylic acid is formed with high (>90%) selectivity [182]

$$HC\equiv CH + H_2O \xrightarrow[150\ °C/30\ atm]{Ni(CO)_4} CH_2=CHCO_2H.$$

When the synthesis is carried out in aqueous alcohol, ROH, the corresponding acrylic ester, $CH_2=CHCO_2R$, is formed. With methyl acetylene as the alkyne substrate, and aqueous methanol as the reaction medium, the major product (>80%) is methyl methacrylate

$$CH_3C\equiv CH + CO + MeOH \longrightarrow CH_2=C(Me)CO_2Me.$$

Although other Group VIII metal carbonyl complexes, e.g. $Fe(CO)_5$, will also catalyse these reactions the preferred metal, in terms of both rate and selectivity, is undoubtedly nickel. On a commercial scale, the nickel is usually added as either nickel bromide or nickel iodide, both of which are rapidly converted to nickel carbonyl species under the reaction conditions, typically ±200 °C and 100 atm. When water is the sole co-reactant, i.e. the free acid is the required product, tetrahydrofuran is frequently used as solvent [53]. A generalized catalysis cycle for the process is shown in Fig. 2.12.

The active catalytic species is thought [183,184] to be a hydrido nickel carbonyl species of the type (2.81), formed by

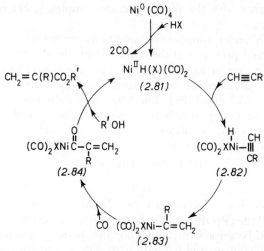

Fig. 2.12 *Catalysis cycle for the nickel catalysed alkyne carboxylation reaction*

oxidative addition of HX to the nickel tetracarbonyl complex. Since $Ni(CO)_4$ is an 18 valence electron species, the oxidative addition reaction is almost certainly preceded by dissociation of at least one carbonyl ligand. When the nickel is added in the form of a nickel(II) halide the HX is the corresponding hydrogen halide formed during the initial reduction of NiX_2 to $Ni(CO)_4$

$$NiX_2 + 5CO + H_2O \longrightarrow Ni(CO)_4 + CO_2 + 2HX.$$

In the absence of added halide (*2.81*) can be formed, albeit more slowly, from either the water or alcohol co-reactant in which case X = OH or OR. As far as the rate of reaction is concerned, the best results are obtained when X = I [185] although in commercial operation X = Br is generally preferred mainly due to the corrosion problems associated with HI. Once formed, the hydrido carbonyl species can co-ordinate the alkyne substrate RC≡CH to give the 18 valence electron nickel(II) intermediate (*2.82*). Markovnikov addition of the nickel–hydride ligand to the co-ordinated alkyne, probably via a four-centre transition state similar to that described for hydrogenation catalysis in Section 2.1, gives the σ-alkene complex (*2.83*). In coming ligand, i.e. CO, assisted ligand migration, i.e. C-alkene to carbonyl, converts (*2.83*) into (*2.84*) and either hydrolysis, R′ = H, or alcoholysis, R′ = alkyl, to give either the acrylic acid or the corresponding R′

ester, together with the nickel–hydrido complex (*2.81*) completes the cycle.

Although under commercial conditions temperatures in excess of 200 °C and pressures of the order of 100 atm are generally used [53], in the case of methyl methacrylate selectivities of over 80% have been obtained at temperatures and pressures as low as 132 °C and 13.5 atm [184]. The major by-product of the reaction is methyl crotonate which is formed by the anti-Markovnikov addition of hydride to alkyne in (*2.82*) to give the linear σ-alkene intermediate (*2.85*), R = Me

$$(CO)_2XNi-CH=CHR.$$

(*2.85*)

Most work has concentrated on the use of nickel catalysts, however $PdBr_2[P(OPh)_3]_2$ in the presence of perchloric acid, is reported to be an active catalyst for the methoxycarboxylation of acetylene giving methyl acrylate with 95% selectivity under mild conditions [186]. As with nickel, it appears likely that a hydrido intermediate is the active catalytic species. In the absence of acid, but in the presence of thiourea, NH_2CSNH_2, and a trace of oxygen, palladium chloride catalyses the reaction between acetylene, carbon monoxide and methanol to give dimethyl maleate (*2.86*) with 90% selectivity under ambient conditions of temperature and pressure [187]

$$CH\equiv CH + CO + MeOH \xrightarrow[O_2]{PdCl_2/NH_2CSNH_2} MeO_2CCH=CHCO_2Me.$$

(*2.86*)

The proposed catalysis cycle for this system is shown in Fig. 2.13. The active species is thought to be a palladium(II) carbonyl cation, (*2.87*), containing two *trans*-thiourea ligands. Nucleophilic attack by methanol, or methoxy anion, on the co-ordinated carbonyl gives the 16 valence electron carboxy species (*2.88*). Acetylene co-ordination followed by carboxy ligand migration, probably via the ubiquitous four-centre transition state, gives the σ-carboxyalkene intermediate (*2.89*) which can then co-ordinate a carbon monoxide before undergoing a further ligand migration reaction, carboxylalkene to carbonyl carbon, to form (*2.90*). Nucleophilic attack by methanol on the metal bonded carbonyl carbon liberates the product maleate and leaves the palladium-hydrido species (*2.91*). In the presence of oxygen, HCl

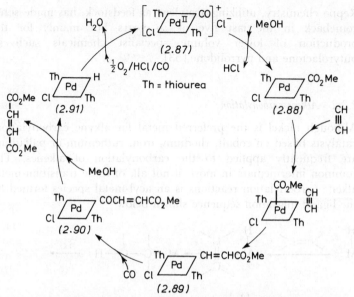

Fig. 2.13 *Catalysis cycle for the palladium catalysed synthesis of dimethyl maleate from acetylene, carbon monoxide and methanol*

and carbon monoxide (2.91) is somehow converted into the active species (2.87). The mechanism for this latter step is not yet clear and the production of water shown in the last step of the cycle is merely speculative.

Little, if any, commercial use is presently made of these palladium catalysed alkyne carbonylations although they begin to show the potential versatility of this type of chemistry. The nickel catalysed system for the production of acrylic acid and esters, especially methylmethacrylate, is still used on a commercial scale, especially in Germany and Eastern Europe, but this route is declining in importance in the path of alternative processes based on cheaper feedstocks such as propylene and acetone. Most of the western world's requirement of methyl methacrylate, an important monomer for the production of acrylic sheet, i.e. a clear polymeric material extensively used as an alternative to glass, is now produced from acetone via the cyanohydrin route:

$$CH_3COCH_3 + HCN \longrightarrow (CH_3)_2C(OH)CN$$

$$(CH_3)_2C(OH)CN + CH_3OH + H_2SO_4 \longrightarrow$$
$$CH_2=C(CH_3)COOCH_3 + NH_4HSO_4.$$

Reppe chemistry, utilizing acetylene as feedstock, has made some comeback in the last five or so years but mainly for the production of lower volume specialist chemicals such as butyrolactone and pyrrolidone [53].

2.3.2 *Alkene carbonylation*

Although nickel is the preferred metal for alkyne carbonylation, catalysts based on cobalt, rhodium, iron, ruthenium or palladium are frequently applied to the carbonylation of alkenes. The common intermediate in most, if not all, of these transition-metal alkene carbonylation reactions is an acyl-metal species formed by the ligand migration sequence shown below

$$
\begin{array}{c}
\text{H} \\
| \\
\text{M}
\end{array}
\xrightarrow{\text{>C=C<}}
\begin{array}{c}
\text{H} \quad \text{C} \\
| \quad \diagdown \\
\text{M} \leftarrow \| \\
\quad \diagup \\
\text{C}
\end{array}
\longrightarrow
\text{M}-\overset{|}{\underset{|}{\text{C}}}-\overset{|}{\underset{|}{\text{C}}}-\text{H}
\xrightarrow{\text{CO}}
$$

$$
\underset{\text{M}-\overset{\overset{\text{O}}{\|}}{\underset{|}{\text{C}}}-\overset{|}{\underset{|}{\text{C}}}-\overset{|}{\underset{|}{\text{C}}}-\text{H}}{} \longrightarrow \text{M}-\overset{\overset{\text{O}}{\|}}{\underset{|}{\text{C}}}-\overset{|}{\underset{|}{\text{C}}}-\overset{|}{\underset{|}{\text{C}}}-\text{H}.
$$

Product formation can be visualized as occurring via nucleophilic attack at the carbonyl carbon of the acyl-metal species with, in the case of a nucleophile of the type HNu, concurrent regeneration of a metal hydride species

$$\text{M}-\overset{\overset{\text{O}}{\|}}{\underset{|}{\text{C}}}-\overset{|}{\underset{|}{\text{C}}}-\overset{|}{\underset{|}{\text{C}}}-\text{H} \xrightarrow{\text{HNu}} \text{M}-\text{H} + \text{H}-\overset{|}{\underset{|}{\text{C}}}-\overset{|}{\underset{|}{\text{C}}}-\text{C}\overset{\diagup \text{O}}{\diagdown \text{Nu}} \quad (2.13)$$

The type of product obtained from such a reaction clearly depends to a large extent on the nature of the nucleophilic species, e.g. with HNu = HOH, acids are formed, with HNu = HOR, esters with HNu = H_2NR, amides are the principal products. When HNu is hydrogen, aldehydes are formed, however we shall reserve consideration of this latter type of reaction for the section on hydroformylation (Section 2.4).

The alkene carbonylation reaction is not always quite so

straightforward as suggested by Equation (2.13). Thus carbonylation of ethene in the presence of an alcohol, ROH, with a palladium or nickel catalyst gives, in addition to the expected product $CH_3CH_2CO_2R$, the keto-ester product $CH_3CH_2C(O)CH_2CH_2CO_2R$ incorporating two ethene and two carbon monoxide molecules [188].

$$CH_2=CH_2 + CO + ROH \longrightarrow CH_3CH_2CO_2R + CH_3CH_2\overset{O}{\underset{\|}{C}}CH_2CH_2CO_2R$$

The formation of the γ-ketocaproic ester (*2.93*) can be rationalized on the basis of the following ligand migration sequence, which is closely analogous to that previously described for the palladium catalysed dimethyl maleate synthesis (Fig. 2.13).

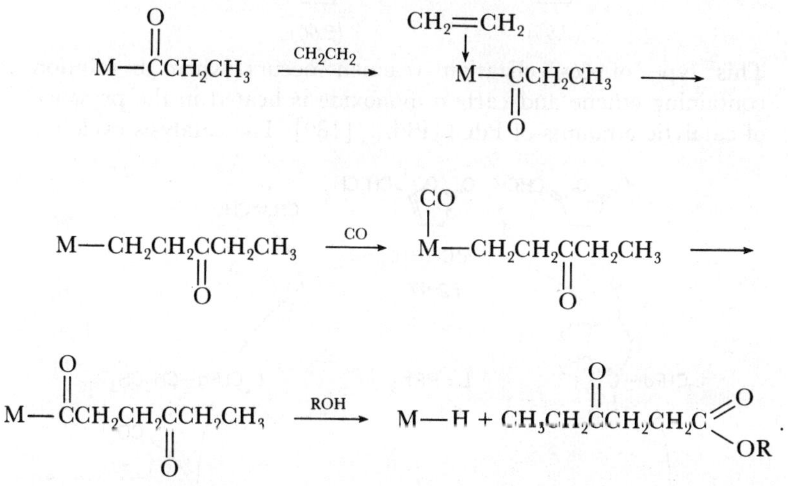

$$M-\overset{O}{\underset{\|}{C}}CH_2CH_3 \xrightarrow{CH_2CH_2} \overset{CH_2=CH_2}{\underset{\downarrow}{}} M-\underset{\underset{O}{\|}}{C}CH_2CH_3 \longrightarrow$$

$$M-CH_2CH_2\underset{\underset{O}{\|}}{C}CH_2CH_3 \xrightarrow{CO} M-\underset{\underset{O}{\|}}{\overset{CO}{\underset{|}{C}}}CH_2CH_2CCH_2CH_3 \longrightarrow$$

$$M-\overset{O}{\underset{\|}{C}}CH_2CH_2\underset{\underset{O}{\|}}{C}CH_2CH_3 \xrightarrow{ROH} M-H + CH_3CH_2\overset{O}{\underset{\|}{C}}CH_2CH_2C\overset{O}{\underset{OR}{\diagdown}}$$

(*2.92*)   (*2.93*)

In the absence of an external nucleophile the di-keto complex (*2.92*) can undergo an internal nucleophilic attack, or cyclization reaction, to give the lactone species (*2.94*). When M = Pd the cyclization, which results from nucleophilic attack by the carbonyl oxygen in the 4 position on the palladium acyl carbon, is accompanied by metal migration from carbon 1 to carbon 4. The metal substituted lactone species (*2.94*) can breakdown via beta hydrogen elimination, from either carbon 3 or carbon 5, to give the two unsaturated lactone products (2.95) and (2.96)

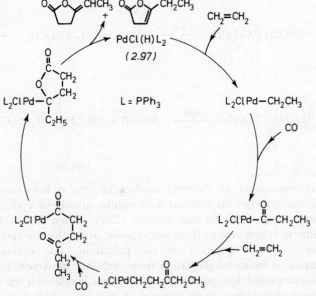

This type of co-cyclization reaction occurs when a solution containing ethene and carbon monoxide is heated in the presence of catalytic amounts of $PdCl_2(PPh_3)_2$ [189]. The catalysis cycle for

Fig. 2.14 *Catalysis cycle for the palladium catalysed co-cyclization of ethene and carbon monoxide*

the latter system is shown in Fig. 2.14. The active catalytic species is thought to be a palladium-hydrido complex (2.97), initially formed from the reaction between $PdCl_2(PPh_3)_2$ and ethene [190]

$$PdCl_2(PPh_3)_2 + CH_2CH_2 \longrightarrow PdCl(H)(PPh_3)_2 + CH_2{=}CHCl$$

(2.97)

Similar co-cyclization reactions have also been observed with both nickel and cobalt carbonyl catalysts [190]. As far as large scale industrial application is concerned the major Reppe type alkene carbonylation process is that for the production of propionic acid from ethene using nickel carbonyl as catalyst under an operating pressure of 200–250 atm [53].

2.3.3 *Alcohol carbonylation*

Undoubtedly the most significant industrial development of Reppe type chemistry in recent years has been the discovery by Paulik and Roth of Monsanto that, in the presence of iodide, certain soluble carbonyl complexes of rhodium or iridium are capable of catalysing the carbonylation of methanol to form acetic acid under low (30–40 atm) pressure conditions

$$CH_3OH + CO \longrightarrow CH_3COOH.$$

An alternative process, using an iodide-promoted cobalt carbonyl catalyst, had previously been developed by BASF, however, the Monsanto process, using rhodium as catalyst, has the advantage of, working under considerably milder conditions (~180 °C/35 atm as compared to ~230 °C/600 atm for cobalt), requiring a lower catalyst concentration (~10^{-3} M as compared to ~10^{-1} M), and giving greater product selectivity (>99% compared to ~90%) [45]. In a world of expensive energy, where we need to make the most efficient use of raw material, the lower energy requirements, i.e. milder operating conditions, and greater selectivity of the rhodium system balance, if not outweigh the greater cost of rhodium compared to cobalt. A similar situation is arising with respect to the cobalt and rhodium hydroformylation catalyst which we shall be considering in the next section. Monsanto started large scale production of acetic acid using the rhodium system in 1970 and suggest that by the mid-eighties over 1 million tonnes per annum will be produced using their process [45].

The active catalytic species is the dicarbonyldiiodorhodium(I) anion, (2.98). When the rhodium is added in the form of the trihalogeno salt, RhX_3, the anion can be rapidly formed under the reaction conditions via the reductive carbonylation sequence [192,193]

$$RhX_3 + 3CO + H_2O \longrightarrow [RhX_2(CO)_2]^- + CO_2 + 2H^+ + X^-.$$

(2.99)

If X is other than iodo then (2.99) converts to (2.98) by halide exchange with the iodide promotor present in solution. Iodide is usually added to the system as methyl iodide, aqueous hydroidic acid or iodine. Adding the rhodium in the form of a tertiary-phosphine complex, e.g. $RhX(CO)(PPh_3)_2$, does not essentially alter the nature of the active catalytic species since, under the reaction conditions, the rhodium–phosphorus bond is ruptured and the tertiary-phosphine ligand rapidly converted to a non-bonding quaternary-phosphine salt, e.g. $[P(Ph)_3Me]^+I^-$. The suggested [45] catalysis cycle for the process is shown in Fig. 2.15. As far as the rhodium is concerned methyl iodide rather than methanol is the substrate and the first stage of the cycle

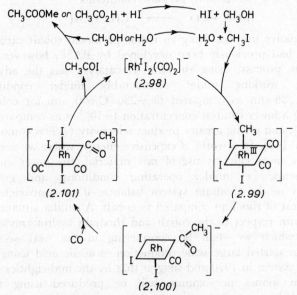

Fig. 2.15 *Catalysis cycle for the rhodium catalysed carbonylation of methanol*

involves oxidative addition of methyl iodide to the 16 valence electron rhodium(I) species (*2.98*) to form the 18 valence electron octahedral rhodium(III) complex (*2.99*). Ligand migration, methyl to carbonyl carbon, converts (*2.99*) into the 16 valence electron acyl species (*2.100*) which quickly picks up a carbon monoxide molecule to return to an 18 valence electron rhodium(III) species (*2.101*). Reductive elimination of acetyl iodide from (*2.101*) completes the cycle. The acetyl iodide is rapidly solvolysed by either the methanol or the water, which are present in the reaction medium, to give either methyl acetate or acetic acid together with hydrogen iodide. The hydrogen iodide thus formed reacts with the methanol substrate to give methyl iodide which keeps the rhodium happy. Under the carbonylation conditions the equilibrium lies well to the right so that under operating conditions the concentration of free HI is very low. This

$$CH_3OH + HI \rightleftharpoons CH_3I + H_2O$$

is important from the practical standpoint of reactor/plant design since HI is one of the least loved of substances among chemical engineers, owing to its corrosive properties.

Kinetic data obtained from the rhodium catalysed carbonylation system [194] indicate that the rate-determining step in the catalysis cycle is the oxidative addition of methyl iodide to the rhodium(I) species (*2.98*). As would be expected on the basis of this finding, bromide and chloride are much less efficient at promoting the reaction than iodide since, in general, the rate of oxidative addition of RX to d^8 metal–halide complexes decreases in the order I > Br > Cl [33]. In the case of this particular system the rate of formation of the bromide analogue of (*2.99*), by oxidative addition of methyl bromide to $[RhBr_2(CO)_2]^-$, is at least an order of magnitude slower than the iodide reaction [45]. Further evidence for the addition of methyl iodide being the rate-determining step has been obtained [195] by measuring the infrared spectrum of the reaction solution under typical operating conditions (100 °C/6 atm) with the aid of a high temperature, high pressure spectrophotometric cell. The spectrum contained two strong bonds at 1996 and 2067 cm^{-1} characteristic of the $[RhI_2(CO)_2]^-$ anion indicating that this is the major metal species present in the catalysis solution and strongly suggesting that the oxidative addition reaction is rate determining. The application of infrared spectroscopy to this particular reaction provides yet another example of the amenability of homogeneous catalyst systems to detailed spectroscopic study.

Iridium carbonyl complexes are also effective catalysts for the carbonylation of methanol under relatively mild conditions [191]. However, compared to rhodium, the behaviour of such systems is somewhat more involved [45]. There appear to be two possible catalysis cycles open to the system: which the iridium follows depends to a large extent on the iodide concentration which is the principal factor in determining the position of equilibrium (2.14);

$$\text{IrI(CO)}_3 \underset{+\text{CO}-\text{I}^-}{\overset{+\text{I}^--\text{CO}}{\rightleftharpoons}} [\text{IrI}_2(\text{CO})_2]^- \quad (2.14)$$

$$(2.102) \qquad\qquad (2.103)$$

The two catalysis cycles proposed for the iridium system are shown in Fig. 2.16. Although (2.103), favoured by high iodide concentrations, is completely analogous to the dicarbonyldi-iodorhodium(I) species (2.98), the catalysis cycle, A, followed

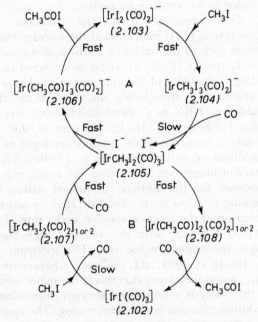

Fig. 2.16 *Catalysis cycles for the indium catalysed carbonylation of methanol*

by the iridium complex differs markedly from that traversed by its rhodium counterpart, cf. Fig. 2.15. Addition of methyl iodide to (*2.103*) to give the octahedral iridium(III) methyl complex (*2.104*) is a fast reaction, cf. rhodium where it is rate-determining. However, conversion of (*2.104*) to the acyl complex (*2.106*) is slow and is thought to go via the 18 valence electron tricarbonyldi-iodo methyl intermediate (*2.105*). (*2.105*) has not been detected spectroscopically and its proposed presence, albeit at extremely low concentration, is based on kinetic data. Whatever the route from (*2.104*) to (*2.105*), there is no doubt that, in marked contrast to the analogous rhodium transformation, it is a slow process. Under the reaction conditions (*2.104*) is the predominant species detected by infrared spectroscopy and in the absence of a positive carbon monoxide pressure (~5 atm) it shows little inclination to rearrange to an acyl complex. Furthermore, the presence of an excess of iodide, needed to favour formation of (*2.103*), inhibits the conversion of (*2.104*) into (*2.106*). Once formed the acyl species (*2.106*) undergoes a fast reductive elimination to give acyl iodide and complete the catalysis cycle. At low iodide levels the predominant iridium species observed spectroscopically is the neutral tricarbonyliodo complex (*2.102*) and the carbonylation reaction follows the lower, B, catalysis cycle. In this case, addition of methyl iodide with concurrent loss of CO is rate-determining. Independent experiments, i.e. not under catalytic carbonylation conditions, have shown [45] that $IrI(CO)_3$ reacts with methyl iodide to give, as major product, the dimeric species $[IrCH_3I_2(CO)_2]_2$; however, it is not clear whether, under carbonylation conditions (~10 atm, 125 °C) (*2.107*) is present as a monomer or a dimer. A similar situation exists for the acyl species (*2.108*), which is rapidly formed from (*2.107*) almost certainly via the 18 valence electron tricarbonyl species (*2.105*). Rearrangement of (*2.107*) to an acyl species prior to co-ordination of carbon monoxide would entail a 16 to 14 valence electron transformation which according to Tolman's 16 and 18 electron rule (see Section 1.6.1) has little to recommend it. Once formed the acyl species reductively eliminates acyl iodide, co-ordinates a molecule of CO and completes the cycle.

Although overall, the rhodium and iridium catalysts give comparable rates under commercially feasible conditions, the complexity of the iridium system makes it unlikely that it will, in the absence of other considerations, significantly replace rhodium as a homogeneous catalyst for the carbonylation of methanol to acetic acid.

We began this section by pointing out that homogeneous rhodium catalyst are making major inroads into the cobalt based processes available for the manufacture of acetic acid. We now move on to consider another process, or class of reaction, where rhodium systems are beginning to challenge conventional cobalt catalysts, namely, hydroformylation.

Further reading on homogeneous carbonylation

General texts and reviews
Bird, C. W. (1967), *Transition Metal Intermediates in Organic Synthesis*, Ch. 7 and 8, New York: Academic Press.
Falbe, J. (1970), *Carbon Monoxide in Organic Synthesis*, Berlin: Springer-Verlag.
Ryang, M. (1970), Organic synthesis with metal carbonyls. *Organometal Chem. Rev.*, Section A, **5**, 67.
Thompson, D. T. and Whyman, R. (1971), in *Transition Metals in Homogeneous Catalysis* (ed. G. N. Schrauzer), p. 147, New York: Marcel Dekker.
Maitlis, P. M. *The Organic Chemistry of Palladium*, Vol. II, p. 18, London: Academic Press.
Heck, R. F. (1974), *Organotransition Metal Chemistry, A Mechanistic Approach*, Ch. 9, New York: Academic Press.
Henrici-Olivé, G. and Olivé, S. (1976), Reactions of carbon monoxide with transition metal-carbon bonds. *Transition Met. Chem.*, **1**, 77.
Davidson, P. J., Hignett, R. R. and Thompson, D. T. (1977), in *Catalysis*, Vol. I, p. 369, London: Chemical Society.
Forster, D. (1979), Mechanistic pathways in the catalytic carbonylation of methanol by rhodium and iridium complexes. *Adv. Organometal Chem.*, **17**, 255.

2.4 Hydroformylation

In terms of product volume hydroformylation – the addition of the units H and CHO to a double bond – represents one of the largest industrial applications of soluble transition-metal catalysts. Presently between 3.6 and 4.5 million tonnes of aldehydes or derivatives are produced annually using processes based on either homogeneous cobalt or rhodium catalyst systems, and there are several plants, using cobalt catalysts, with individual capacities in excess of 300 000 tonnes per annum [196]. As mentioned in the introduction to Section 2.3, the reaction was discovered in 1938 by O. Roelen, while working for Ruhrchemie in Germany [179,180]. He found that when an olefinic Fischer–Tropsch fraction was recycled over the heterogeneous, cobalt based, Fischer–Tropsch catalyst in the presence of synthesis gas, a mixture of carbon monoxide and hydrogen, there was a marked increase in the concentration of aldehydic products in the recycle stream [185]. Originally, it was thought that the reaction was due to the heterogeneous cobalt catalyst but later work revealed that the actual catalyst was a soluble cobalt carbonyl species with the reaction taking place in a homogeneous liquid phase. The process is frequently referred to as the 'oxo' process, with oxo being short for oxonation, i.e. the adding of oxygen to a double bond. However, the term hydroformylation is descriptively more accurate and probably more useful in characterizing this type of reaction.

The primary product of hydroformylation is an aldehyde containing one more carbon atom than the alkene substrate

$$RCH=CH_2 + CO + H_2 \longrightarrow RCH_2CH_2CHO + RCH(CH_3)CHO \quad (2.15)$$

$$\text{'normal'} \quad \text{'iso'}$$
$$\text{product} \quad \text{product}$$

In the absence of isomerization there are, depending on the orientation of H—CHO addition across the double bond, two possible aldehyde products from an unsymmetrical alkene substrate. With a terminal alkene as substrate, Equation (2.15), the two are referred to as the 'normal' and 'iso' product. In all but a few specialized cases the normal isomer is the preferred commercial product and the normal–iso ratio is an important parameter in industrial hydroformylation processes: generally speaking, the higher it is the better. A major feedstock for the hydroformylation is propene, of the over 5 million tonnes of propene used for the manufacture of chemicals in 1976 in Europe, nearly 1 million tonnes went as hydroformylation feedstock [53] and in the western world some 2.7 million tonnes of butanal are produced annually using this process [1]. Most of the normal butanal is either converted to 2-ethyl hexanol via an aldol condensation/hydrogenation sequence, the ALDOX process, or directly hydrogenated to n-butanol. Since the iso-isomer has considerably fewer end uses and hence a much lower commercial value, high normal–iso ratios are particularly important in propene hydroformylation. Similarly, in the production of higher, C_{7+}, alcohols by hydroformylation and subsequent, or concurrent, hydrogenation, selectivity to the n-isomer is an important consideration in process design since the normal products are especially useful in the manufacture of biodegradable detergents and high-grade plasticizers (plasticizers are chemicals added to synthetic resins, e.g. polyvinylchloride, to improve among other things their moulding properties).

Three types of homogeneous transition-metal complexes are used in industrial hydroformylation processes. In order of both their present commercial importance and historic development, these are:

(i) simple cobalt carbonyl, or rather hydrido cobalt carbonyl, complexes

(ii) hydrido cobalt carbonyl complexes having tertiary phosphine ligands, and

(iii) tertiary phosphine hydrido rhodium carbonyl species.

In the following sub-sections we will consider each of these three in turn, then go on to compare and contrast rhodium and cobalt based systems, briefly describe some other transition-metal species which catalyse this reaction and finally touch on the development of asymmetric hydroformylation catalysts.

Since the reaction is of such commercial significance, in the first four sub-sections some space is allotted to describing the industrial application of the process.

2.4.1 *Cobalt catalysts*

(a) *Unmodified cobalt carbonyl systems*

Processes using unmodified, i.e. no added inert ligands, cobalt carbonyl catalysts were first operated commercially in the late 1940s and still constitute some 80%, in terms of capacity, of today's hydroformylation plants. Typically temperatures in the range 110 °C to 180 °C with pressures in the range 200 to 350 atm are employed and the cobalt is intially fed into the reactor as cobalt salts, (e.g. cobalt acetate), cobalt carbonyls (e.g. $Co_2(CO)_8$) or even cobalt metal. Irrespective of how the cobalt is introduced, it is converted under the reaction conditions to the true catalyst precursor $CoH(CO)_4$, (*2.109*). Before entering the catalysis cycle [197] shown in Fig. 2.17, this 18 valence electron species loses a carbon monoxide ligand to give the four

HOMOGENEOUS CATALYST SYSTEMS IN OPERATION

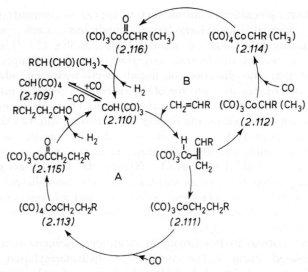

Fig. 2.17 *Catalysis cycles for cobalt catalysed hydroformylation; cycle A leading to 'normal' product and cycle B to 'iso' product*

co-ordinate 16 valence electron species, $CoH(CO)_3$ (*2.110*), which is the active catalytic species

$$CoH(CO)_4 \rightleftharpoons CoH(CO)_3 + CO \quad (2.16)$$

$\quad\quad$ (*2.109*) $\quad\quad\quad\quad\quad$ (*2.110*)

The next stage in the catalysis cycle involves co-ordination of the alkene substrate (16 valence electron to 18 valence electron transformation) followed by rapid hydride migration to give a cobalt(I)-alkyl species. In the absence of any double bond migration, we will return to the question of alkene isomerization under hydroformylation conditions a little later, the metal hydrogen can add to either the α-carbon atom (Markovnikov addition) or to the β-carbon atom (anti-Markovnikov addition) to give either the iso alkyl product (*2.112*) or the normal alkyl product (*2.111*). Electronically we might expect metal hydrogen addition to occur at the α-carbon giving the branched product, the metal being an electron rich centre. However, on steric grounds the large metal group will clearly prefer to become attached to the least substituted carbon atom giving rise to the linear alkyl product. At low temperatures (0 °C), in the absence of carbon monoxide, electronic factors appear to dominate while under normal hydroformylation conditions steric effects predominate [198] and the normal alkyl complex (*2.111*) is the major product of the hydride migration step. It is tempting to assume that the direction of initial metal hydrogen addition totally determines the outcome of the reaction in terms of product aldehyde linearity, i.e. the normal–iso ratio. That this is not the case is demonstrated by the finding that when an internal alkene, such as hept-3-ene, is used as substrate under normal hydroformylation conditions the major product is normal octanal. Similarly, at 110 °C with pCO ~ 90 atm, the normal–iso ratios obtaining using pentene as substrate are not too different with either pent-1-ene or pent-2-ene as feed (4.5 and 3.1, respectively). At a low partial pressure of carbon monoxide (1.7 atm) the ratios are identical, i.e. 1.3 [153].

Under normal hydroformylation conditions, isomerization of a co-ordinated alkene is fast compared to hydroformylation, thus the direction of initial hydride addition is of only marginal importance in determining the normal–iso ratio. We use the expression 'co-ordinated alkene' deliberately since it seems, from studies using deuterated alkenes [1], that complete isomerization can occur within the cobalt–alkene complex, probably by a series of 1,2-hydrogen shifts (see Section 2.4), without elimination of free, isomerized, alkene. As far as the cobalt is concerned, on steric grounds, the preferred π-alkene complex is that involving the terminal isomer of the alkene substrates thus, under conditions where isomerization is rapid, it is always the α and β aldehyde products which predominate.

Once formed, the 16 valence electron alkyl tricarbonyl cobalt species, either (*2.111*) or (*2.112*), picks up a molecule of carbon monoxide to give the 18 valence electron tetracarbonyl intermediate, (*2.113*) or (*2.114*). Alkyl to carbonyl migration forms the acyl species, (*2.115*) or (*2.116*). It is this last step which is thought to be most influential in determining the product linearity [199].

For unmodified cobalt carbonyl catalyst systems, the most influential single reaction parameter in determining the normal–iso ratio for a given alkene is the carbon monoxide partial pressure, pCO. With propene as substrate, increasing the pCO from 2.5 atm to 90 atm at 100 °C results in almost a tripling of the normal–iso ratio, i.e. from 1.6 to 4.4 [153]. Hydrogen partial pressure has only a small effect on product linearity and the normal–iso ratio is essentially independent of the reaction temperature for temperatures above ±50 °C. The dependence of product selectivity on carbon monoxide partial pressure can be rationalized in terms of steric crowding in the alkyl to acyl transition state [199]. In the catalysis cycles illustrated in Fig. 2.17 the alkyl to acyl transformation is shown as a 16 to 18 valence electron sequence

$$(CO)_3Co^ICH_2CH_2R \xrightarrow{CO} (CO)_4Co^ICH_2CH_2R \longrightarrow$$

$$\qquad (2.111) \qquad\qquad\qquad (2.113)$$

NVE   16   18

$$(CO)_3Co^I\overset{O}{\overset{\|}{C}}CH_2CH_2R$$

$$(2.115)$$

16

The transition state for the latter step in this process, i.e. (2.113) to (2.115), may be visualized as shown below.

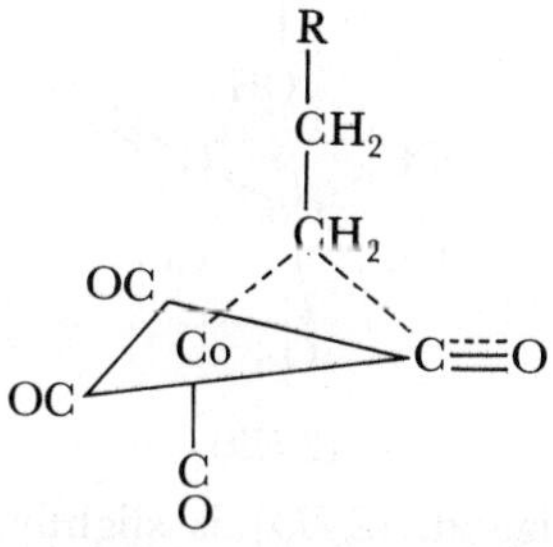

(2.117)

Being co-ordinately unsaturated 16 valence electron species both (2.111) and (2.115) have a marked tendency to pick up any stray donor molecules, thus increasing the carbon monoxide partial pressure will not only favour conversion of (2.111) into (2.113) but also conversion of (2.115) into the 18 valence electron pentaco-ordinate complex $(CO)_4Co^ICOCH_2CH_2R$, (2.118).

$$(CO)_4Co^ICOCH_2CH_2R \longrightarrow (CO)_3Co^I\overset{O}{\overset{\|}{C}}CH_2CH_2R \xrightarrow{CO}$$

(2.113)　　　　　　　(2.115)

NVE　　18　　　　　　　16

$$(CO)_4Co^I\overset{O}{\overset{\|}{C}}CH_2CH_2R$$

(2.118)

18

Under high pCO the independent existence of the 16 valence electron species (2.115) would seem unlikely and the concerted transformation (2.113) to (2.118) can be visualized as occurring via a transition state of the type (2.119), where cobalt–carbonyl bond making occurs concurrently with cobalt–alkyl bond breaking and alkyl–carbonyl bond making. By virtue of its extra, albeit

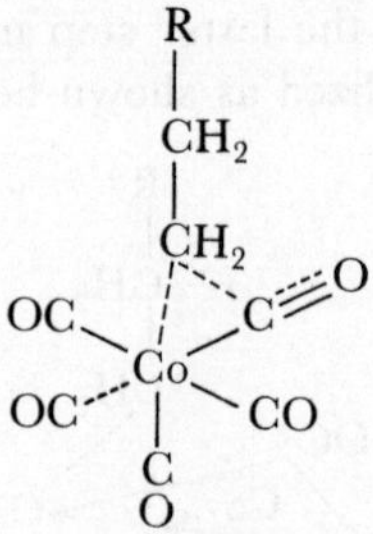

(2.119)

incipient, carbonyl ligand, (2.119) is slightly more crowded than (2.117) and therefore of somewhat higher energy. This energy difference will clearly be more marked when the alkyl unit is di-alkyl substituted as in the complexes leading to the branched product shown in cycle B of Fig. 2.17. Thus, increasing the CO pressure will favour cycle A over cycle B and thus increase the selectivity to linear product.

Having formed the acyl-cobalt species, be it a tricarbonyl complex, (2.115) or (2.116), or a tetracarbonyl complex of the type

(2.118), the next step in the catalysis cycle is hydrogenolysis to form the product aldehyde. Under hydroformylation conditions this hydrogenolysis almost certainly occurs via oxidative addition of hydrogen to a cobalt(I)-acyl complex.

$$(CO)_n Co^I COR' + H_2 \longrightarrow (CO)_n Co^{III}(H)_2 COR'$$

$$(2.120) \hspace{4cm} (2.121)$$

where $R' = CH_2CH_2R$ or $CHR(CH_3)$. When $n = 3$ this presents no problem, being a 16 to 18 valence electron process between two readily available oxidation states. However, with $n = 4$, as in (2.118), we would be dealing with an 18 to 20 valence electron sequence which has little to recommend it. With a tetracarbonyl acyl complex, carbonyl dissociation must precede hydrogen addition

$$(CO)_4 Co^I COR' \rightleftharpoons (CO)_3 Co^I COR + CO. \hspace{1cm} (2.17)$$

Under high pressure $(CO:H_2 \simeq 1:1)$ hydroformylation conditions Equilibrium (2.17) lies very much to the left, and the rate-determining step in the catalysis cycle occurs during the hydrogenolysis of the acyl complex. In fact, acyl tetracarbonyl complexes of the type (2.118) with $R = C_6H_{13}$ have been detected using infrared spectroscopy in a cobalt/oct-1-ene system at 150 °C/250 atm [200]. An alternative hydrogenolysis sequence involves reaction between a tricarbonylcobalt acyl species and $CoH(CO)_4$:

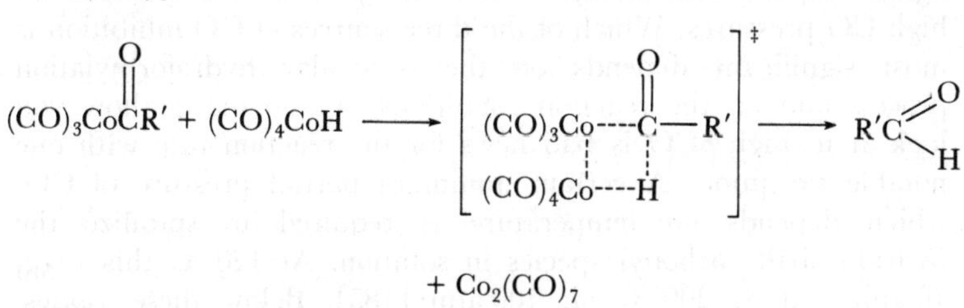

$$(CO)_3 Co\overset{O}{\overset{\|}{C}}R' + (CO)_4 CoH \longrightarrow \left[\begin{array}{c} (CO)_3 Co \cdots \overset{O}{\overset{\|}{C}} - R' \\ (CO)_4 Co \cdots H \end{array} \right]^{\ddagger} \longrightarrow R'C\overset{O}{\underset{H}{\diagdown}}$$

$$+ Co_2(CO)_7$$

with the hydridocobalt carbonyl being regenerated via either

$$Co_2(CO)_7 + H_2 \longrightarrow CoH(CO)_3 + CoH(CO)_4$$

or

$$Co_2(CO)_7 + CO \longrightarrow Co_2(CO)_8$$
$$Co_2(CO)_8 + H_2 \longrightarrow 2 CoH(CO)_4.$$

Although $CoH(CO)_4$ reduction of acyl cobalt has been shown to occur under stoichiometric reaction conditions [201], kinetic considerations make it unlikely that such a sequence plays a major role in the catalytic process.

As we have already seen, increasing the carbon monoxide partial pressure has a beneficial effect on the product linearity. Such an increase does, however, have a detrimental effect on the overall rate of the hydroformylation reaction. Kinetic measurements [202] have shown that the rate is inversely proportional to the CO partial pressure:

$$d[\text{aldehyde}]/dt \approx k_{\text{obs}}[\text{alkene}][\text{cobalt}]p(H_2)p(CO)^{-1} \qquad (2.18)$$

The retarding influence of increased CO pressure can be understood by reference to the equilibria shown in Equations (2.16) and (2.17). In both cases the tetracarbonyl species is catalytically inactive and in both cases its formation is favoured by high CO partial pressure. A further source of CO inhibition arises through the sequence

$$CoH(CO)_3 + CoH(CO)_4 \rightleftharpoons Co_2(CO)_7 + H_2$$

$$Co_2(CO)_7 + CO \rightleftharpoons Co_2(CO)_8. \qquad (2.19)$$

$$(2.122)$$

In order to regenerate active hydridocarbonyl species the octacarbonyl-dicobalt(0) complex (*2.122*) must lose a carbonyl ligand, Equilibrium (2.19). Clearly this process is not helped by high CO pressures. Which of the three sources of CO inhibition is most significant depends on the particular hydroformylation process and on the reaction conditions but no matter how you look at it, high pCO is bad news for the reaction rate with one notable exception. A certain minimum partial pressure of CO, which depends on temperature is required to stabilize the hydridocobalt carbonyl species in solution. At 120 °C this is *ca.* 10 atm and at 200 °C *ca.* 100 atm [185]. Below these values, metallic cobalt will precipitate and that is even worse news for the reaction rate! It is the crucial need to maintain the cobalt in solution which dictates the high pressures used in commercial hydroformylation processes having unmodified cobalt catalysts.

We have already briefly mentioned (Section 2.1.2*d*) that $CoH(CO)_4$ can function as a catalyst for the hydrogenation of aldehydes to alcohols. It is not a particularly active catalyst and

HOMOGENEOUS CATALYST SYSTEMS IN OPERATION 111

at temperatures below *ca.* 180 °C only some 10% of the aldehyde product is converted to alcohols. Similarly, only a small proportion (±1%) of the alkene feed is hydrogenated to alkane. With temperatures in excess of 250 °C (CO/H_2 pressure *ca.* 250 atm) more efficient conversion to alcohols can be obtained. The active species in the hydrogenation cycle, Fig. 2.18, is probably, as in the hydroformylation cycle, the hydridotricarbonyl complex (*2.110*). Inital co-ordination of the aldehyde to the metal centre is suggested [203] to take place via a side-on π-bonding interaction, (*2.123*). Hydride ligand migration to the carbonyl carbon of the aldehyde, via a four-centre transition state such as (*2.126*) can then form the σ-bonded alkoxy species (*2.124*).

$$\left[\begin{array}{c} RHC\!=\!\!=\!\!O \\ | \quad\quad | \\ H\text{----}Co(CO)_3 \end{array} \right]^{\ddagger}$$

(*2.126*)

Oxidative addition of hydrogen to this 16 valence electron d^8 complex gives the 18 valence electron d^6 dihydridocobalt(III) intermediate (*2.125*) which can then break down, by reductive

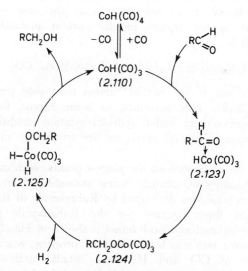

Fig. 2.18 *Catalysis cycle for the cobalt catalysed hydrogenation of aldehydes to alcohols*

elimination, to form the alcohol product and complete the catalysis cycle. In view of the acidity of the cobalt hydride, in aqueous solution $CoH(CO)_4$ is as strong an acid as HCl, we might possibly expect addition of the H—Co unit across the aldehyde C=O unit to occur in the opposite direction, i.e. to give hydroxyalkylcobalt complex of the type (2.127).

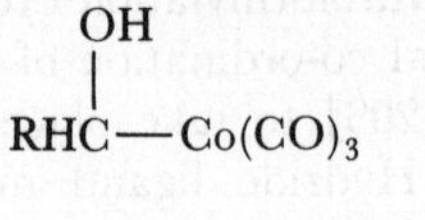

$$\begin{array}{c} OH \\ | \\ RHC-Co(CO)_3 \end{array}$$

(2.127)

Oxidative addition of hydrogen to (2.127) followed by hydride migration and reductive elimination of product alcohol, as described for the alkoxy cycle (Fig. 2.18), would complete the catalysis cycle. On steric grounds (2.124) is clearly preferred over (2.127). This observation, taken together with the finding [204] that formate esters can be produced from aldehydes under hydroformylation conditions, strongly suggests that the alkoxy route is the preferred cycle [190].

As with the hydroformylation reaction the cobalt catalysed hydrogenation of aldehydes is subject to carbon monoxide inhibition. However, in the latter case the rate is inversely proportional to the square of the carbon monoxide partial pressure [203]:

$$d[\text{alcohol}]/dt = k_{obs}[\text{aldehyde}][\text{cobalt}]p(H_2)(CO)^{-2} \qquad (2.20)$$

rather than just to the partial carbon monoxide presssure, cf. Equation (2.18). This accounts, to some extent, for the low hydrogenation activity under hydroformylation conditions where carbon monoxide partial pressures are frequently in excess of 100 atm.

In commercial operations the largest plants, with capacities of ca. 325 000 tonnes per annum, using unmodified cobalt catalysts are based on processes developed by Ruhrchemie or BASF [196]. A simplified flow diagram for the Ruhrchemie process to n-butanal, n-butanol and iso-butanol is shown in Fig. 2.19. Vessel 1 is the reactor into which the substrate propene, reactant syngas (a mixture of CO and H_2) and cobalt catalyst are fed. Temperatures in the range 130–175 °C and total pressures in the range 200–300 atm are used and the catalyst concentration, in

HOMOGENEOUS CATALYST SYSTEMS IN OPERATION

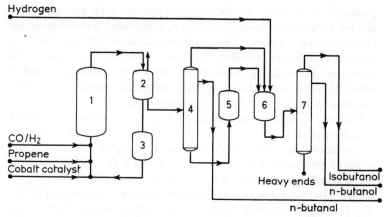

Fig. 2.19 *Simplified flow diagram for the Ruhrchemie hydroformylation process using unmodified cobalt catalysts*

terms of the percentage of metal compared to alkene feed is between 0.1 and 1%. The reaction products, including cobalt catalyst, are drawn off at the top of reactor 1 and pass into a separator, 2, in which the mixture is treated thermally and/or chemically to yield a cobalt free organic phase and a cobalt rich inorganic phase. This part of the process is known as 'decobaltation'. In reactor 3, the cobalt rich phase is further treated to convert it into a form suitable for recycling to the hydroformylation reactor 1.

A number of methods have been developed for cobalt removal and recycling, and most involve at some stage an acid treatment. One scheme particularly suitable for the Ruhrchemie process involves initial neutralization of the acidic hydridocobaltcarbonyl with aqueous sodium bicarbonate

$$CoH(CO)_4 + NaHCO_3 \longrightarrow Na[Co(CO)_4] + H_2O + CO_2$$

followed by acidification of the separated aqueous inorganic phase with sulphuric acid to regenerate the hydridocobaltcarbonyl species

$$2Na[Co(CO)_4] + H_2SO_4 \longrightarrow 2CoH(CO)_4 + Na_2SO_4.$$

The cobalt free organic phase, containing mainly *n*- and iso-butanal in the approximate ratio of 4:1, is fed to a distillation column 4 where the aldehydes are separated from the other organic reaction products, e.g. alcohols and aldol condensation products. If alcohols are the required product then the aldehydes are fed to a hydrogenation reactor 6, containing a heterogeneous hydrogenation catalyst. The resulting alcohols then go on to a second distillation column 7. The 'bottoms', i.e. the aldol condensation products, from 4 can be fed to a cracking unit 5 where they are either thermally or catalytically broken down into constituent aldehydes prior to hydrogenation in 6.

One of the major disadvantages of processes using unmodified cobalt catalysts is the high energy requirement in terms of temperature and pressure. Also if, as is usual, the n-isomer is the required product, a normal–iso ratio of 3 or 4 to 1, which is about the best achievable with such processes, leaves room for improvement. The first important breakthrough in overcoming these disadvantages occurred in the early sixties when Slaugh and Mullineaux [205–207], working in Shell Oil's laboratories in California, discovered that adding tertiary phosphine ligands to cobalt cabonyl systems led to complexes which did not depend on high CO pressures for their stability and which were able to catalyse the conversion of alk-1-enes to normal alcohols with over 90% selectivity.

(b) Tertiary-phosphine modified cobalt carbonyl systems

Compared to the unmodified systems, cobalt carbonyl hydroformylation catalysts [208] containing alkyl tertiary-phosphines, e.g. PBu_3, are more stable – H_2/CO pressures as low as 5 to 10 atm may be used for reaction temperatures in the range 100 to 180 °C cf. 100 to 320 atm for $CoH(CO)_4$. They have greater hydrogenation activity thus giving rise to alcohol rather than aldehyde products under hydroformylation conditions. With linear α-olefins as substrate, normal–iso ratios are typically twice those obtained using unmodified catalysts, e.g. 8:1 compared to 4:1. However, somewhat on the negative side, their activity is considerably lower than that encountered with cobalt carbonyl catalysts not containing phosphine ligands – at 145 °C hydroformylation with an unmodified catalyst is some five times faster than that obtained with a modified one at 180 °C.

Mechanistically, the reaction is closely related to that described in the previous sub-section for $CoH(CO)_n$. The active species is $CoH(CO)_2L$, (*2.127*), formed by dissociation of a carbonyl ligand from $CoH(CO)_3L$, (*2.128*):

HOMOGENEOUS CATALYST SYSTEMS IN OPERATION

$$CoH(CO)_3L \rightleftharpoons CoH(CO)_2L + CO \quad (2.21)$$
$$(2.128) \qquad\qquad (2.127)$$

cf. Equation (2.16). The cobalt is often added in the form of the stable dicobaltoctacarbonyl complex, $Co_2(CO)_8$, in which case (2.128) can be formed by the following sequence

$$Co_2(CO)_8 + 2L \rightleftharpoons Co_2(CO)_6L_2 + 2CO$$
$$Co_2(CO)_6L_2 + H_2 \rightleftharpoons 2CoH(CO)_3L.$$

The double catalysis cycle, i.e. leading to aldehyde and then on to alcohol product, is shown in Fig. 2.20. As for the unmodified cobalt catalysts the cycles consist of a series of alternating 16/18 valence electron processes as indicated by the inner cycles in Fig. 2.20.

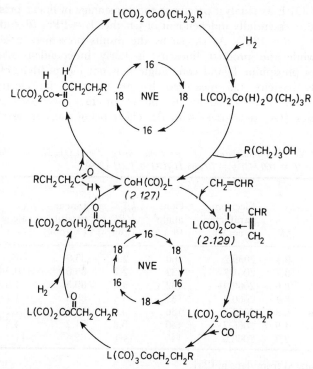

Fig. 2.20 *Catalysis cycles for the hydroformylation/hydrogenation process catalysed by tertiary phosphine modified cobalt carbonyl complexes*

Owing to the increased steric crowding, resulting from the presence of the bulky tertiary-phosphine ligand (the cone angle for PBu$_3$, a frequently used ligand in hydroformylation systems, is 132° compared to ~95° for a carbonyl ligand [26]) CoH(CO)$_2$L shows a marked preference for terminal over internal alkenes when it comes to forming π-complexes of the type (2.129). Similarly, in all the subsequent ligand migration reactions in the cycle, transition states leading to the least substituted alkyl or acyl complex are much favoured. In these modified cobalt systems it is the steric effect of the tertiary ligands which is mainly responsible for the higher normal–iso ratios. Having said this, the maximum influence is apparently achieved with a ligand cone angle of ca. 130° and there is little or no improvement in the normal–iso ratio on going to more bulky trialkylphosphine ligands. Thus in the catalytic hydroformylation of hex-1-ene using CoH(CO)$_3$L as catalyst precursor the percentage of linear product formed is essentially independent of whether L = PPr$^i_3$ (θ = 160°) or PPr$^n_3$ (θ = 132°). As shown by the results presented in Table 2.2, while the product linearity is fairly independent of the tertiary phosphine ligand cone angle it is not independent of the ligands basicity (as measured by pK_a), or of its electronic parameter. The selectivity to linear product decreases as v increases (pK_a decreases) with the effect being most marked for

Table 2.2 *Hydroformylation of hex-1-ene using $Co_2(CO)_8/2L$ as catalyst precursor, T = 160 °C; p ≃ 70 atm H_2/CO = 1.2/1* [209]

L	pK_a	Electronic parameter* (v cm^{-1})	Cone angle* ($\theta°$)	k_r† x 10^3 (min^{-1})	Linear product (% total)	Aldehyde to alcohol ratio†
PPr$^i_3$	9.4	2059.2	160	2.8	85.0	—
PEt$_3$	8.7	2061.7	132	2.7	89.6	0.9
PPr$^n_3$	8.6	2060.9‡	132	3.1	89.5	1.0
PBu$^n_3$	8.4	2060.3	132	3.3	89.6	1.1
PEt$_2$Ph	6.3	2063.7	136	5.5	84.6	2.2
PEtPh$_2$	4.9	2066.7	140	8.8	71.7	4.3
PPh$_3$	2.7	2068.9	145	14.1	62.4	11.7

*Calculated from data in [26]
†calculated from data in [209]
‡estimated

HOMOGENEOUS CATALYST SYSTEMS IN OPERATION 117

the arylphosphine ligands, PEt_2Ph, $PEtPh_2$ and PPh_3. Concurrent with this decrease in linear selectivity there is a decrease in the hydrogenation activity of the system, as reflected in the increasing aldehyde to alcohol ratio, but an increase in the overall hydroformylation activity, as reflected in the increasing k_r values (see Table 2.2). These latter three effects of ligand basicity or electronic parameter can be rationalized in terms of the Equilibrium (2.22):

$$CO + CoH(CO)_3L \rightleftharpoons CoH(CO)_4 + L \qquad (2.22)$$
$$(2.128) \qquad\qquad\qquad (2.109)$$

The equilibrium concentration of $CoH(CO)_4$ increases with decreasing tertiary phosphine basicity and increasing ligand electronic parameter. This is exactly what we would expect since ligand basicity essentially measures the σ-donor properties of the phosphorus and, as discussed in Section 1.2.3b, the electronic parameter v attempts to quantify the electron donor/acceptor properties of the ligand. Both are closely related to ligand lability thus a tertiary phosphine ligand having either a low pK_a or a high v will be more likely to dissociate from (2.128), and thus be replaced by CO to give (2.109), than one having these two parameters at the opposite end of the scale. Since $CoH(CO)_3$ is a more active, but less selective, hydroformylation catalyst than $CoH(CO)_2L$ the activity of the system will increase, and its selectivity decrease as we go from $L = PEt_3$ to $L = PPh_3$. Similarly, the tertiary-phosphine substituted complex is a more active hydrogenation catalyst than its carbonyl counterpart which goes some way towards explaining the decrease in hydrogenation activity encountered as v increases.

As far as hydrogenation activity is concerned, the superior activity of cobalt/tertiary-phosphine systems compared to the simple cobalt carbonyl system has been ascribed to an increase in the hydridic character of the cobalt-hydride on substituting carbonyl by tertiary phosphine. Being certainly a better σ-donor, and probably a poorer π-acceptor ligand, than CO the trialkyl phosphine ligand will increase the electron density at the metal centre and thus at the hydride hydrogen centre. Enhancing the negative charge at this centre will facilitate the nucleophilic ligand migration, hydride to acyl carbon. Similarly, increasing the electron density at the metal centre will favour oxidative addition of H_2. Both of these effects would aid the hydrogenation cycle.

However, possibly more important [210] is the resultant effect of being able to operate under lower carbon monoxide pressures.

As is the situation with the unmodified system, increasing the pressures decreases the reaction rate. This is not surprising since, as we have seen, many of the active, 16 valence electron, catalytic species are formed by CO dissociation. However, as can be appreciated by comparing the kinetic rate Equations (2.18) and (2.20), the inhibiting effect of increasing CO partial pressure is more strongly felt by the hydrogenation reaction, where the rate is inversely proportional to the square of pCO, than by the hydroformylation reaction, where the rate dependence involves only pCO^{-1}.

Although it is not clear just how important the electronic properties of the tertiary-phosphine ligand are in increasing the hydrogenation activity of the system, they undoubtedly play a major role in increasing its stability which permits low pressure operation. It has been suggested [211] that the enhanced stability is due to the tertiary-phosphine being a poorer π-acceptor than CO and thus leaving more π-acceptor electron density available to the carbonyl ligands. At least equally important must be the greater σ-donor properties of the tertiary-phosphine ligand. Whether the increased electron density at cobalt results from the phosphine ligand taking less back or giving more to the metal centre is not clear. Whatever the cause there is more electron density available for the metal–carbonyl bonding orbitals and the complexes are considerably more stable on tertiary-phosphine substitution.

The increased steric bulk of the tertiary-phosphine ligand, in addition to increasing the selectivity while decreasing the hydroformylation activity, alters the nature of the rate-determining step. Whereas with unmodified cobalt catalysts hydrogenolysis of the cobalt–acyl species appears to be rate-determining, with the modified system initial olefin co-ordination, or possibly hydride ligand transfer to co-ordinated olefin, is the slowest step in the catalysis cycle. This is principally due to the steric bulk of the tertiary-phosphine ligand and the modified systems show a marked preference for terminal (unsubstituted) over internal alkenes. Since the complexes are reasonably good isomerization catalysts the selectivity to normal products, although not necessarily the overall reaction rate, is frequently independent of whether internal or terminal alkenes are fed to the catalyst system. Thus with $Co_2(CO)_8/2PBu_3$ as catalyst precursor hydroformylation of either hex-1-ene or hex-2-ene at 195 °C/27 atm results in essentially the same (80 and 81%, respectively) selectivity to normal products although the overall reaction rate is a factor 2 slower with the hex-2-ene substrate [212]. Provided the cobalt to tertiary-phosphine ratio is above one, i.e. so that complexes of the type $CoH(CO)_nPBu_3$ can be effectively formed, both the selectivity and the activity of the system are unaffected by further additions of tertiary-phosphine ligand. Below Co:P ratios of 1 the selectivity of the system decreases, due to the presence of $CoH(CO)_4$, and at low CO pressures the activity decreases due to decomposition to cobalt metal [118,119].

If alcohols are the desired reaction products, then using the modified cobalt catalyst systems is advantageous in that the whole sequence can be carried out in one reactor and there is no need to incorporate a second hydrogenation stage. Unfortunately the systems are not totally selective towards aldehyde reduction and some, typically of the order of 15%, of the alkene feed is also hydrogenated. In both economic and energy terms this represents a net loss since alkanes are of considerably lower commercial value than alkenes and conversion of alkanes into alkenes requires expenditure of energy.

In Fig. 2.21 a simplified flow diagram for the Shell hydroformylation process using modified cobalt catalysts is shown. Typically the reaction is carried out in the temperature range 160–200 °C, with a CO/H_2 pressure in the range 50–100 atm. Alcohols are the major (~80%) products and normal–iso ratios generally exceed 8:1. Compared to the process using unmodified cobalt catalysts, cf. Fig. 2.20, less hardware, i.e. capital investment, is involved, although, owing to the lower catalytic activity of the modified systems, it is necessary to employ larger reactor volume to achieve the same conversion per pass. However, since the catalyst is considerably more stable, there is no need to include a decobaltation step and catalyst separation can occur in the first distillation column after the reaction mixture, from the top of reactor 1, has passed through the gas separation unit 2. The catalyst solution is returned to the reactor via a catalyst make up unit, 4, where any spent catalyst component is replenished. Final product separation takes place in the distillation train represented by columns 5, 6 and 7.

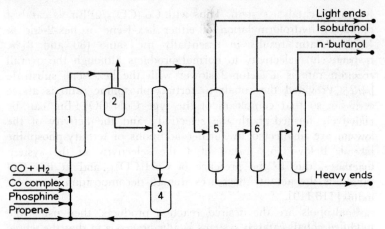

Fig. 2.21 *Simplified flow diagram for the Shell hydroformylation process using tertiary phosphine modified cobalt catalysts*

In spite of being able to effectively function under milder conditions than their unmodified counterparts, cobalt/tertiary-phosphine catalysts do suffer from low activity and the increased hydrogenation activity is not always advantageous.

The next development in industrial hydroformylation catalysis occurred with the introduction of tertiary-phosphine modified rhodium catalysts.

2.4.2 *Rhodium catalysts*

The potential advantages of rhodium based hydroformylation catalysts over those involving cobalt were first recognized in the mid-fifties when it was disclosed [213] that rhodium is a much more active hydroformylation metal than cobalt capable of effectively operating under much milder temperature and pressure conditions. As with the unmodified cobalt systems the rhodium can be introduced into the reaction medium in a number of forms [46], e.g. rhodium on a support, as a chloro or carboxylic salt or as a metal-carbonyl complex such as $Rh_4(CO)_{12}$ or $Rh_2(CO)_4Cl_2$, all of which under the hydroformylation conditions (70–150 °C and 50–150 atm total pressure H_2:CO = 1:1) form the catalytic active species, $RhH(CO)_3$. This is completely analogous to the cobalt complex (*2.110*) and the overall catalysis cycle is essentially the same as that described for $CoH(CO)_3$, Fig. 2.17.

As found with cobalt, the rate-determining step is hydrogenolysis of the metal-acyl species, $Rh(COR)(CO)_3$, cf. complex (2.115). Unlike cobalt the rhodium systems exhibit little or no aldehyde hydrogenation activity and thus the final reaction mixture contains almost no alcohols. Unmodified rhodium carbonyl hydroformylation catalysts are between 10^2 and 10^4 times more active than their cobalt analogues [196]. They are also highly active isomerization catalysts, but give rise to disappointingly low normal–iso ratios which seldom exceed 1:1, even with simple alkenes such as propene (cf. Co ~ 4:1). This low selectivity together with the high price of rhodium, ca. 3500 times that of cobalt, far outweighs the activity advantage of rhodium and no industrial hydroformylation process using unmodified rhodium carbonyl catalysts are presently in operation.

The major breakthrough in the commercial application of rhodium based hydroformylation systems occurred in the early sixties when Slaugh and Mullineaux [214] of Shell Oil's Emeryville laboratories disclosed the use of trialkyl-phosphine and -arsine complexes of rhodium as alkene hydroformylation catalysts. Independently, Osborn, Young and Wilkinson (1965) working at Imperial College in London discovered [215] that under a H_2/CO atmosphere, in benzene solution, $RhCl_3(PPh_3)_3$ gave rise to a catalyst system capable of catalysing the hydroformylation of pent-1-ene or hex-1-ene under essentially ambient conditions.

Several tertiary-phosphine rhodium complexes, e.g. $RhCl(CO)L_2$, L = PBu_3 or PPh_3, $RhH(CO)(PPh_3)_3$, $RhCl(PPh_3)_3$ have been used as catalyst precursors [206]. Under hydroformylation conditions they all seem to give rise to the hydridodicarbonyldi(tertiaryphosphine)rhodium species, $RhH(CO)_2L_2$, (2.130). This is one of the key intermediates in these systems. In the case of the chloro-containing precursors, formation of this intermediate is a slow process and the hydroformylation reaction is frequently preceded by an induction period which can be removed by adding an organic base such as triethylamine to the reaction mixture. When $RhCl(CO)L_2$ is the catalyst precursor, (2.130) can be formed by the oxidative addition/reductive elimination sequence shown below

$$Rh^I Cl(CO)L_2 + H_2 \longrightarrow Rh^{III}Cl(H)_2(CO)L_2 \longrightarrow Rh^I H(CO)L_2 + HCl$$

$$Rh^I H(CO)L_2 + CO \longrightarrow Rh^I H(CO)_2L_2.$$

The triethylamine removes the HCl so that the overall reaction becomes [152]

$Rh^I Cl(CO)L_2 + H_2 + CO + Et_3N \longrightarrow Rh^I H(CO)_2 L_2 + Et_3 NHCl$.

When $RhH(CO)L_3$ is used as catalyst precursor there is no induction period and addition of Et_3N has no effect on the rate of hydroformylation. Infrared spectroscopy, carried out under hydroformylation conditions, has shown that with both $RhH(CO)(PPh_3)_3$ and $RhCl(CO)(PPh_3)_2/Et_3N$ the same, albeit not yet fully characterized, complex is formed under a positive CO/H_2 pressure [217].

Two catalyst cycles have been suggested for the hydroformylation reaction catalysed by modified rhodium systems [152]. These are shown in Figs. 2.22 and 2.23. The first, somewhat similar to the modified cobalt catalysis cycle leading to aldehydes, lower cycle in Fig. 2.20, involves initial dissociation of a triphenylphosphine ligand from (*2.131*) to give the 16 valence electron species (*2.132*). Co-ordination of alkene followed by rapid

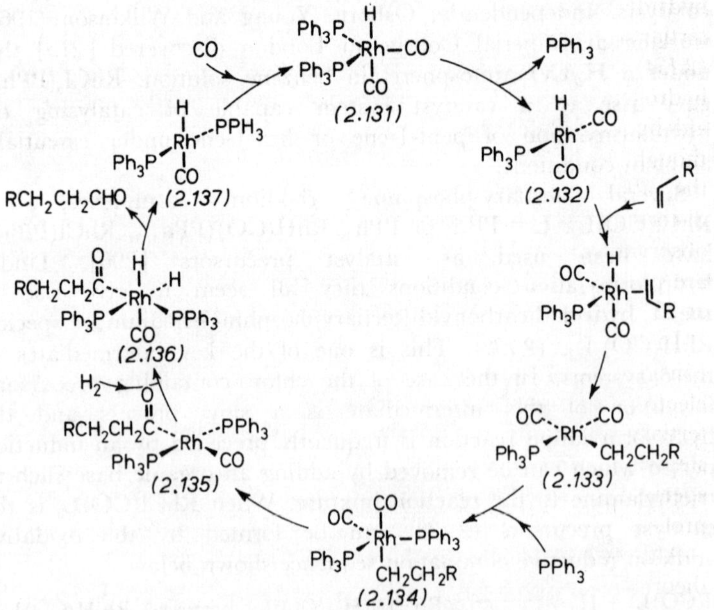

Fig. 2.22 *Dissociative catalysis cycle for the alkene hydroformylation reaction catalysed by $RhH(CO)_2(PPh_3)_2$*

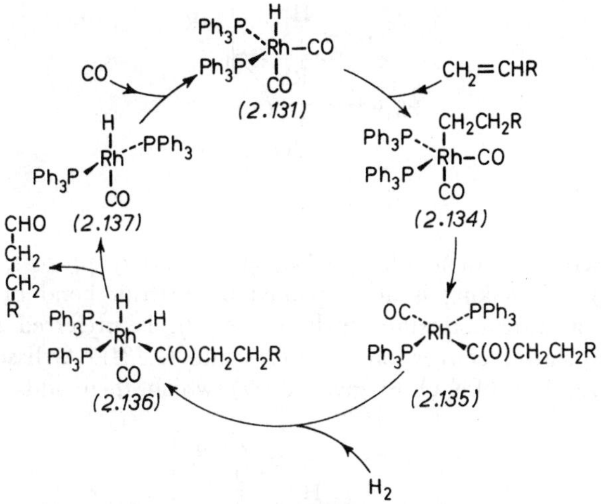

Fig. 2.23 *Associative catalysis cycle for the alkene hydroformylation reaction catalysed by* $RhH(CO)_2(PPh_3)_2$

hydride ligand migration gives (*2.133*) which quickly picks up a tertiary phosphine ligand to give the 18 valence electron species (*2.134*). Alkyl ligand migration followed by oxidative addition of hydrogen gives (*2.136*) which can reductively eliminate product aldehyde and pick up CO to complete the cycle. The slow step is thought to be the oxidative addition of H_2 to (*2.135*). Addition of the alkene to a dicarbonyl, i.e. (*2.132*), rather than to a di-tertiary-phosphine species, i.e. (*2.137*), was substantiated by the observation, via ^{1}H n.m.r. spectroscopy, that if (*2.137*) and (*2.131*) are present together in solution only the latter reacts with ethene at 25 °C, 1 atm [218].

In the associative cycle shown in Fig. 2.23 the alkene is considered to add directly to penta-co-ordinate 18 valence electron species (*2.131*) without prior dissociation of either a tertiary-phosphine or a carbonyl ligand. If this occurs via a stepwise process involving the hexa-co-ordinate rhodium(I) species (*2.138*) it would constitute an 18 to 20 valence electron process.

Although it should be pointed out that there are, as yet, no theoretical reasons to exclude such a process, experimental experience, which forms the basis of Tolman's 16/18 electron rule, mitigate against such a sequence. In the absence of direct

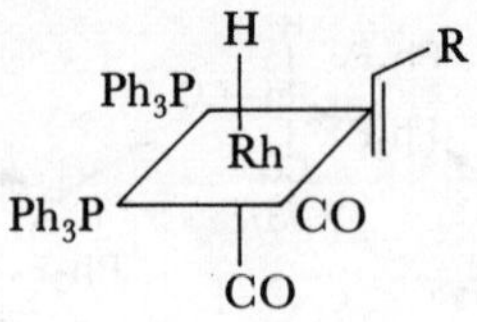

$$\begin{array}{c} \text{Ph}_3\text{P} \quad \overset{\text{H}}{\underset{\text{Rh}}{|}} \quad \overset{R}{\nparallel} \\ \text{Ph}_3\text{P} \diagup \quad \diagdown \text{CO} \\ \text{CO} \end{array}$$

(*2.138*)

spectroscopic evidence for (*2.138*) a concerted process, where alkene bond making is accompanied by hydride bond breaking, i.e. via a transition state such as (*2.139*), is preferred for the non-dissociative conversion of (*2.131*) into (*2.134*). A dissociative route, via loss of CO to give (*2.137*) which then adds alkene,

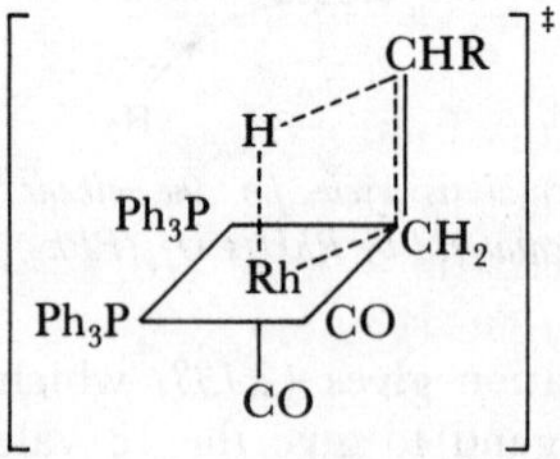

$$\left[\begin{array}{c} \text{CHR} \\ \text{H}\cdots\overset{|}{\underset{\text{Rh}}{|}}\text{CH}_2 \\ \text{Ph}_3\text{P}\diagup\quad\diagdown\text{CO} \\ \text{CO} \end{array} \right]^{\ddagger}$$

(*2.139*)

undergoes ligand migration and reassociates CO to form (*2.134*), has been ruled out on the basis of the n.m.r. experiment described above, although on the basis of the 16/18 electron rule it would be preferred to either the stepwise or concerted schemes proposed.

The cardinal difference between the associative and dissociative hydroformylation catalysis cycles is that in the associative cycle the rhodium never has fewer than two tertiary-phosphine ligands. The associative cycle is sterically more demanding and would, therefore, be expected to more strongly favour linear product formation. This is borne out by the experimental observation that increasing the tertiary-phosphine to rhodium ratio in the catalyst solution increases the selectivity of the system towards linear aldehyde formation. Thus with $RhH(CO)(PPh_3)_3$ as catalyst precursor in the hydroformylation of propene at 100 °C, ~35 atm $CO:H_2$ = 1:1, the normal–iso ratio is only a little over 1:1. With a ten-fold excess of triphenylphosphine, selectiveness of the order of 70% (normal–iso ≃2:1) were obtained [219] and when the reaction was run with triphenylphosphine as solvent, corresponding to a 600-fold excess of ligand over complex, a straight to branched chain ratio of 15.3:1 was obtained at 125 °C/~12.5 atm [220]. In the absence of added tertiary-phosphine, decreasing the partial pressure of carbon monoxide also increases the selectivity of the reaction. Both of these effects can be rationalized in terms of the equilibrium sequence shown below, Equation (2.24) [221,222]. High ligand concentrations and/or low CO partial pressures favour a predominance of species substituted by more than one tertiary phosphine ligand

$$\text{RhH(CO)(PPh}_3\text{)}_3 \underset{-\text{PPh}_3}{\overset{+\text{CO}}{\rightleftharpoons}} \text{RhH(CO)}_2\text{(PPh}_3\text{)}_2 \underset{-\text{PPh}_3}{\overset{+\text{CO}}{\rightleftharpoons}} \text{RhH(CO)}_3\text{(PPh}_3\text{)}$$

$$\underset{-\text{PPh}_3}{\overset{+\text{CO}}{\rightleftharpoons}} \text{RhH(CO)}_4. \qquad (2.24)$$

Although increasing the PPh_3:Rh ratio increases the selectivity of the system it unfortunately decreases the rate of hydroformylation – with propene as feed the rate of butanal formation of 100 °C, ~35 atm decreases from 13 g/min to 2.5 g/min as the PPh_3:Rh ratio is increased from 5:1 to 50:1 [219]. Thus in practical operation it is frequently necessary to reach a compromise between reaction rate and reaction selectivity. In addition to its beneficial influence on selectivity, increasing the PPh_3:Rh ratio also has a beneficial effect, from the standpoint of terminal alkene hydroformylation, on the isomerization activity of the catalyst in that it markedly reduces it. At 150 °C/35 atm, with $RhCl(CO)(PPh_3)_2$ as catalyst precursor, the isomer distribution in the product aldehyde is essentially independent of whether hex-1-ene, hex-2-ene or hex-3-ene is used as feed with some 20% of the product being 2-ethylpentanal. Whereas with a 20-fold excess of PPh_3 and hex-1-ene as feed no 2-ethylpentanal was detected and the selectivity to 1-heptanal was 76.2% [216].

As far as the nature of the tertiary-phosphine ligand is concerned, of those tested, triphenylphosphine appears to be optimal in terms of reaction rate, selectivity and also of cost. Trialkylphosphines generally give rise to catalysts having lower activity and poor selectivity. Thus under similar operating conditions a rhodium catalyst containing tributylphosphine was some six times less active than one containing triphenylphosphine and ten percentage points less selective [221]. This contrasts sharply with the situation found with the modified cobalt systems, see above. However, under the lower CO pressure used in the rhodium case the triphenylphosphine can effectively compete with CO for co-ordination sites and even in the absence of excess PPh_3 little or no $RhH(CO)_n$ is formed from $RhH(CO)(PPh_3)_3$ at low (<10 atm) pCO. The decreased activity obtained with the Rh/PBu_3 system is not fully understood although it may be due to the tributylphosphine ligand, which is a better σ-donor than triphenylphosphine, stabilizing the dihydrido–acyl complexes of the type (2.136) to the detriment of the overall reaction rate.

One of the disadvantages of the modified cobalt systems is their propensity to hydrogenate the alkene substrate. Potentially this is also a problem with the rhodium systems since, in the absence of carbon monoxide, $RhH(CO)(PPh_3)_3$ is an efficient alkene hydrogenation catalyst. However in the presence of excess triphenylphosphine ligand and/or with increasing carbon monoxide partial pressure, the hydrogenation activity is effectively depressed. Furthermore although these factors also decrease the hydroformylation rate their effect is an order of magnitude less than on the hydrogenation rate. Thus, with $RhCl(CO)(PPh_3)_2$ as catalyst precursor, a 50-fold excess of triphenylphosphine at 100 °C, 35 atm, H_2/CO 1:1, and hex-1-ene as feed, only trace amounts of hexane were formed over 17 h. Similarly in the commercial process, which is described below, when propene is used as feed it is claimed [223] that less than 2% is hydrogenated to propane.

Commercial development of a hydroformylation process using soluble rhodium catalysts, generally $RhH(CO)(PPh_3)_3$ is the catalyst precursor, has been carried out jointly by Union Carbide Corporation, Davy International Ltd. and Johnson Matthey. The first plants, one in Puerto Rico with a capacity of 130 000 tonnes/annum n-butanal from propene, and two in Texas City with a capacity of 45 000 tonnes/annum propanal from ethene, came on stream in the mid-seventies and construction of a 100 000 tonnes/annum butanal plant in Sweden is well advanced. In broad outline, the simplified flow diagram for the process is identical to that described for the process using tertiary-phosphine modified cobalt catalyst (Fig. 2.21). Because of the high catalyst cost, addition of a feedstock (propene, carbon monoxide and hydrogen) purification step is needed to reduce the concentration of sulphur containing catalyst poisons in the feed. As a result of the considerably greater activity of the rhodium system (ca. a factor of 10^3), greater conversion per pass can be achieved which allows use of either a smaller reactor vessel or total omission of substrate recycle. Similarly, the lower temperature and pressure requirements (*ca.* 100 °C/20 atm versus *ca.* 180 °C/80 atm) simplify plant design and reduce capital cost.

Before going on to consider hydroformylation catalysts based on metals other than cobalt and rhodium, let us briefly pause to compare and contrast the three processes we have described up to now.

2.4.3 *Cobalt versus rhodium*

In Table 2.3, some data relating to the industrial operation of the three, i.e. unmodified cobalt, tertiary phosphine modified cobalt and Rh/PPh$_3$, catalyst systems are shown. Although actual operating conditions and selectivities are a well-guarded commercial secret some general conclusions can be drawn from the data presented in Table 2.3. On going from left to right there is a marked decrease in the energy requirement, in terms of both milder operating conditions and greater product selectivity. The relatively high alkane make in the Co/L systems is compensated, to some extent, by the feasibility of a one-step process to alcohols. On the other hand, if the aldehyde is the desired product, e.g. butanal for the manufacture of 2-ethylhexanol, then a one-step process to the alcohol can be a disadvantage. The lower energy

Table 2.3 *Operating data for cobalt and rhodium hydroformylation processes*

	Unmodified cobalt	Ligand modified cobalt	Ligand modified rhodium
Temperature °C	140–180	160–200	80–120
Pressure, atm	250–350	50–100	15–25
Cobalt concentration, % metal/olefin	0.1–1.0	0.5–1.0	10^{-2}–10^{-3}
Normal:iso ratio	3–4:1	6–8:1	10–14:1*
Aldehydes, %	*ca.* 80	—	*ca.* 96
Alcohols, %	*ca.* 10	*ca.* 80	—
Alkanes, %	*ca.* 1	*ca.* 15	*ca.* 2
Other products, %	*ca.* 9	*ca.* 5	*ca.* 2

*With alk-1-enes as feed

requirements of the modified systems has aided their growth in recent years when industry has been subjected to rapidly escalating energy costs. There is, however, a price to pay and this is reflected in the relative prices of rhodium and cobalt metal which are in the approximate ratio of 3500:1. This is compensated for to some extent by the greater (10^2–10^3) activity of the rhodium system which allows lower catalyst concentrations to be used. Nevertheless, if severe economic penalties are to be avoided catalyst loss must be kept to an absolute minimum. Even a loss of only 3–5 p.p.m. would correspond to a price penalty of some 10 to 15 cents per kg in the manufacture of butanal from propene, assuming a rhodium price of $30 000 per kg, which, for a 100 000 tonne plant operating at 85% loading would correspond to a catalyst loss of between 85 000 and 127 000 dollars per year. There is also a problem of physical availability. In 1974 the total world production of rhodium amounted to 6 tonnes compared to 22 000 tonnes of cobalt in the western hemisphere alone [224] and in 1976 of the 4.29 tonnes produced nearly 25% came from the Soviet Union.

In spite of these difficulties, homogeneous processes based on rhodium are prospering [225,226], with the lower plant capital costs and energy requirements apparently outweighing the high catalyst investment. The high stability of the rhodium/triphenylphosphine system frequently allows the product to be distilled directly from the catalyst solution. For butanal production a 'gas-sparged' reactor has been described in which gaseous propene, hydrogen and carbon monoxide are 'sparged' through the catalyst solution and the butanal product is continuously removed in the vapour effluent [227].

2.4.4 *Other metal catalysts*

A comparison of the relative hydroformylation activity of the transition metals which readily form metal carbonyl complexes was published in 1977 [228]. As expected rhodium emerged supreme:

	Rh	>	Co	>	Ru	>	Mn	>	Fe	>	Cr, Mo, W, Ni
relative activity	10^3–10^4		1		10^{-2}		10^{-4}		10^{-6}		0

The most studied ruthenium hydroformylation catalyst system is that formed from $Ru(CO)_3(PPh_3)_2$ [229]. Under fairly rigorous conditions (120 °C, 100 atm, $CO:H_2 = 1:1$) alkene to aldehyde selectivities of some 99% have been achieved although the normal–iso ratio is somewhat disappointing in that it seldom exceeds 3:1. The active catalytic species is thought to be the dihydrido complex (*2.141*) formed by oxidative addition of hydrogen to (*2.140*) with concurrent, or prior, loss of CO. The rate determining step in the hydroformylation cycle, shown in Fig. 2.24, appears to be this oxidative addition of hydrogen and, unless (*2.141*) is added as such to the reaction medium, an induction period is found (cf. that observed in the rhodium system when $RhCl(CO)(PPh_3)_2$ is used as catalyst precursor). The ruthenium cycle differs from both the cobalt and the rhodium cycles in that alkene addition occurs subsequent to hydrogen addition. The alkene prefers to co-ordinate to the formally ruthenium(II) species rather than to the electron rich ruthenium(0) complex (*2.140*). Like the modified cobalt and rhodium catalysts the ruthenium system shows a marked preference for terminal alkenes although the final normal–iso ratio obtained is considerably lower (*ca.* 2:1 as compared to *ca.* 8:1 and 10:1 for Co and Rh, respectively). Adding excess tertiary-phosphine improves the normal–iso ratio, up to 5:1, but markedly decreases the rate (~10% conversion in 20 h with molten triphenylphosphine as solvent). In the absence of added

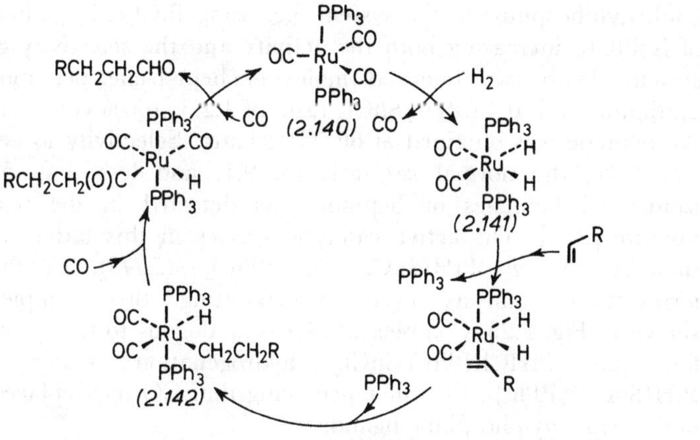

Fig. 2.24 *Catalysis cycle for the alkene hydroformylation reaction catalysed by $Ru(CO)_3(PPh_3)_2$. Only that cycle leading to linear products is shown*

tertiary-phosphine the normal–iso selectivity is essentially the same as that found with $RhH(CO)_2(PPh_3)_2$ in the absence of excess triphenylphosphine. In both cases it is likely that the steric course of the reaction is decided in the hydride ligand step, i.e. the step forming (2.133) in the rhodium cycle shown in Fig. 2.22 and that forming (2.142) in Fig. 2.24. At this stage in the cycles only one tertiary-phosphine ligand is attached to the metal centre and hence steric hindrance, which is probably the most important factor in determining the normal–iso ratio, is at a minimum.

A recent addition to the arsenal of homogeneous hydroformylation catalysts is the platinum–tin combination reported in the mid-seventies [203,231]. In the absence of tin chloride, platinum chloride, and complexes of platinum chloride containing tertiary-phosphine or tertiary-arsine ligands, show negligible hydroformylation activity. However, when a tin(II) halide, preferable $SnCl_2$, is added to the system, some highly active and highly selective hydroformylation catalysts are produced. The situation is reminiscent of that found in hydrogenation where addition of $SnCl_2$ to $PtHCl(PPh_3)_2$ converts a fairly inert complex into an active hydrogenation catalyst [19].

K_2PtCl_4 shows negligible hydroformylation activity. Addition of a five-fold excess of $SnCl_2$ produces a system capable of hydroformylating hept-1-ene to a mixture of octanal (60%), isomerized heptenes (13%) and n-heptene (10%) at 66 °C and 100 atm, CO/H_2 1:1. Adding two mole equivalents of triphenylphosphine to the system, i.e. using $PtCl_2(PPh_3)_2$ in place of K_2PtCl_4, increasing both the activity and the selectivity of the system. With two hundred moles of hept-1-ene per mole of platinum, and at $Pt:PPh_3:SnCl_2$ ratio of 1:2:5, 100% conversion of the heptene was obtained at 66 °C/100 atm. Selectivity to octanal was 85%, the normal–iso ratio $ca.$ 9:1, and less than 5% of isomerized heptenes or heptane was detected in the reaction mixture [231]. The active catalytic species in this latter case is thought to be $PtH(SnCl_3)(CO)(PPh_3)$, (2.143). A hydroformylation catalysis cycle incorporating this complex is shown in Fig. 2.25. Complex (2.143) is analogous to that proposed for the $PtHCl(PPh_3)_2SnCl_2$ hydrogenation system, i.e. $PtH(SnCl_3)(PPh_3)_2$, the difference being that CO has replaced one of the triphenylphosphine ligands.

As far as hydroformylation selectivity to n-octanal is concerned, there appears to be no significant trend as a function of either the

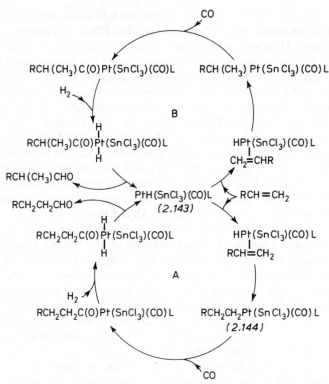

Fig. 2.25 *Catalysis cycles for the alkene hydroformylation reaction catalysed by* $PtH(SnCl_3)(CO)(PPh_3)$ *for both n, cycle A, and iso, cycle B, product formation.*
$L = PPh_3$

steric (cone angle θ) or electronic (electronic parameter ν) properties of the phosphorus donor ligand provided it is basic enough to remain co-ordinated to the platinum during the course of the reaction. With extremely weak basic ligands such as triphenylphosphite ($pK_a \simeq 0$) the selectivity drops to that found with $K_2PtCl_4/SnCl_2$ indicating that the phosphorus ligand is no longer bonded to the platinum centre (cf. PPh_3 versus PBu_3 in the ligand modified cobalt system). As with the other catalyst systems discussed, under non-isomerization conditions, the rate of alkene hydroformylation decreases with increasing alkene substitution, i.e.

$$RCH{=}CH_2 > RR^1C{=}CH_2 > RCH{=}CHR^1$$

reflecting the relative ease of alkene co-ordination. Similarly the hydroformylation rate decreases as the PPh$_3$ concentration is increased. However, in the case of the platinum–tin system, this could be due either to complete displacement of the tin ligand to form inactive PtCl$_x$(PPh$_4$)$_{4-x}$ or to PPh$_3$ association with one of the platinum/tin species in the catalysis cycle (Fig. 2.25) to give a sterically crowded 18 valence electron species of the type Pt^{II}(H)$_x$(SnCl$_3$)A(CO)(PPh$_3$)$_2$, where A = alkene and x = 1, or A = an alkene derivative, e.g. RCH$_2$CH$_2$ or RCH$_2$CH$_2$CO and x = 0. In either case lower catalytic activity would be expected.

The effectiveness of tin chloride, and to a much lesser extent germanium and lead chloride, in promoting both the hydroformylation and hydrogenation activity of these platinum complexes is not yet fully understood. Group IVB ligands such as SnCl$_3^-$, are generally recognized to be weak σ-donors and strong π-acceptors. They have empty orbitals of the correct symmetry and energy to accept electron density from filled platinum 5d or hybrid dp orbitals. Thus they can effectively lower the electron density at the platinum centre which, it has been suggested, favours both initial platinum hydride formation [232] and subsequent attack by alkene and carbon monoxide [233] while stabilizing the complexes against reduction to platinum(0) metal. SnCl$_3^-$ exerts a strong *trans*-labilizing influence, i.e. *trans*-effect, of the same order of magnitude as H$^-$. Thus, assuming the correct stereochemistry in the reaction intermediates, it would be expected to facilitate the ligand migration steps implicit in the hydroformylation cycle. The relative importance of these two closely related electronic effects is not yet clear. In addition to exerting an electronic effect on the course of the reaction SnCl$_3^-$ would also be expected to exert a steric influence since it is a relatively bulky ligand having a cone angle in excess of 120° (cf. PCl$_3$ θ = 124°). Thus the presence of SnCl$_3$ in the catalyst system will, sterically at least, favour linear products. Again the relative importance of steric and electronic effects in these systems remains to be resolved although it is quite clear, from this work and the earlier hydrogenation work, that SnCl$_3^-$ is a ligand of considerable potential in homogeneous catalysis.

In terms of activity, the platinum–hydrido complex PtH(SnCl$_3$)(CO)(PPh$_3$) is some five times more active than the modified cobalt system under analogous conditions [230] and some recent work [234] has shown that a platinum dichloride–tertiary-phosphine–tin(II) chloride system, where the tertiary-phosphine is 1,2-bis(diphenylphosphino-methyl)-cyclobutane, catalyses the hydroformylation of pent-1-ene at a rate which exceeds that obtained with $RhH(CO)(PPh_3)_3$. At 100 °C, 100 atm, $CO:H_2$ = 1:1, the platinum complex is 1.3 times as active as the rhodium species. With pent-1-ene as feed the product, at 100% conversion, consisted of hexanal (79%), n-pentane (6.4%), pent-2-ene (13.4%) and polymeric material (0.2%). Remarkably, the normal–iso ratio was 99:1 which is attributed to the steric constraint imposed by the bulky chelating di-phosphine ligand. Finally, before leaving the subject of hydroformylation, just a few words about asymmetric hydroformylation.

2.4.5 *Asymmetric hydroformylation*

As in catalytic hydrogenation (see Section 2.1.3) the stereoselective addition of an AB unit, HCHO in the case of hydroformylation, across a prochiral double bond offers the possibility of generating optical activity in the product, e.g.

$R^1R^2C{=}CH_2 + HCHO \longrightarrow R^1R^2\overset{*}{C}HCH_2CHO +$

or

$\phantom{R^1R^2C{=}CH_2 + HCHO \longrightarrow}R^1R^2\overset{*}{C}(CHO)CH_3$

$R^1CH{=}CHR^2 + HCHO \longrightarrow R^1CH_2\overset{*}{C}H(CHO)R^2 +$

$\phantom{R^1CH{=}CHR^2 + HCHO \longrightarrow}R^1\overset{*}{C}H(CHO)CH_2R^2.$

Involving as they do more ligand migration steps, hydroformylation catalysis cycles offer more opportunity for racemization than their hydrogenation counterparts and optical yields in asymmetric hydroformylation reactions have not yet reached those achieved by asymmetric catalytic hydrogenation.

Some of the best results have been obtained using styrene as substrate and, with $Rh_2Cl_2(CO)_4/4L$, where L = *(2.145)*,

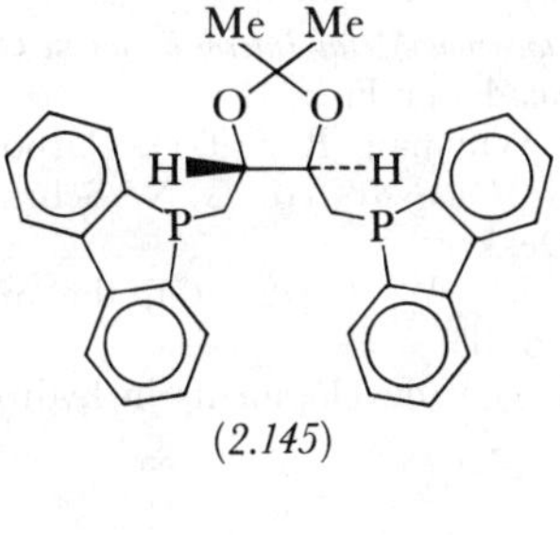

(2.145)

optical yields in excess of 44% have been reported [235]. With non-aryl substituted prochiral alkenes, e.g. but-2-ene, optical yields are considerably lower, seldom exceeding 10%, and it has been suggested [236] that the higher stereoselectivity associated with styrene may be due to formation of π-benzyl intermediates of the type (2.146). This type of bonding would increase the conformational rigidity in the diastereoisomeric intermediate thus making asymmetric induction more likely. The platinum/tin

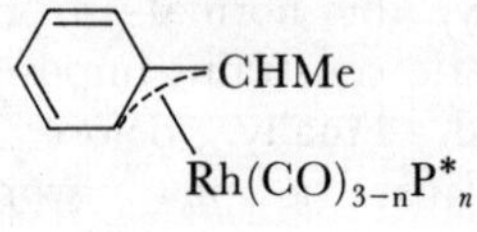

$$\text{C}_6\text{H}_5\text{—CHMe}$$
$$\text{Rh(CO)}_{3-n}\text{P}^*_n$$

(2.146)

system with (-)diop as ligand also catalyses the asymmetric hydroformylation of n-butenes [237]. With this system the products and optical yields are independent of the butene used suggesting that asymmetric induction occurs after formation of the platinum-alkyl intermediate, e.g. (2.144) in the catalysis cycle shown in Fig. 2.25.

Catalytic asymmetric hydroformylation has considerable potential in the synthesis of biologically or pharmacologically active molecules. However, commercial application will probably require the development of more stereoselective catalyst systems. Having raised the subject of biologically active or naturally occurring molecules we now move on to another area of homogeneous catalysis which is proving very useful in the synthesis of natural products, namely transition-metal alkene oligomerization.

Further reading on homogeneous hydroformylation

General texts and reviews

Bird, C. W. (1967), *Transition Metal Intermediates in Organic Synthesis*, Ch. 6, p. 117, New York: Academic Press.

Thompson, D. T. and Whyman, R. (1971), Carbonylation, in *Transition Metals in Homogeneous Catalysis* (ed. G. N. Schrauzer), Ch. 5, p. 147, New York: Marcel Dekker.

Orchin, M. and Rupilius, W. (1972), On the mechanism of the oxo reaction. *Catal. Rev.*, **6**, 85.

Paulik, F. E. (1972), Recent developments in hydroformylation catalysis. *Catal. Rev.*, **6**, 49.

Heck, R. F. (1974), *Organotransition Metal Chemistry, A Mechanistic Approach*, Ch. 9, p. 201, New York: Academic Press.

Marko, L. (1974), Hydroformylation of olefins with carbonyl derivatives of the noble metals as catalysts, in *Aspects of Homogeneous Catalysis* (ed. R. Ugo), Vol. 2, Ch. 1, Dordrecht: Reidel.

Falbe, J. (1975), Technical reaction of carbon monoxide with metal carbonyls as catalysts. *J. Organometal. Chem.*, **94**, 213 (text in German).

Henrici-Olivé, G. and Olivé, S. (1976), Reactions of carbon monoxide with transition metal–carbon bonds. *Transition Met. Chem.*, **1**, 77.

Davidson, P. J., Hignett, R. R. and Thompson, D. T. (1977), Homogeneous catalysis involving carbon monoxide, in *Catalysis*, Vol. 1, p. 369, London: Chemical Society.

Pino, P., Piacenti, F. and Bianchi, M. (1977), in *Organic Synthesis via Metal Carbonyls* (eds. I. Wender and P. Pino), Vol. 2, p. 43, New York: Wiley.

Pruett, R. L. (1979), Hydroformylation. *Advan. Organometal. Chem.*, **17**, 1.

2.5 Oligomerization

Oligomerization is a general term applied to that class of reactions in which a substrate, A, or a mixture of substrates, A and B, is converted into a higher molecular weight product by a series of mutal coupling reactions

$$nA \longrightarrow A_n$$
$$xA + yB \longrightarrow A_xB_y.$$

The simplest type of oligomerization is dimerization where $n = 2$. When n is large (>100) we move into the area of polymerization which we will be considering in the next section. As far as oligomerization is concerned, for most practical purposes we are interested in systems where n, or $x + y$, is less than 15 or 20. Oligomerization reactions involving two different substrate molecules are referred to as co-oligomerization or hetero-oligomerization reactions. A number of factors are important in describing catalytic oligomerization reactions; for instance the degree of oligomerization, i.e. the value of n or $x + y$, will be determined by the relative rates of the propagation and termination steps; unless the substrate is perfectly symmetrical the way in which it is incorporated into the growing chain will be important in deciding the stereochemistry of the final product; with certain substrates, e.g. dienes, there is the possibility of forming linear or cyclic oligomers; with two different substrates the order and the manner in which they are incorporated will be of prime importance in determining the nature of the oligomeric product; finally, as we are dealing with a catalytic bond making process, we have the possibility of producing asymmetric, optically active, molecules.

In the following five sub-sections we have chosen the individual systems with the express aim of illustrating the above points. We begin by considering two of the simplest alkene substrates, i.e. ethene and propene, and one of the most active homogeneous alkene oligomerization catalysts – nickel.

2.5.1 *Nickel complexes as homogeneous alkene oligomerization catalysts*

Numerous catalyst systems involving transition-metal complexes are known which convert alkenes into oligomers [238]. However, one of the most active – catalytic reactions have been carried out at temperatures as low as $-100\ °C$ – and certainly one of the most extensively studied is that involving nickel [23,239].

One of the simplest methods of preparing such catalysts is via the so-called Ziegler method (see Section 2.6) in which a nickel(II) compound of the type NiX_2 or NiX_2L_2 (X = anionic ligand; L = donor ligand, such as a tertiary-phosphine) is treated with an excess of an alkyl-aluminium or alkyl-aluminium halide species such as $Al(C_2H_5)_xCl_{3-x}$ (x = 1, 2 or 3) or $(C_2H_5)_3Al_2Cl_3$. Alternatively, an organo nickel compound of the type $NiR_xX_{2-x}L_n$ (where x and n are generally equal to one or two) may be used to provide the nickel component of the catalyst system. Of the latter type, dimeric π-allyl nickel halides (*2.147*) or their mono-tertiary-phosphine adducts (*2.148*) have been frequently used [240] and the subsequent reactions with the alkyl-aluminium compounds well characterized, as shown below.

A single crystal X-ray structural determination has confirmed the stereochemistry of (*2.150*) with $L = P(cyclohexyl)_3$ and $AlR_xX_{3-x} = AlMe_3$ [23]. While alkyl-aluminium compounds usually give rise to the most active catalysts, other Lewis acids such as BF_3, $TiCl_4$ or SbF_5 can also be used in combination with nickel as indeed can Brønsted acids such as H_2SO_4 or CF_3COOH [23]. In all cases, irrespective of the detailed method of preparation, it appears that the active catalytic species is a hydrido-nickel complex of the type 'NiHY(L)', (*2.151*), where Y is either a complex Lewis acid type anion, e.g. $XAlR_xX_{3-x}$ in (*2.149*) and (*2.150*), or a Brønsted anion such as $CF_3CO_2^-$. The presence

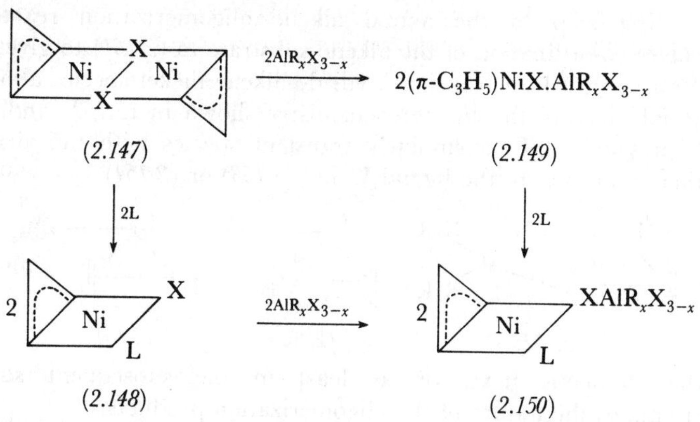

HOMOGENEOUS CATALYST SYSTEMS IN OPERATION

[Scheme showing complexes (2.147) → (2.149) with 2AlR$_x$X$_{3-x}$ giving $2(\pi\text{-}C_3H_5)\text{NiX·AlR}_x X_{3-x}$; both convert downward with 2L to (2.148) and (2.150) respectively, with (2.148) → (2.150) via 2AlR$_x$X$_{3-x}$, yielding XAlR$_x$X$_{3-x}$]

of the donor ligand, L, is not strictly essential although, as we shall see shortly, it can play an influential role in directing the course of the oligomerization reaction.

The precise way in which 'NiHY(L)' is formed depends on the nature of the catalyst precursor. When a Ni(0)/Brønsted acid combination is used oxidative addition of the Brønsted acid (HY) to the Ni(0) complex will directly form (2.151), e.g. [154, 241]

$$\text{NiL}_4 + \text{HY} \longrightarrow \text{HNi(L)}_{4-n}\text{Y} + n\text{L}$$

$$n = 0 \text{ or } 1.$$

When a π-allyl nickel complex is used in combination with a Lewis acid the π-allyl unit frequently ends up in the alkene substrate, with the nickel hydride species being formed by a sequence such as that shown below, with cyclo-octene as substrate [23]:

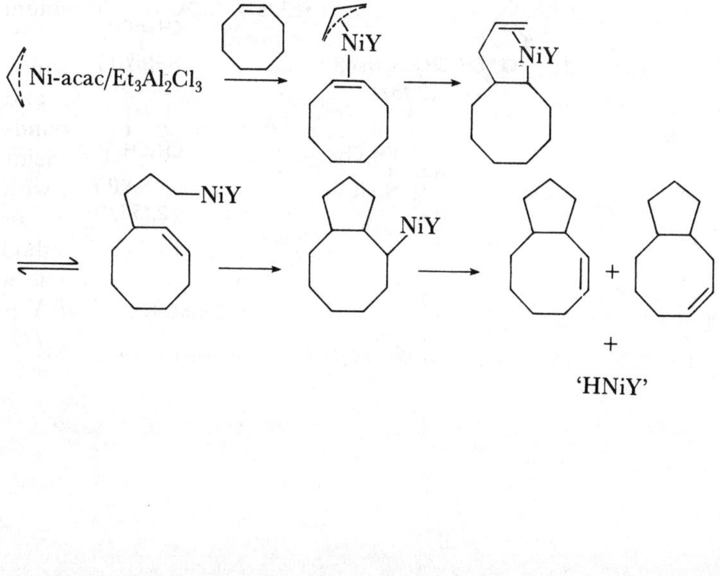

[Reaction sequence: Ni-acac/Et$_3$Al$_2$Cl$_3$ with cyclooctene forming various allyl-NiY intermediates, finally giving cyclooctene products + 'HNiY']

The first step in the actual alkene oligomerization reaction involves co-ordination of the alkene substrate to (2.151) to give, in the absence of L, a transient hydrido-alkene nickel species almost certainly having the *cis* stereochemistry shown in (2.152) and, in the presence of L, a similarly transient species with the alkene either *cis* or *trans* to the ligand L, i.e. (2.153) or (2.154).

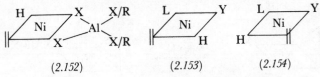

(2.152)   (2.153)   (2.154)

What happens next, or at least in the subsequent steps, determines the nature of the oligomerization products.

(*a*) *Ethene oligomerization; dimerization versus further oligomerization*
A set of catalysis cycles for the nickel catalysed ethene

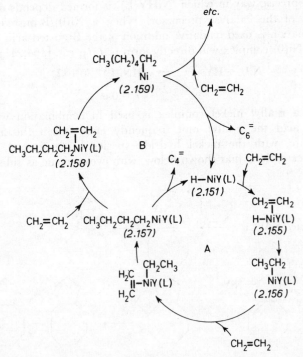

Fig. 2.26 *Catalysis cycles for the nickel catalysed oligomerization of ethene*

oligomerization reaction is shown in Fig. 2.26. After formation of the hydrido-alkene species the next step involves a hydride to alkene ligand migration reaction to form the ethyl–nickel intermediate (2.156). Further ethene co-ordination followed by ethyl migration gives the butyl–nickel species (2.157). At this stage of the cycle (2.157) can either undergo a β-hydrogen elimination reaction to give product butene and the nickel hydrido intermediate (2.151), thus completing the dimerization cycle A, or co-ordinate a further molecule of ethene to give (2.158), thus entering the trimerization cycle B. Similarly (2.159) can either undergo a β-hydrogen elimination reaction to complete cycle B or co-ordinate a further molecule of ethene and enter a tetramerization cycle (shown in Fig. 2.26 as etc.). An important factor in determining the degree of oligomerization, or in the extreme polymerization, is the relative rate of the elimination, or displacement, step and the alkyl migration to co-ordinated alkene, or insertion, step. Using the simplified scheme:

$$RCH_2CH_2Ni \xrightarrow{CH_2CH_2} \begin{array}{c} \xrightarrow{k_i} RCH_2CH_2CH_2CH_2Ni \\ \xrightarrow{k_d} RCH=CH_2 + HNi-\|\begin{array}{c}CH_2\\CH_2\end{array} \end{array}$$

the degree of oligomerization will depend on the k_i/k_d ratio. If $k_d \gg k_i$ then we will be dealing predominantly with cycle A and butenes will be the major products of the reaction. If $k_d \ll k_i$ then we will be into cycle B and higher cycles and higher oligomers or polymers will result. We can influence the k_i/k_d and thus steer the reaction by modifying the nature of the donor ligand L. As in so many homogeneous catalytic reactions, the most dramatic effects are obtained when L is a tertiary-phosphine ligand as evidenced by the results shown in Table 2.4 [23].

These results indicate that the k_i/k_d ratio increases with the steric bulk of the tertiary-phosphine ligand, i.e. somewhat surprisingly greater steric constraint in the nickel system favours chain growth over product elimination. This result has been rationalized [23] by assuming that the elimination reaction proceeds via a penta (or higher) co-ordinated nickel hydride intermediate having two alkene molecules co-ordinated to the metal centre, whereas alkyl to alkene migration (insertion)

Table 2.4 *Influence of tertiary-phosphine ligands on the dimerization and oligomerization of ethene**

L	Electronic parameter ν (cm^{-1})	Steric parameter θ (degrees)	butenes (%)	hexenes (%)	Octenes and higher oligomers (%)
P(CH$_3$)$_3$	2064.1	118	>98	~1	—
PPh$_3$	2068.9	145	90	10	—
P(cyclohexyl)$_3$	2056.4	170	70	25	5
P(Bu^t)$_2$Pr^i	2057.1	~175†	25	25	50
PBu$^t_3$‡	2056.1	182	—	—	polyethylene

*Catalyst precursor (π-C$_3$H$_5$)NiBrL/Et$_3$Al$_2$Cl$_3$ at -20 °C 1 bar ethene
†estimated
‡four-fold excess of PBu$^t_3$ over Ni

proceeds by a less sterically hindered tetra co-ordinate intermediate. Adopting this explanation the transition states for the two processes can be envisaged as (*2.160*) and (*2.161*), respectively, wherein the differing steric requirements are clear.

$$\left[\begin{array}{c} \text{CH}_2 \\ \| \\ \text{CH}_2 \end{array}\!\!\!\begin{array}{c} \text{H} \text{----} \text{CHR} \\ \vdots \quad \| \\ \text{Ni} \text{----} \text{CH}_2 \\ /\backslash \\ \text{Y} \quad \text{L} \end{array}\right]^{\ddagger} \quad \left[\begin{array}{c} \text{Y} \\ \diagdown \\ \text{L} \end{array}\!\!\!\begin{array}{c} \text{C}\!=\!\text{C} \\ \diagdown \\ \text{Ni} \text{----} \text{CH}_2 \\ \diagup \\ \quad \text{CH}_2\text{R} \end{array}\right]^{\ddagger}$$

(*2.160*) (*2.161*)

This would imply that in the catalysis cycle shown in Fig. 2.26 (*2.157*) transforms directly into (*2.155*) without passing through the intermediate nickel-hydride species (*2.151*). This would seem probable since, in the presence of excess alkene and in the absence of other strongly co-ordinating molecules, it is unlikely that the co-ordinatively unsaturated species (*2.151*) would be capable of independent existence.

With a symmetric alkene, such as ethene, the direction of either H or alkyl addition is, in the absence of isomerization, immaterial. However, when we move to an unsymmetric alkene, such as propene, it becomes an important factor in determining the oligomer product distribution.

(b) Propene oligomerization; regioselectivity

During the nickel catalysed oligomerization of propene, or indeed any alkene of the type $RCH=CH_2$, the initial hydride migration step can occur to give either a linear alkyl product (H to C_2, Ni to C_1) or a branched alkyl product (H to C_1, Ni to C_2). Similarly the alkyl to alkene migration, implicit in the next step of the oligomerization cycle, can also occur in two distinct directions. These dual migration possibilities clearly give rise to a range of isomeric products and in Fig. 2.27 we have shown routes to the six possible isomeric products arising from the dimerization of propene [239]. The situation is akin to, although somewhat more complex than, that described in hydroformylation when we were discussing the normal–iso ratio (see Section 2.4) and as was found there, the course of the reaction can be significantly influenced by the nature of the donor ligand present in the catalyst system. In Table 2.5 some representative data relating to the effect of tertiary-phosphine ligands on the nickel catalysed dimerization of propene are given [23,242]. Assuming that the reactivity of the nickel-n-propyl species, (*2.162*), towards propene is the same as that of the nickel-iso-propyl species (*2.163*) and assuming that no isomerization occurs during the course of the oligomerization it is possible, by determining the product distribution to calculate the ratio or $Ni \rightarrow C_1$, and $Ni \rightarrow C_2$ addition in both step 1 and step 2 of Fig. 2.27. Thus for step 1:

$$\frac{\% \; Ni \rightarrow C_1}{\% \; Ni \rightarrow C_2} = \frac{\%(\text{2-methylpent-1-ene} + \text{hex-1-ene} + \text{hex-2-ene})}{\%(\text{4-methylpent-1-ene} + \text{4-methylpent-2-ene} + \text{2,3-dimethylbut-1-ene})}$$

and for step 2

$$\frac{\% \; Ni \rightarrow C_1}{\% \; Ni \rightarrow C_2} = \frac{\%(\text{2-methylpent-1-ene} + \text{2,3-dimethylbut-1-ene})}{\%(\text{4-methylpent-1-ene} + \text{4-methylpent-2-ene} + \text{hex-1-ene} + \text{hex-2-ene})}$$

The data in Table 2.5 indicate that, with the exception of $PBu^t_2Pr^i$, the nature of the tertiary-phosphine has little influence on the direction of the first ligand (H) migration step. The predominant addition product being the iso-product (*2.163*) formed by hydride migration to the α-carbon of the propene unit,

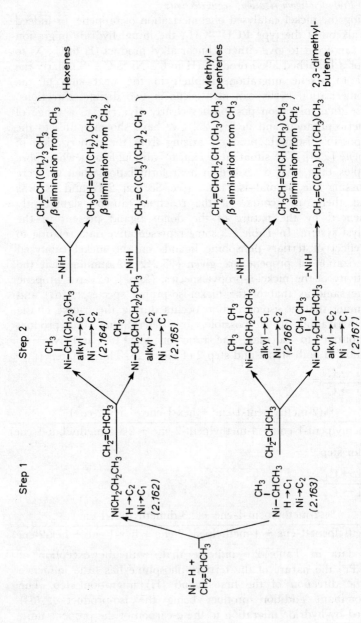

Fig. 2.27 Routes to the six possible isomeric products from the dimerization of propene

Table 2.5 *Effect of tertiary-phosphine ligands on the nickel catalysed dimerization of propene**

Tertiary phosphine L	Electronic parameter ν (cm^{-1})	Steric parameter θ (degrees)	Σ Hexenes	Σ Methyl pentenes	Σ 2,3-dimethyl-butene	First-step Ni→C$_1$: Ni→C$_2$	Second-step Ni→C$_1$: Ni→C$_2$
PMe$_3$	2064.1	118	9.9	80.3	9.8	15:85	15:85
PEt$_3$	2061.7	132	9.2	69.7	21.1	17:83	29:71
PBu$_3$	2060.3	132	7.1	69.6	23.3	18:82	34:66
PPr$^i_3$	2059.2	160	1.8	30.3	67.9	20:80	86:14
P(cyclohexyl)$_3$	2056.4	170	3.3	37.9	58.8	26:74	83:17
P(Bu^t)$_2$Pr^i	2056.1	~175†	0.6	70.1	29.1	69:31	98:2

*Catalyst precursor (π-C$_3$H$_5$NiX)$_2$/Et$_3$Al$_2$Cl$_3$/L at −20 °C 1 atm
†Estimated

i.e. Markovnikov addition. Only with the very bulky tertiary-phosphine, $PBu^t_2Pr^i$, is the direction of this addition significantly altered to favour the less sterically demanding n-propyl product (*2.162*). It is during the second step in the oligomerization sequence that the tertiary-phosphine ligands exert their steering influence [23]. Increasing the size of L strongly favours formation of the less sterically demanding alkyl intermediates having the methyl substituent at C_2 (relative to Ni) rather than at C_1, i.e. (*2.165*) and (*2.167*) rather than (*2.164*) and (*2.166*). This is particularly marked for tertiary-phosphine of the type PBu^t_2R where the ratios are 85:15, 97:3 and 98:2 for R = Me, Et and Pr^i, respectively [243] and with $PBu^t_2Pr^i$ the yield of *n*-hexene products is virtually zero. Studies [23,239] using a large range of tertiary-phosphine ligands have confirmed that the effect on the direction of nickel-alkyl migration to propene is largely steric in origin with electronic properties, as reflected in say v, playing only a minor role.

Up to now we have concentrated on substrates containing only one double bond, an equally important class of oligomerization reactions, which offer the possibility of forming cyclic as well as linear products, are those involving diene molecules. To illustrate this latter type of reactions we have chosen one of the simplest, and commercially most readily available of dienes – buta-1,3-diene.

2.5.2 *Butadiene oligomerization*

Depending on the catalyst system and the reaction conditions a number of different products can be obtained from the catalytic oligomerization of butadiene as exemplified in Fig. 2.28. There are basically two types of products, i.e. cyclic and open chain oligomers. In the absence of isomerization the cyclic oligomers can be viewed as simple combinations of butadiene units as indicated by the dashed lines in Fig. 2.28. An examination of the structure of the open chain products reveals that they cannot be broken down into simple butadiene units and that a hydrogen atom transfer reaction must be involved in their synthesis. As with alkene oligomerization, several catalysts systems are known to be effective for butadiene oligomerization [244]. However, once again it is the nickel systems which have received most attention [55] and which are among the best characterized [239], at least for cyclo-oligomerization. As far as oligomerization to open chain

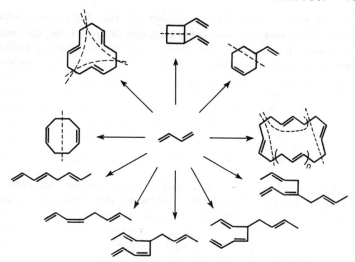

Fig. 2.28 *Range of products obtainable from the catalytic oligomerization of butadiene*

products is concerned, palladium based catalyst systems have proved extremely useful [245].

(a) Nickel(0) systems: cyclo-oligomerization
In contrast to the situation found in the nickel catalysed alkene oligomerization, where the active catalytic species is a nickel(II) complex *vide supra*, in the butadiene oligomerization reactions the active species is a nickel(0)-butadiene complex. This can be formed, either by the reduction of a nickel(II) salt, e.g. nickel acetylacetonate, in the presence of butadiene, or by ligand exchange between butadiene and a zero-valent nickel compound, e.g. Ni(cyclo-octadiene)$_2$, Ni(π-C$_3$H$_5$)$_2$ or Ni(CH$_2$=CHCN)$_2$. In the absence of donor ligands, such as tertiary phosphine, and of hydrogen donor solvents, such as alcohols or secondary amines, the main product of the nickel catalysed oligomerization of butadiene is the cyclo-trimer *trans,trans,trans*-cyclo-octadodecatriene (*2.168*). Small amounts of the *trans,trans,cis*-(*2.169*) and the *trans,cis,cis*-(*2.170*) isomer together with some cyclic dimers and higher oligomers are also produced. Generally, however, the selectivity towards (*2.168*) is greater than 80% and with some systems, e.g. Ni(acac)$_2$/(C$_2$H$_5$)$_2$AlOC$_2$H$_5$, exceeds 90% [239].

Detailed insight into the mechanism of this butadiene cyclo-trimerization reaction has been obtained from an elegant and painstaking investigation by Wilke and his co-workers [239]. They isolated the key intermediate (2.171) formed by the coupling

(2.168)　　　　　(2.169)　　　　　(2.170)

of three butadiene units at a nickel atom. Further addition of butadiene to this intermediate results in the formation of the cyclo-dodecatriene (2.168) and the regeneration of (2.171).

(2.171)

A possible catalysis cycle incorporating (2.171) is shown in Fig. 2.29.

It seems likely that the first step involves co-ordination of two or more butadiene molecules to the nickel centre to give a nickel(0)–alkene complex possibly of type (2.172). Reductive coupling of the two diene units gives the di-(π-allyl)nickel(II) species (2.173) which can exist in equilibrium with the π-allyl, σ-allyl isomer (2.174). Co-ordination of third butadiene in the *cis*-configuration gives (2.175) in which the co-ordinated butadiene molecule can 'insert' into the π-allyl nickel bond to give (2.171) or one of its isomeric analogues such as (2.176). Oxidative coupling of the C_{12} chain in either (2.171) or (2.176) with concurrent reduction of the nickel(II) to nickel(0) gives the product(s) and completes the cycle. From its stereochemistry (2.171) would be expected to give *trans,cis,cis*-cyclododecatriene, (2.170). (2.176), which contains a *syn,syn* arrangement of the π-allyl groups, should form the *trans,trans,trans*-isomer (2.168). Since the *trans,trans,trans*-isomer is the major product of the reaction it would seem that (2.171) can rapidly isomerize under the reaction conditions, i.e. in the presence of excess butadiene, to give (2.176) or an equivalent *syn,syn* isomer. By adding suitable

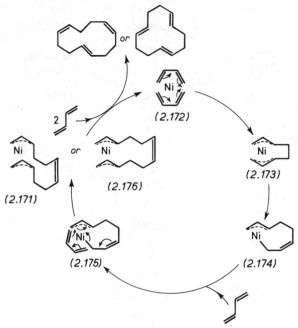

Fig. 2.29 *Catalysis cycle for the nickel(0) catalysed butadiene trimerization*

ligands, e.g. P(OC₆H₄-*o*-C₆H₅)₃ and P(cyclohexyl)₃, it is possible to isolate ligand stabilized adducts of both (*2.173*) and (*2.174*), i.e. (*2.177*) and (*2.178*):

(*2.177*)   (*2.178*)

Furthermore, on adding triethylphosphine to (*2.171*) it is possible to mimic the catalytic reaction stoichiometrically as shown below:

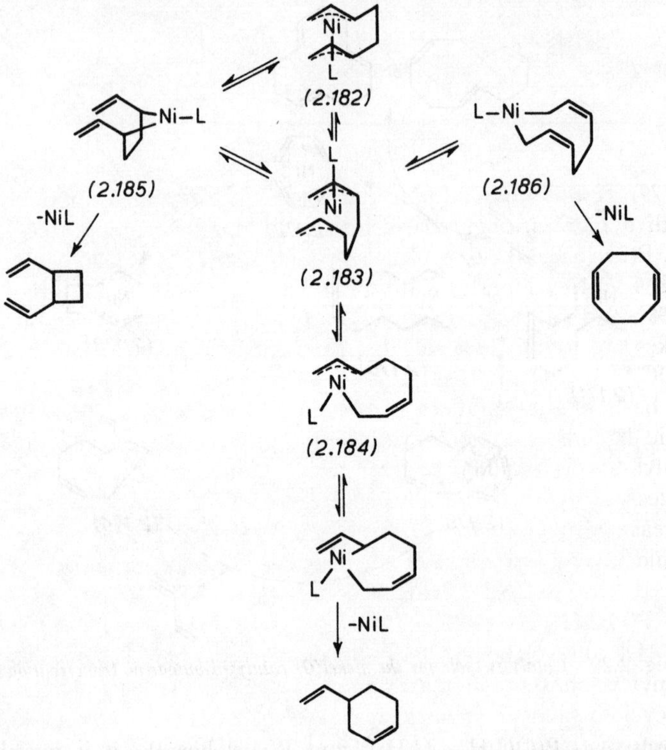

Fig. 2.30 *Sequence of reactions resulting in the formation of 1,5-cyclo-octadiene, 4-vinylcyclohexene, and* cis-*1,2-divinylcyclobutane from the cyclodimerization of butadiene*

In addition to providing a means of stabilizing reaction intermediates, such that they can be isolated, tertiary-phosphine ligands also provide a means of steering the course of the nickel catalysed oligomerization reaction.

Addition of a tertiary-phosphine or -phosphite ligand to the butadiene trimerization catalyst systems described above, generally in the ligand:Ni ratio of 1:1, produces a system capable of selectively catalysing the cyclodimerization of butadiene to a mixture of 1,5-cyclo-octadiene (*2.179*) 4-vinylcyclohexene (*2.180*) and *cis*-1,2-divinylcyclobutane (*2.181*). The relative distribution of these products is dependent upon the nature of the phosphorus ligand. With bulky, weakly basic, ligands such as $P(OC_6H_4\text{-}o\text{-}C_6H_5)_3$, $\theta = 152°$, $\nu = 2085$ cm^{-1}, 1,5-cyclo-octadiene

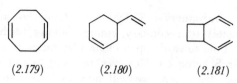

(2.179)   (2.180)   (2.181)

(2.179) is the predominant (96%) product of the dimerization. With a more strongly basic ligand, of roughly comparable size, e.g. P(cyclohexyl)$_3$, $\theta = 170°$, $\nu = 2056.4$ cm^{-1}, 1,5-cyclo-octadiene and 4-vinylcyclohexene are produced in about equal amounts (41% and 40%, respectively) and cis-1,2-divinylcyclobutane makes up some 14% of the product mixture (cf. 0.2% with P(OC$_6$H$_4$-o-C$_6$H$_5$)$_3$ [246]. These results have been rationalized on the basis of the sequence of reactions shown in Fig. 2.30 [239]. Basic ligands, i.e. low ν value, favour formation of the π-allyl, σ-allyl species (2.184) whereas less basic ligands, i.e. high ν values, favour the bis-π-allyl complexes (2.182) and (2.183). Thus increasing the basicity, electron donor properties, of the ligand should favour formation of the vinylcyclohexene product. This is indeed the case as evidenced by the results obtained with L = P(OC$_6$H$_5$)$_3$, $\nu = 2085.3$ cm^{-1}, L = P(C$_6$H$_5$)$_3$ $\nu = 2068.9$ cm^{-1} and L = P(cyclohexyl)$_3$, $\nu = 2056.4$ cm^{-1}, where the yield of 4-vinylcyclohexene increases from 7.4 to 27 to 40% while the 1,5-cyclo-octactiene yield decreases from 81 to 64 to 41%. Also, as previously mentioned, by working at lower temperature (the dimerizations reactions are usually carried out at ±80 °C) it has proved possible to isolate (2.183), with L = P(OC$_6$H$_4$-o-C$_6$H$_5$)$_3$, and (2.184) with L = P(cyclohexyl)$_3$. As well as the electronic factors already discussed, the steric properties of the ligand also play a role in directing the course of the reactions. With weakly basic ligands, i.e. tertiary-phosphites, the steric bulk influences the relative concentrations of (2.185) and (2.186) and hence the relative yields of 1,5-cyclo-octadiene and cis-1,2-divinylcyclobutane. Increasing the size of L favours formation of the less sterically demanding intermediate (2.186) and hence increases the yield of cyclo-octadiene. Thus with P(OC$_6$H$_4$-o-C$_6$H$_5$)$_3$ and P(OC$_6$H$_5$)$_3$, which both have essentially the same electronic parameter ν (i.e. 2085 and 2085.3 cm^{-1}, respectively), the relative yields of cyclo-octadiene and divinylcyclobutane are 96/81% and 0.2/9.2% reflecting the smaller bulk of P(OC$_6$H$_5$)$_3$ (cone angles 152° and 128°, respectively).

A further modification of the nickel(0) catalyst systems,

involving the introduction of hydrogen donor co-catalyses, e.g. alcohols, phenols or secondary amines, allows further reaction steering, this time towards open chain oligomeric products of the type shown in Scheme 2.4. However, in order to add a little more variety to the discussion, we will now look at some butadiene linear oligomerization catalysts involving one of nickels nearest neighbours in the periodic table, namely palladium.

(b) *Palladium(0) systems; linear oligomerization* [245,247,248]

In an aprotic solvent such as benzene or tetrahydrofuran at *ca.* 100 °C, palladium(0) complexes of the type PdL_nA, where L = PPh_3 or PEt_3, n = 2 or 3 and A = a weakly co-ordinating ligand such as maleic anhydride or benzoquinone, catalyse the linear dimerization of butadiene to 1,3,7-octatriene (*2.187*). When π-allyl-palladium acetate dimer, $(\pi\text{-}C_3H_5PdOAc)_2$, is used as the catalyst precursor the major product (79%) of the reaction is the linear trimer, 1,3,6,10-dodecatetraene (*2.188*). In the presence of

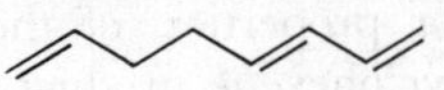

(*2.187*) 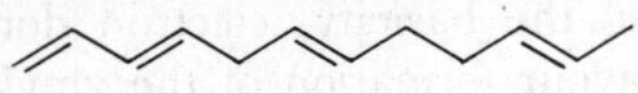   (*2.188*)

protic solvents, e.g. alcohols, phenols, carboxylic acids or amines, with $Pd(PPh_3)_2$(maleic anhydride) as catalyst precursor, the hydrogen substituted product *trans*-2,7-octadiene (*2.189*) is formed.

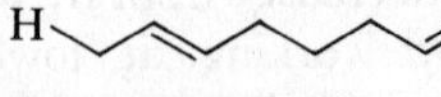

(*2.189*)

Two mechanisms have been proposed for the palladium catalysed dimerization reaction. In the first it is suggested [240,249] that, in the initial stages, the reaction is essentially the same as that found for nickel (see Fig. 2.30), i.e. two butadiene molecules couple around the metal centre to give the *bis*-π-allyl species (*2.190*), where M = Ni or Pd. In the case of nickel this species undergoes further carbon–carbon coupling to give the cylic dimer products. When M = Pd (*2.190*) undergoes a rapid hydrogen transfer reaction, effectively from C-4 to C-6, to give the linear

'ML' + 2 ⟶ L—M ⟶ + 'ML'

(*2.190*)

HOMOGENEOUS CATALYST SYSTEMS IN OPERATION 151

octatriene product. The difference in behaviour between nickel and palladium is then ascribed to the greater atomic size of palladium compared to nickel. It is suggested that the carbon atoms, which must couple to give the cyclic products, cannot get close enough in the palladium species so an alternative reaction course, i.e. hydrogen transfer, ensues. If this is indeed the case then this reaction would constitute one of the few examples of the metal size, rather than that of the associated ligands, directing the course of a catalytic reaction. In line with this mechanistic proposal it has been suggested [250,251] that, in the case of the palladium acetate catalysed oligomerization where trimers are the major products, bimetallic acetate bridged dimers of the type (*2.191*) are involved in the catalysis cycle. In the second

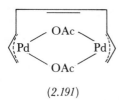

(*2.191*)

mechanistic proposal [247] the active species is suggested to be a palladium(II) hydrido species and the step-wise catalysis cycle shown in Fig. 2.31, which is closely analogous to the nickel catalysed alkene oligomerization cycle, Fig. 2.26, is proposed to account for the linear products.

The presently available experimental data do not allow a definite choice, assuming that such a distinction will be universally applicable, to be made between the two mechanisms although deuteration studies [252], in which mono C-6 deuterated 1,3,7-octatriene was obtained from the $Pd(PPh_3)_2$(maleic anhydride) catalysed dimerization of butadiene in C_3H_7OD, tend to favour the latter step-wise sequence in protic solvents.

On leaving hydroformylation to move on to look at catalytic oligomerization, we stated that this latter reaction could prove useful in the synthesis of natural products. This is particularly true when the diene substrate is isoprene since regiospecific hand-to-tail dimerization or telomerization, i.e. the inclusion of a hetero-atom substituent in the dimer molecule, offers the possibility of synthesizing a range of mono-terpenes. Indeed, natural rubber itself consists of a linear head-to-tail polymer of isoprene units. This particular facet of catalysed diene

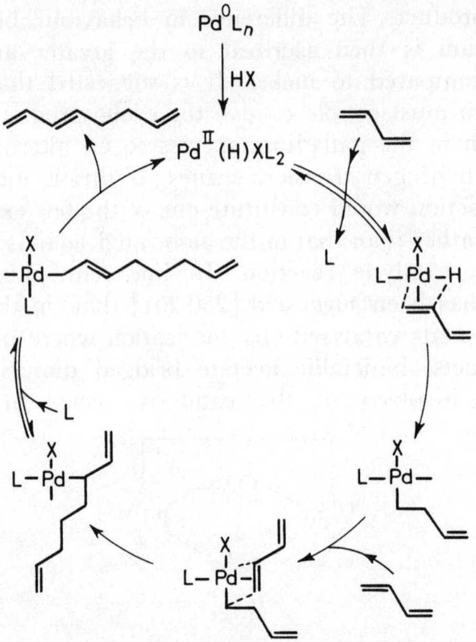

Fig. 2.31 *Catalysis cycle for the palladium catalysed step-wise dimerization of butadiene*

dimerization is at an early stage of development [245] and there are, as yet, few direct catalytic routes to naturally occurring mono-terpenes although the natural product citronellol (*2.192*) has been prepared in reasonable yields by the sequence shown below [253]:

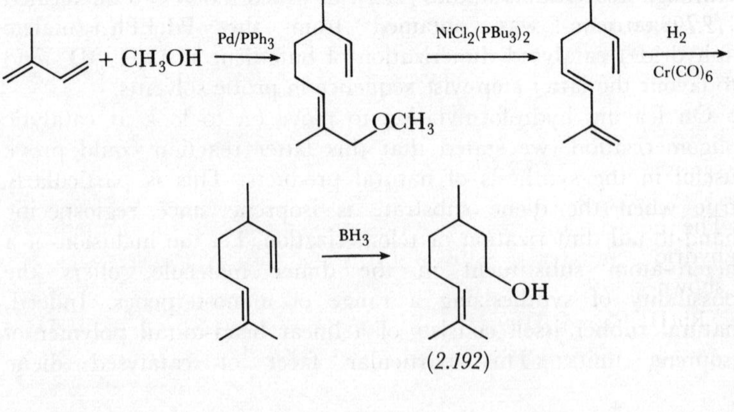

Although this type of reaction shows considerable potential, a dimerization reaction which has already clearly demonstrated its usefulness, at least in a commercial sense, is the codimerization of ethene and butadiene to form *trans*-hexa-1,4-diene.

2.5.3 *Ethene/butadiene codimerization*

Trans-hexa-1,4-diene is manufactured on a commercial scale (*ca.* 2000 tonnes per annum), mainly in the United States, using a soluble rhodium catalyst, in a process developed by DuPont. The major use of this material is as a comonomer in the production of ethene–propene–diene synthetic rubber of which some 100 000 tonnes was produced in 1975. The introduction of the diene molecule into the ethene/propene copolymer is necessary since otherwise the resulting polymers would be completely saturated which, although being an advantage as far as resistance to oxidation and photodegradation is concerned, would make them extremely difficult to vulcanise, i.e. cross-link the chains.

The rhodium catalysed codimerization of ethene and butadiene [254–256] is quite a triumph for catalytic selectivity, with selectivities towards *trans*-hexa-1,4-diene of the order of 80% being readily achieved with $RhCl_3$ as catalyst precursor in ethanol. The other products of the reaction are *cis*-hexa-1,4-diene (*ca.* 10%) and hexa-2,4-diene (*ca.* 10%) together with minor quantities of 3-methylpenta-1,4-diene and higher ethene oligomers. The selectivity, which is somewhat surprising since, in the absence of butadiene, the same catalyst system is active for ethene dimerization is attributed [254] to thermodynamic control through a relatively stable π-crotyl-rhodium complex of the type (*2.193*), where L = a weak donor ligand or solvent molecule.

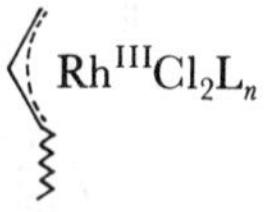

$$\begin{array}{c} \diagdown \\ \diagup \end{array} Rh^{III}Cl_2L_n \qquad n \text{ probably} = 2$$

(*2.193*)

The active catalytic species is thought to be a rhodium(III) hydride complex possibly formed via the sequence of reactions shown below

$Rh(III)Cl_3 + C_2H_4 + H_2O \longrightarrow Rh(I)Cl + 2HCl + CH_3CHO$

$Rh(I)Cl + HCl \longrightarrow Rh(III)(H)Cl_2.$

Support for this suggestion comes from the observation that, when $RhCl_3.3H_2O$ is used as catalyst precursor, an induction period occurs which can be eliminated by using a rhodium(I) complex, e.g. $Rh(C_2H_4)_2(acac)$, in the presence of HCl [11]. A set of catalysis cycles leading to both *trans-* and the *cis-*hexadiene products is shown in Fig. 2.32 [256]. Since the reaction is generally carried out either in a potential donor solvent, e.g. ethanol, or in the presence of added donor ligands, and since such donors are thought to be influential in moderating both the activity and selectivity of the catalyst system, they have been included in the catalysis cycles. Their inclusion also allows us to express the cycle as a series of 16/18 valence electron steps.

The first step in both catalysis cycles is co-ordination of a

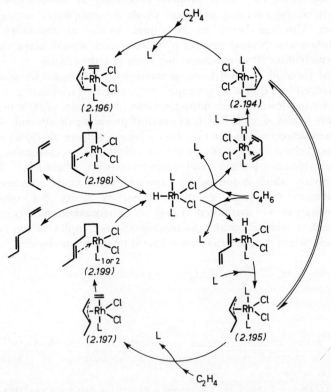

Fig. 2.32 *Catalysis cycles for the rhodium catalysed codimerization of ethene and butadiene to cis- and trans-hexa-1,4-diene (L = a weak ligand or solvent molecule)*

butadiene molecule followed by rapid hydrogen ligand migration to form the π-crotyl species, either as the *anti* isomer (*2.194*) or the *syn* isomer (*2.195*). As indicated in the cycles, isomerization between these two isomers is possible and the *trans* to *cis* ratio of the final product hexa-1,4-dienes is determined both by the ratio of (*2.195*) to (*2.194*) initially formed, and by the degree of isomerization of (*2.194*) to the more stable (*2.195*) prior to co-ordination of ethene in the next step of the cycle. The *trans* to *cis* ratio is sensitive to the presence and nature of the donor molecule. In the absence of other than substrate donors, i.e. in a non- or extremely weak co-ordinating medium such as toluene, the *trans* to *cis* ratio is less than 5; in the presence of a large excess, *ca*. 1000-fold, of alcohol (MeOH or EtOH) it rises to *ca*. 20. With a strong donor such as tributylphosphine oxide, Bu_3PO, or $\{(CH_3)_2N\}_3PO$ a donor/rhodium ratio of as little as 5:1 is sufficient to bring about the same increase in *trans*-selectivity. By co-ordinating to the metal centre, the donor molecule(s) will increase the steric crowding in the catalyst sphere, thus favouring initial transoid co-ordination of butadiene and formation of the less sterically demanding *syn*-π-crotyl isomer (*2.195*). Furthermore, by effectively competing with ethene, i.e. slowing the formation of either (*2.196*) or (*2.197*), the donor ligand allows more time for isomerization of (*2.194*) to (*2.195*). This latter competition can be a mixed blessing since, if it is too effective the donor molecule will prevent formation of (*2.197*) and thus kill the codimerization reaction or at least considerably slow it down. Thus, addition of pyridine kills the catalytic activity and addition of more than a *ca*. two-fold excess of Bu_3PO decreases the codimerization rate.

Under normal operating conditions, i.e. both ethene and butadiene present in large excess, the rate-determining step in the sequence is the allyl to alkene ligand migration (ethene insertion) to form either (*2.198*) or (*2.199*). β-hydrogen elimination, with loss of product hexa-1,4-diene and reformation of the hydrido-rhodium(III) species completes the cycles.

As mentioned above, the high selectivity towards hexa-1,4-diene formation is attributed to the relative stability of the π-crotyl complex, either (*2.194*) or (*2.195*), compared to the other possible alkyl-rhodium species that might conceivably be formed under the reaction conditions. Since alkene insertion is rate determining the product composition will be dependent on the relative concentration of intermediates of the type (*2.200*). The

R = π-crotyl species is much preferred over the R = ethyl species and similarly, insertion of ethene into the π-crotyl bond is much faster than insertion of butadiene. Hence, hexa-1,4-diene is the thermodynamically controlled product. Further evidence for the

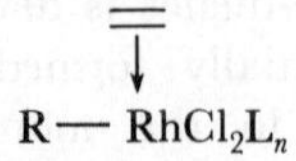

$$R — RhCl_2L_n$$

(2.200)

relative stability of the π-crotyl species has been obtained by the isolation of (2.201) from the reaction between butadiene and ethyl-rhodium(III) chloride [254]. Although hexa-1,4-dienes are

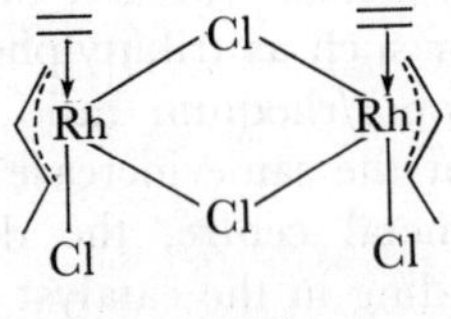

(2.201)

the initial products of the codimerization they are not the most thermodynamically stable hexadiene isomers and, unless there is an extremely large excess of butadiene present together with an adequate amount of ethene, the rhodium system will catalyse isomerization of the 1,4-isomers to the conjugated 2,4-isomers. Commercially this is undesirable and reaction conditions are chosen to minimize this isomerization and keep it below ca. 10%.

Nickel, palladium, cobalt and iron catalyst systems have also been developed for this codimerization reaction [256]. However, as far as selectivity to *trans*-hexa-1,4-diene is concerned the rhodium system is still about the best although some remarkable selectivities of the order of 100% at almost 100% conversion, to *cis*-hexa-1,4-diene have been reported with cobalt based catalysts.

We concluded the last section on hydroformylation with a brief reference to the application of that particular catalytic process to asymmetric synthesis. The metal catalysed oligomerization systems we have considered here also have considerable potential in the field of catalytic chiral synthesis and before going on to discuss the logical extension of oligomerization, namely polymerization, we will pause to mention two examples.

2.5.4 Asymmetric oligomerization

Most success, as regards optical purity, has been obtained via codimerization reactions between ethene and a strained carbocyclic alkene such as norbornene or norbornadiene [23,239,257]:

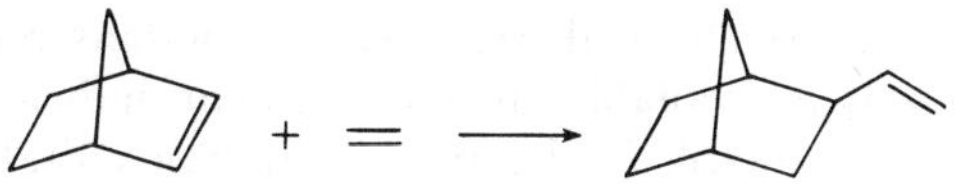

(2.202)

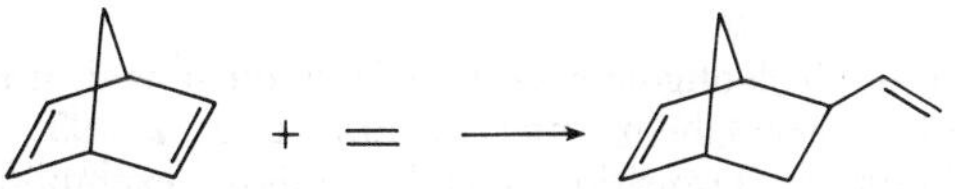

(2.203)

With $(\pi\text{-allyl})\text{NiX}_2/(\text{C}_2\text{H}_5)_3\text{Al}_2\text{Cl}_3/\text{L}$ as catalyst precursor and $\text{L} = (-)\text{-(menthyl)}_2\text{P}(iso\text{-propyl})$, i.e. (2.204), optical yields of 80.6% at $-97\,°\text{C}$ and 77.5% at $-65°$ have been obtained for (2.202) and (2.203), respectively. The optical yield improves

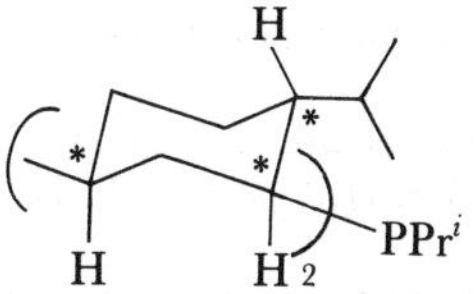

(2.204)

dramatically as the temperature is reduced thus, while the enantiomeric excess obtained in the synthesis of (2.202) is only 30% at 20 °C at −97 °C it is 80.6%. The inherent high activity of the nickel systems makes operating at such low temperatures a practical proposition which can be further enhanced by using reactive strained alkenes. The synthesis is, however, not limited to such strained alkenes and an optical purity of 64% has been achieved in the nickel catalysed codimerization of ethene and butene to yield $(+)\text{-(S)-CH}_2\!\!=\!\!\text{CHCH}(\text{CH}_3)\text{CH}_2\text{CH}_3$ at −40 °C with $(-)\text{-(menthyl)}_2\text{PCH}_3$ as the asymmetric phosphorus ligand. Similarly an optical yield of 37% has been obtained in the codimerization of ethene and styrene to form (2.205) at −10 °C using the same catalyst system.

$$\text{(2.205)}$$

Although the full scope of these types of reactions remains to be explored the data available at the moment indicate they will prove extremely valuable in the area of catalytic chiral synthesis.

Further reading on homogeneous oligomerization

General texts and reviews

Wilke, G. (1963), Cyclooligomerisation of butadiene with transition metal π-complexes. Angew Chem. Internat. Edit. Eng., **2**, 105.

Cramer, R. (1968), Transition metal catalysis exemplified by some rhodium-promoted reactions of olefins. *Acc. Chem. Res.*, **1**, 186.

Bogdanovic, B., Henc, B., Karmann, H. G., Nussel, H. G., Walter, D. and Wilke, G. (1970), Olefin transformations catalysed by organonickel compounds. *Ind. Eng. Chem.*, **62**, 34.

Heimbach, P., Jolly, P. W. and Wilke, G. (1970), Allylnickel intermediates in organic synthesis. *Adv. Organometal. Chem.*, **8**, 29.

Keim, W. (1971), π-allyl systems in catalysis, in *Transition Metals in Homogeneous Catalysis* (ed. G. N. Schrauzer), p. 59, New York: Marcel Dekker.

Tolman, C. A. (1971), Role of transition metal hydrides in homogeneous catalysis, in *Transition Metal Hydrides* (ed. E. L. Muetterties), p. 271, New York: Marcel Dekker.

Maitlis, P. M. (1971), *The Organic Chemistry of Palladium. Vol. II Catalytic Reactions*, p. 33, London: Academic Press.

Bogdanovic, B., Henc, B., Lösler, A., Meister, B., Pauling, H. and Wilke, G. (1973), Asymmetric synthesis with the aid of homogeneous transition-metal catalysts. *Angew. Chem. Internat. Edit. Engl.*, **12**, 954.

Heck, R. F. (1974), *Organotransition Metal Chemistry, A Mechanistic Approach*, Ch. 5 and 8, New York: Academic Press.

Jolly, P. W. and Wilke, G. (1975), *The Organic Chemistry of Nickel. Vol. II Organic Synthesis*, New York: Academic Press.

Bogdanovic, B. (1979), Selectivity control in nickel-catalysed olefin oligomerisation. *Adv. Organometal. Chem.*, **17**, 105.

Chiusoli, G. P. and Salerno, G. (1979), Synthetic applications of organonickel complexes in organic chemistry. *Adv. Organometal. Chem.*, **17**, 195.

Su, A. C. L. (1979), Catalytic codimerisation of ethylene and butadiene. *Adv. Organometal. Chem.*, **17**, 269.

Tsuji, J. (1979), Palladium catalysed reactions of butadiene and isoprene. *Adv. Organometal. Chem.*, **17**, 141.

2.6 Polymerization

Polymerization processes represent one of the single most important class of reactions used in today's petrochemical industry. Some thirty different basic types of polymers, formed by the linking together of a single type of reactant molecule-monomer, or copolymers, formed by the linking together of two or more different monomers, are now commercially available with end uses ranging from the manufacture of plastic buckets to the production of artifical kidneys. In 1976 the world consumption (excluding the USSR, eastern Europe and China) of polymerization products, i.e. plastics, resins, synthetic rubbers and non-cellulose synthetic fibres, exceeded 45 million tonnes [258] and production of synthetic rubber, mainly styrene–butadiene copolymer and poly-butadiene, was more than double that of the natural material (8 million tonnes compared to 3.5 million tonnes) [53]. Two of the largest volume polymers are polyethylene (1976 world production *ca.* 12.5 m million tonnes) and polypropylene (1976 world production *ca.* 3 million tonnes) and it is for these two products that the discoveries of Professor Ziegler in Germany and Professor Natta in Italy have proved most significant.

2.6.1 *Ziegler–Natta catalysts*

While investigating the multiple insertion of ethene into an aluminium alkyl bond to give long chain aluminium-alkyls of the type (*2.206*), K. Ziegler and his co-workers discovered that small amounts of nickel inhibit this polymerization reaction.

$$Al-R + nCH_2CH_2 \longrightarrow Al(CH_2CH_2)_nR \quad (n \leq 100 \text{ when } R = CH_2CH_3)$$

(*2.206*)

Termination occurs after one insertion step with but-1-ene being the major organic product [239]. Not only did this observation herald the development of many of the nickel oligomerization catalysts discussed in the previous section but it also stimulated an extensive study of the effect of other transition metals on the aluminium–alkyl polymerization system. These studies led, in the early 1950s, to the discovery [259] that certain transition-metal complexes, e.g. zirconium acetylactonate, in the presence of aluminium-alkyl compounds, e.g. AlEt$_2$Cl, will catalyse the polymerization of ethene under mild conditions, e.g. 50 °C/10 atm, to give a material having a molecular weight in excess of 50 000. This discovery was of tremendous technical significance since previously ethene had been considered a very difficult molecule to polymerize, needing pressures of over 1000 atm and temperatures of the order of 200 °C. In 1952 Ziegler granted a licence on his low pressure process to the Italian chemical company Montecatini. G. Natta of the Milan Polytechnic Institute, who was a consultant for Montecatini, extended Ziegler's work to the polymerization of propene and higher α-olefins [260]. In 1955 Natta announced [261,262] the preparation and, equally important, the characterization of stereoregular polymers of α-olefins produced using Ziegler type catalysts based on titanium. He established that these highly crystalline polymers were made up of long sequences of monomer units all having the same stereochemistry. It is this latter discovery which has led to the modern processes for the production of polypropylene.

Since these original discoveries many metal combinations have been found to be capable of catalysing alkene polymerization and the definition of a Ziegler catalyst has been extended to include any combination of alkyl, hydride or halide of groups I–III with a transition-metal salt or complex of groups IV–VIII, including Sc, Th and U, capable of catalysing the polymerization of ethene via a co-ordination, rather than an ionic or free radical, mechanism [263–265]. Although some truly homogeneous Ziegler–Natta catalyst systems are known it is, up to now, the heterogeneous systems based on titanium which have proved the most attractive and, industrially, most widely used, especially in the isotactic (see Section 2.6.3) polymerization of propene. Being heterogeneous they do not, strictly speaking, qualify for inclusion in our present discussions. However, in view of their very close mechanistic relation to homogeneous systems and their commercial and scientific importance it would be churlish to disqualify them on such a technicality.

Commercial Zeigler–Natta catalysts are generally prepared by treating either titanium tetrachloride or titanium trichloride with an aluminium-alkyl or an aluminium-alkyl-chloride, e.g. $AlEt_3$ or Et_2AlCl. $TiCl_3$ can exist in four different crystalline modification, α, β, δ and γ, and the exact form of the catalyst depends on the detailed method of preparation, e.g. the solvent, or rather slurry medium used, the rate of addition of aluminium alkyl, the rate of stirring, the reaction temperature etc. Since both activity and selectivity are closely related to the physical form of the catalyst, detailed methods of preparation are a closely guarded commercial secret. The polymerization reaction is thought to occur on the exposed edges of small $TiCl_3$ crystallites and the mechanism of Cossee and Arlman, first proposed [266,267,268] in 1964 is still widely accepted in its essential details.

The active catalytic centre is suggested to be an octahedral titanium of the type (*2.207*) having four chloro ligands, being part of the crystal lattice, an alkyl group, initially resulting from chloro-alkyl exchange between the $TiCl_3$ and the aluminium-alkyl species, and a vacant site. The first stage of the reaction involves alkene co-ordination at the vacant site. This is followed by alkyl to alkene migration, via a four-centre transition state of the type (*2.208*), to give an elongated alkyl chain species, (*2.209*), which now contains the vacant site in the apical position. Further co-ordination of ethene followed by a repetition of the sequence constitutes the polymerization process. Termination can occur via a β-elimination reaction to give a long chain α-alkene, and a titanium-hydrido centre which can either reinitiate polymerization or interact with an aluminium-alkyl compound to reform the titanium-alkyl site (*2.207*).

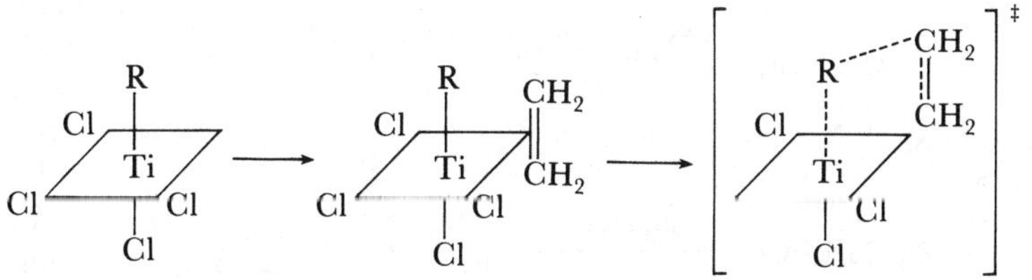

(*2.207*)   (*2.208*)

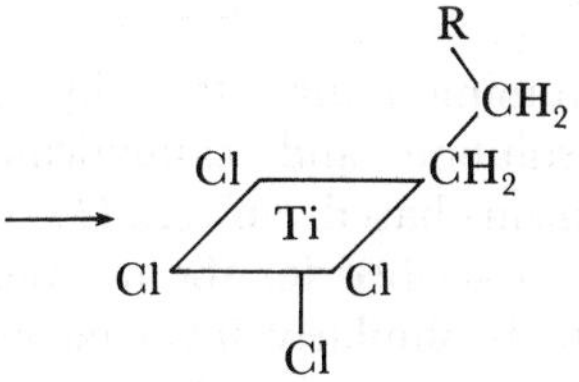

(*2.209*)

The polymer growth reaction is thus an alkyl migration reaction with the growing alkyl chain alternating between two *cis* octahedral sites at the titanium centre. In this mechanism the aluminium alkyl plays no essential role in the propagation step, nor is it necessarily associated with the titanium centre during the catalytic reaction. Its key roles are seen as alkylating the transition metal, possibly acting as a chain transfer centre and a scavenger during the reaction, and functioning as a reducing agent when titanium tetrachloride is used as catalyst precursor. Chain growth is seen as taking place exclusively at the titanium centre. Alternative mechanisms involving two metal centres, either two titanium or a titanium and an aluminium, have been proposed [263] but on the basis of the published data the monometallic mechanism is generally preferred.

Several homogeneous catalyst systems based on titanium and aluminium are known and in these the aluminium is closely associated with the titanium centre, as a ligand rather than a reactant, throughout the course of the alkene oligomerization/polymerization. One such system is that prepared by treating $(\eta^5\text{-}C_5H_5)_2TiCl_2$ with $EtAlCl_2$ [269]. The active catalyst is thought to be the titanium(IV) complex (*2.211*) formed by the sequence of reactions shown below

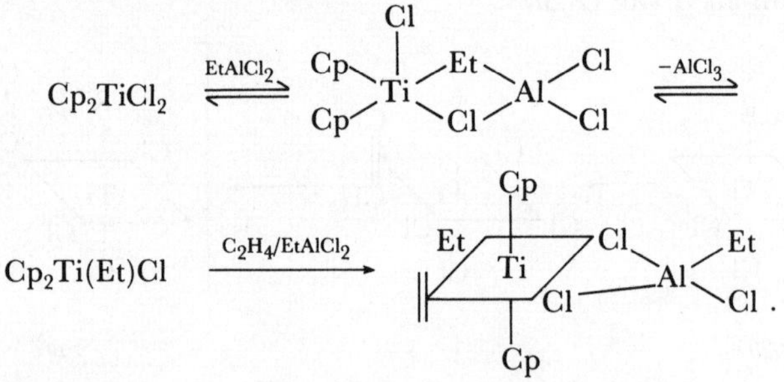

$$Cp_2TiCl_2 \xrightleftharpoons{EtAlCl_2} \begin{array}{c} Cp \\ Cp \end{array}\!\!\!Ti\!\!\begin{array}{c} Cl \\ | \\ Cl \end{array}\!\!Al\!\!\begin{array}{c} Et \\ \diagdown \\ Cl \end{array}\!\!\begin{array}{c} Cl \\ \diagup \\ Cl \end{array} \xrightleftharpoons{-AlCl_3}$$

$$Cp_2Ti(Et)Cl \xrightarrow{C_2H_4/EtAlCl_2} \text{(2.211)}$$

(*2.210*)   (*2.211*)

In the absence of the aluminium-alkyl the ethyl–titanium complex (*2.210*) is stable in solution and catalytically inactive. The polymerization mechanism based on (*2.211*) is essentially the same as that previously described for the heterogeneous catalysts. Whether the aluminium is similarly involved as a ligand in the heterogeneous systems is not totally clear although it has been suggested [270] that the presence of $AlCl_3$ in a $TiCl_3$ crystal

HOMOGENEOUS CATALYST SYSTEMS IN OPERATION 163

lattice can have the effect of increasing the exposure, and thus possibly the reactivity, of surface Ti centres.

Although, because of their limited activity, a point to which we will return later, the homogeneous Ti/Al polymerization catalysts reported to data are of negligible commercial significance, they have provided some interesting insights into a possible mechanism for the polymerizations. Calculations [271,272] based on the hypothetical ethene oligomerization system $Ti(OMe)_4/AlEt_3$ – the actual catalyst system $Ti(OC_6H_4\text{-}p\text{-}Me)_3OBu^n/AlEt_3$ proved too expensive in terms of computer time – have indicated that the most stable configuration of the initial catalyst is not an octahedral complex having a vacant site, cf. Cossee and Arlman mechanism, but rather a triagonal-bipyramidal complex of the type (*2.212*), Fig. 2.33. The calculations also show that the titanium unpaired electron is strongly localized, coefficient 0.742,

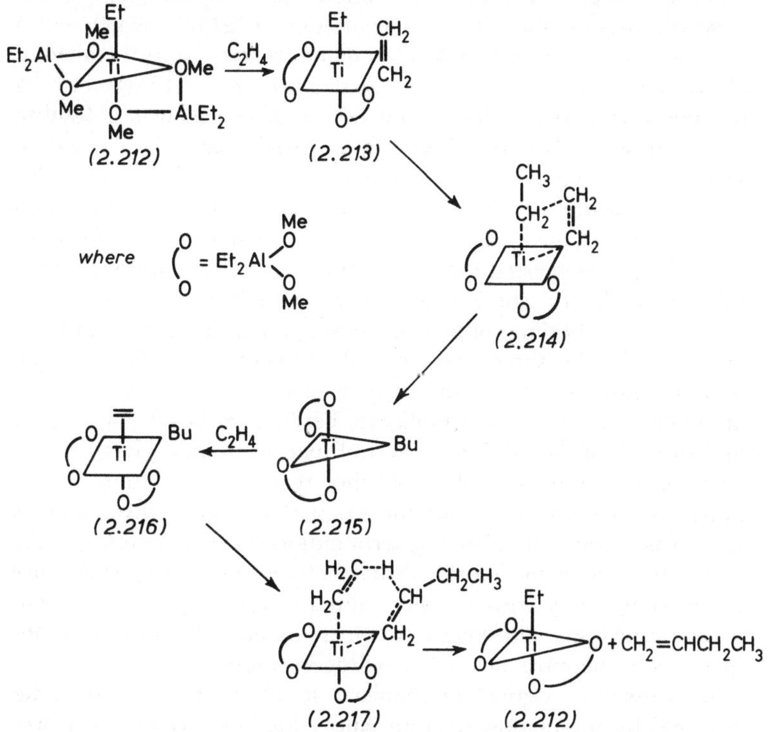

Fig. 2.33 *Dimerization of ethene using $Ti(OMe)_4/AlEt_3$ as catalyst precursor*

in the highest energy occupied molecular orbital (d_{xz}) and makes a significant contribution to the titanium-alkyl bond. Adding the ethene causes the complex to adopt an octahedral configuration. Only a weak bond is formed between the incoming ethene and the titanium with little back bonding into the π^*-antibonding orbitals of the ethene. As soon as the ethene enters the bonding sphere of the titanium a weak bond is also established between the alkyl and the alkene through the titanium d_{xz}-orbital. Thus the system can be visualized as proceeding smoothly to the four-centre transition-state species (2.214) at an early stage along the reaction co-ordinate with the titanium d_{xz}-orbital acting as a 'transfer agent'. This type of mechanism is conceptually very similar to 'direct insertion' of the alkene into the metal alkyl bond in that it implies that alkene-alkyl interaction starts as soon as the alkene enters the bonding sphere of the metal. Completing the ethyl to ethene transfer, or combination reaction, gives the trigonal-bipyramidal butyl complex (2.215) which, on co-ordinating a further molecule ethene, reverts to an octahedral alkene/alkyl species (2.216) analogous to (2.213). In a polymerization system repetition of the sequence would lead to chain growth. However, with this particular oligomerization system termination occurs giving product but-1-ene. The proposed nature of this termination reaction is, by comparison with most of the reactions we have considered up to now, unusual. It is suggested that as the ethene molecule approaches the titanium in (2.215) the ethene carbon atoms become progressively more negative because of their increasing polar interaction with the titanium. At the same time the β-hydrogens in the butyl group became more positive as they approach the ethene unit such that an ionic interaction is established between a butyl β-hydrogen and the C-2 of the ethene — a sort of hydrogen bond giving rise to a transition-state complex of the type (2.217) from which but-1-ene is eliminated and the original titanium-ethyl complex (2.212) is regenerated. Such a termination process is conceptually very attractive in the Ziegler–Natta polymerization sequence since it directly regenerates the initial ethyl species without necessitating the intermediacy of a titanium-hydride complex inherent in the more normal β-hydrogen elimination reaction.

In Cossee's original mechanism stabilization of a titanium d-orbital by interaction with an empty antibonding orbital of the alkene was suggested to play a key role in the catalytic reaction

[266]. The theoretical studies described above are essentially in line with this suggestion with the stabilization of the d_{xz}, the 'transfer' orbital, being an important factor. There is, however, little theoretical evidence for the back donation, $d \rightarrow \pi^*$, being the source of this stabilization as originally suggested by Cossee. The main conclusion, in agreement with Cossee's proposal, is that 'activation' of the Ti-α-C bond, i.e. activation towards migration, comes from the stabilization of the d_{xz}-orbital of the octahedral complex by ethene co-ordination and that this orbital can then act as a 'transfer orbital' to smooth the course of the alkyl to alkene migration [271].

Some earlier calculations [273] using (2.218) as the model catalyst, while introducing the idea that a titanium d-orbital can effectively function as a transfer agent in the formation of the propyl complex (2.219), had discounted the ethene d-orbital stabilizing effect as of negligible importance. However, these earlier authors made no explicit assumptions about the oxidation state of the titanium which could possibly explain the difference. Although it is probably premature to make any definitive statements regarding the relative importance of these orbital

$$\begin{array}{cc} \text{Cl} \diagdown \quad \diagup \text{Cl} \diagdown \overset{\text{Cl}}{\underset{\text{Cl}}{|}} \diagup \text{Me} \\ \text{Cl} \diagup \text{Al} \diagdown \text{Cl} \diagup \text{Ti} \diagdown \hspace{-2pt}/\hspace{-4pt}/ \end{array} \longrightarrow \begin{array}{c} \text{Cl} \diagdown \quad \diagup \text{Cl} \diagdown \overset{\text{Cl}}{\underset{\text{Cl}}{|}} \\ \text{Cl} \diagup \text{Al} \diagdown \text{Cl} \diagup \text{Ti}\text{—Pr} \end{array}$$

(2.218)　　　　　　(2.219)

stabilization effects, it is felt that theoretical calculations of the type referred to above will become increasingly important in gaining a detailed insight into this and many other catalytic systems. With increasing theoretical sophistication and increasing computing power total system orbital calculations will undoubtedly prove of major importance in the future design and elucidation of catalytic systems.

As was mentioned in the introduction to this section, Ziegler–Natta catlaysts have proved particularly important in the production of polyethylene and polypropylene and, before leaving the subject of polymerization, we will consider these two materials and the commercial processes used in their production in a little more detail.

2.6.2 Polyethylene

Approximately half of the total volume of polyethylene presently manufactured is still produced using high pressure processes similar to those originally developed by Imperial Chemical Industries in the late 1930s [53]. Typically the polymerization reaction occurs in a tubular reactor under high pressure (1500 to 3000 atm) with temperatures in the range 100–300 °C. The polymerization, which occurs via a free radical mechanism, is initiated either by including small amounts of oxygen in the ethene feed or by introducing organic peroxides into the reactor. Preferred peroxides are those having a half-life of ca. 1 min at between 110 °C and 210 °C. The reaction is highly exothermic, 106 kJ mol^{-1} at 25 °C, thus efficient cooling and careful temperature control is required. After leaving the high pressure reactor, the product is usually separated in two separation stages operating at 300 atm and 1–3 atm, respectively. The polymer melt is granulated and cooled with water to produce polymer pellets suitable for further processing, e.g. to produce film [274]. In the BASF process [275] it is claimed that for the production of 100 kg of polyethylene 103 kg of ethene is required. Being a free radical process considerable chain branching occurs during the polymerization reaction and thus the density of the resulting polymeric material is somewhat below that of the equivalent, essentially linear chain, material (0.92–0.93 compared to ca. 0.94–0.97 gcm^{-3}) produced using low pressure Ziegler systems. For this reason polyethylene produced using high pressure, free radical, processes is referred to as low density polyethylene (LDPE) while that produced using the lower pressure, non-free radical, i.e. Ziegler catalyst, processes is referred to as high density polyethylene (HDPE). As far as the production of HDPE is concerned, there are basically two types of processes presently in use.

(a) *Solution or slurry processes*

In these types of processes the ethene is continuously fed with catalyst and hydrocarbon diluent into relatively large, compared to the tubular LDPE reactors, polymerization vessels. In solution processes a hydrocarbon diluent such as cyclohexane is used which functions as a solvent for the polyethylene. Polymer content of the solution is limited by the viscosity that can be reasonably handled and thus there is a practical limit to the degree of polymerization that can be allowed in such processes. In the more widely practised slurry processes the hydrocarbon diluent used, e.g. hexane [276] or a light naphtha fraction [277] is a poor solvent for the polyethylene and the reaction takes place in a slurry with the polymer being obtained as a powder. Typical operating conditions employ pressures in the range 10–30 atm and temperatures from 80 °C to 200 °C. High molecular weight material is easily produced using slurry process and hydrogen is often used to control the degree of polymerization. Typically 1.03 to 1.07 tonnes of ethene is required to produce 1 tonne of HDPE [276–278]. The lower pressure requirements of these polyethylene processes considerably simplify the engineering problems associated with the high pressure processes and there is clearly no need for costly compression facilities. On the other hand, larger reactor volumes are generally required and there is also the necessity to include solvent recovery equipment in such plants. A development which is claimed to overcome this latter problem is the HDPE gas phase process announced by Union Carbide Corporation [279].

(b) Gas phase ethylene polymerization
In this type of process ethene and catalyst, in the form of a dry powder, are fed continuously to a fluidized bed reactor where polymerization takes place. No hydrocarbon diluent is included and reaction pressures of the order of 18 atm with temperatures in the range 85–100 °C, depending on the product desired, are employed. Polyethylene with a density of 0.94 to 0.966 gcm^{-3} is produced and Union Carbide claim a polyethylene yield of 98% on ethene [279]. A possible problem with this process is accurate temperature control, with such a highly exothermic reaction, in the absence of any liquid diluent. It is, however, suggested that because there is no need to include solvent recovery and drying equipment, savings of 15% on capital investment and 10% in operating costs are possible [279].

A vital requisite in all of these HDPE processes employing Ziegler type catalysts is high catalytic activity. Ideally polymer yields in excess of 10 000 kg mole^{-1} of titanium are required in order to avoid the costly process of removing the solid catalyst from the polymer after the reaction – a process known as de-ashing. It is in this aspect that homogeneous catalysts fall short, having activities orders of magnitude less than those of their heterogeneous counterparts. Theoretically there seems no reason why, by correct choice of ligands, a highly active homogeneous system should not be designed but, as of now, such a system has not been reported. One general problem with most heterogeneous catalysts is the presence of sites of differing activity; it is extremely difficult to produce a heterogeneous catalyst surface of uniform activity. Furthermore, partially as a result of crystal rupture under polymerization reaction, it is not unusual for the activity of the catalyst to alter during the course of the reaction. Both of these effects can lead to a polymer having a broad molecular weight distribution. The discovery of a highly active homogeneous catalyst would be expected to overcome this problem and allow the selective synthesis of commercially attractive, narrow molecular weight distribution HDPE. A second area where the discovery of a highly active homogeneous system might be expected to make a significant contribution is in the synthesis of isotactic polypropylene.

2.6.3 *Polypropylene*

In the case of polypropylene, produced using Ziegler–Natta, rather than cationic or free radical catalysts, the monomer units can combine to give basically three types of macromolecules, i.e. those in which the methyl groups are randomly orientated in respect to the carbon skeleton, known as atactic; those in which the methyl groups are alternatively placed relative to the carbon skeleton, known as syndiotactic; and those in which the methyl groups all occur on the same side of the carbon backbone, known as isotactic. These three arrangements, which all result from head-to-tail linking of the monomer units, are illustrated in Fig. 2.34. Commercially the stereo-regular, e.g. isotactic, material is the desired product. The high molecular weight isotactic material resembles HDPE in many respects and finds widespread applications in such areas as injection moulding – a sailing dingy with a polypropylene hull is presently on the market, film production, for both packaging and sacking, and fibre manufacture, for use in both carpets and clothing. Recently synthetic paper based on polypropylene fibre has also been developed. The atactic material on the other hand is amorphous, of low strength and of little commercial value. Early polypropylene processes included an atactic removal stage in the

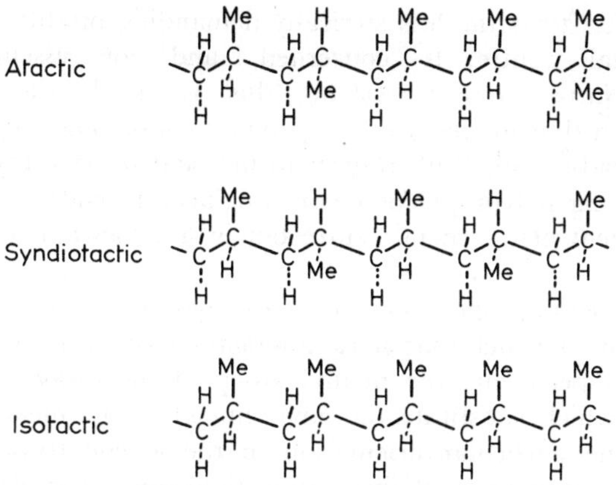

Fig. 2.34 *The three molecular configuration resulting from the linear head-to-tail polymerization of propene*

work-up procedure; now with modern catalysts, selectivities to the isotactic polymer greater than 95% can be achieved, thus eliminating the need to include an atactic removal step. Similarly, the development of highly active catalysts has further allowed the de-ashing stage to be dispensed with, thus considerably simplifying process design and operation. Typical polymerization conditions are temperatures in the range 50–100 °C and pressures of 5 to 30 atm. With the catalyst system, hydrocarbon diluent and monomer being fed continuously to the polymerization reactor, *ca.* 1060 tonnes of propene are required for the production of 1000 tonnes of polypropylene [280,281].

The remarkable stereoselectivity of these propene polymerization catalysts can be attributed to the steric environment of the active metal centre. Given a growing polymer chain at a metal centre, monomer propene insertion, or rather polymer chain, i.e. alkyl, migration to incoming partially or fully co-ordinated alkene, can occur to give either the α-CH_2 product (2.220) or the α-$CH(CH_3)$ product (2.221). These two processes are known as primary and secondary insertion, respectively [282].

$$M-P + CH_2=CH(CH_3) \begin{array}{l} \nearrow M-CH_2CH(CH_3)-P \text{ (primary insertion)} \\ \quad\quad\quad (2.220) \\ \searrow M-CH(CH_3)CH_2-P \text{ (secondary insertion)} \\ \quad\quad\quad (2.221) \end{array}$$

Clearly (*2.220*) is the less sterically demanding product, as far as the metal centre is concerned, and for essentially all Ziegler–Natta catalyst systems this is much the preferred orientation thus giving rise to a polymer containing propene units linked head-to-tail. With respect to the tacticity (see Fig. 2.34) of the resulting polypropylene chain, this also depends on the steric environment of the metal centre but in a somewhat more subtle manner.

For isotactic propagation the steric environment of the catalyst centre must be such that steric interactions between the incoming propene (specifically the methyl group of the propene) and the ligands, which constitute the environment of the metal, strongly favour one particular orientation in the alkene (propene)-alkyl (growing polymer chain) transition-state complex which accompanies the alkyl migration (alkene insertion) chain propagating reaction. One such complex, leading to isotactic polymerization, is shown in Fig. 2.35. The steric bulk of the two sides of the metal centre, represented by the ligands X and Y, must be different so that a given monomer orientation is favoured and furthermore the incoming monomer is sterically constrained to approach the activated metal–carbon bond of the growing polymer chain always from the same side [282]. The inference in Fig. 2.35 is that the side represented by ligand Y is more sterically demanding than that represented by X. As a result of such stereo-orientation, macromolecules of isotactic polypropylene have a helix structure with ternary symmetry (i.e. structural repetition every three monomer units) with an identity period of 6.5 Å [283]. If we consider a given polymer chain orientation then, taking the situation where the alkene has co-ordinated in

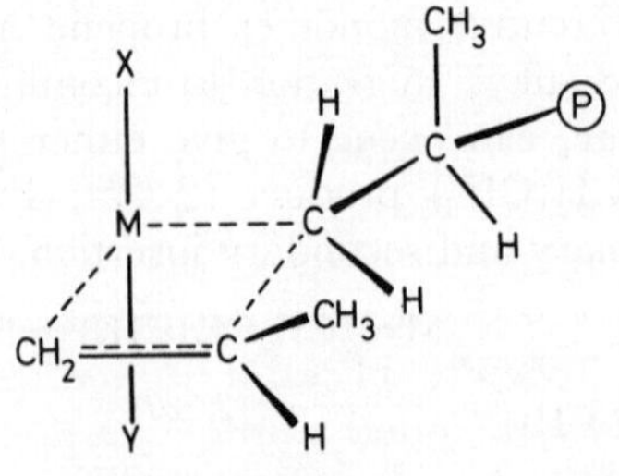

Fig. 2.35 *Possible transition state complex in the isotactic polymerization of propene*

HOMOGENEOUS CATALYST SYSTEMS IN OPERATION

the two possible orientation prior to *cis*-alkyl addition, i.e. (*2.222*) and (*2.223*), for stereoregular polymer formation (containing 98% of say, isotactic material) the free energy difference between the two alkyl to alkene migration steps (i.e. chain propagation) must be of the order of 10.5 KJ mol^{-1} at 25 °C [284]. It is this difference which the steric environment of the catalyst centre must ensure if stereoregular polymerization is to be achieved.

(*2.222*)   (*2.223*)

The homogeneous catalyst system, $VCl_4/AlEt_2Cl$, has been reported to polymerize propene to the syndiotactic material provided the temperature is maintained below *ca.* −40 °C [285]. Stereoregulation in this case has been attributed to strong methyl–methyl interactions which force the incoming propene to always have a configuration which is opposite to that of the last added propene [264]. However, the activity of these systems is too low to make them commercially useful. As far as the highly active isotactic heterogeneous $TiCl_3$ systems are concerned, active sites are created in the surface of the crystal lattice such that the monomer/growing polymer orientation leading to isotactic material is favoured. Just how, and indeed where, such sites are formed in the crystal lattice is still subject to debate [265] although there seems little doubt that the rigidity of the crystal lattice, compared to say the rigidity of the ligand set in a mononuclear titanium species, is of prime importance in enhancing the stereo regulating properties of the heterogeneous systems.

Potentially Ziegler–Natta catalysis should prove a fruitful and rewarding field for homogeneous catalysis. Providing suitable activity (>*ca.* 250 kg polymer per g of catalyst) can be achieved the catalyst can, in most cases, be left in the polymer product thus alleviating one of the normal disadvantages of homogeneous systems, i.e. the difficulty of catalyst/product separation. The uniformity in site activity found with homogeneous systems should allow the manufacture of narrow molecular weight distribution material which would be commercially attractive. However, with regard to stereo-regulation it would seem that there will be a need to design more rigid, but at the same time more open, ligand systems if the stereo-selectivity and activity of the heterogeneous systems is to be matched.

Further reading on polymerization and Ziegler–Natta catalysis

General texts and reviews

Reich, L. and Schindler, A. (1966), *Polymerisation by Organometallic Compounds*, New York: Wiley.

Boor, J. (1967), The nature of the active site in the Ziegler-type catalyst. *Macromol. Rev.*, **2**, 115.

Youngman, E. A. and Boor, J. (1967), Syndiotactic polypropylene. *Macromol. Rev.*, **2**, 33.

Ketley, A. D. (ed.) (1967), *The Stereochemistry of Macromolecules*, Vol. 1, London: Arnold.

Henrici-Olivé, G. and Olivé, S. (1967), The active species in homogeneous Ziegler–Natta catalysts for the polymerisation of ethylene. *Angew. Chem. Internat. Edit.*, **6**, 790.

Bikales, N. A. (1968), Homogeneous catalysis of polymerisation. *Advances in Chemistry Series*, No. 70, p. 233, Washington: American Chemical Society.

Raff, R. A. (1969), Polyethylene, in *Ethylene and its Industrial Derivatives* (ed. S. A. Miller), p. 335, London: E. Benn.

Valvassori, A., Longi, P. and Parini, P. (1973), Polypropylene, in *Propylene and its Industrial Derivatives* (ed. E. G. Hancock), p. 155, London: E. Benn.

Chien, J. C. W. (ed.) (1975), *Co-ordination Polymerisation*, New York: Academic Press.

Cesca, S. (1975), The chemistry of unsaturated ethylene–propylene based terpolymers. *Macromol. Rev.*, **10**, 1.

Tsutsui, M. and Courtney, A. (1977), σ–π rearrangements of organotransition metal compounds. *Adv. Organometal. Chem.*, **17**, 241, and Olefin polymerisation. The Ziegler–Natta process, p. 256.

Caunt, A. D. (1977), Ziegler polymerisation, in *Catalysis*, Vol. 1, p. 234, London: Chemical Society.

2.7 Oxidation

In choosing systems to discuss under the heading of oxidation, we have adopted the classical nineteenth century definition of oxidation, namely 'that oxidation corresponds to gain of oxygen or loss of hydrogen' [286]. Furthermore, in order to restrict the discussion to a manageable length we will consider only those reactions which result in the incorporation of one or more oxygen atoms, thus precluding oxidative coupling reactions, such as the palladium catalysed coupling of benzene [287] or the copper catalysed coupling of phenol [288].

As suggested by Sheldon and Kochi [289], homogeneous metal catalysed oxidation reactions may be conveniently divided into two types which they designated as *homolytic* and *heterolytic*.

Homolytic systems are those which involve free radicals as intermediates and in which the catalysis cycle involves the metal in a series of one electron oxidation or reduction steps. Liquid phase homolytic oxidations, generally referred to as autoxidations, occur via a radical chain mechanism and there is no *a priori* necessity to involve the metal centre at all stages in the reaction. In fact, in most metal catalysed autoxidation reactions the bulk of the chemical transformation occur outside the co-ordination sphere of the metal.

Heterolytic, metal catalysed, homogeneous oxidation processes are much more akin to most of the systems considered in previous sections; the organic substrate and/or oxygen containing reactant is activated by co-ordination at, or addition to, the metal centre. If the catalysis cycle involves the metal undergoing a change in oxidation state such changes occur via a series of two electron steps. Free radicals are not involved as intermediates and the metal remains intimately involved with the substrate/reactant during most, if not all, of the chemical transformations involved in the catalysis cycle.

While the systems discussed below have been chosen with the objective of exemplifying these two basic types of reaction it should be appreciated that oxidation, like most natural processes, does not take too kindly to being categorized. Thus in several systems, especially those involving peroxy intermediates and metals potentially capable of participating in both types of catalysis, it is not always possible, nor indeed desirable, to make a clear distinction between the two types of oxidation processes. Thus cyclohexane oxidation, discussed under homolytic oxidation, may well at high catalyst concentrations involve substantial complex formation between the alkane and the cobalt(III) centre [290]. Similarly, in propene epoxidation, discussed under heterolytic oxidation, radical chain, i.e. homolytic, decomposition of the hydroperoxide reactant can occur to the detriment of the epoxidation selectivity [287].

2.7.1 *Homolytic oxidation*

Reaction between organic compounds and oxygen which proceed via a free radical chain mechanism may be described in terms of three general processes, namely initiation, propagation and termination. The basic autoxidation scheme, incorporating these three processes, is represented by Equations (2.25) to (2.31):

Initiation:
$$\text{In} \longrightarrow \text{In}^{\cdot} \quad (2.25)$$
$$\text{In}^{\cdot} + \text{RH} \longrightarrow \text{InH} + \text{R}^{\cdot} \quad (2.26)$$

Propagation:
$$\text{R}^{\cdot} + \text{O}_2 \longrightarrow \text{RO}_2^{\cdot} \quad (2.27)$$
$$\text{RO}_2^{\cdot} + \text{RH} \longrightarrow \text{RO}_2\text{H} + \text{R}^{\cdot} \quad (2.28)$$

Termination:
$$2\text{R}^{\cdot} \longrightarrow \text{R}_2 \quad (2.29)$$
$$\text{R}^{\cdot} + \text{RO}_2^{\cdot} \longrightarrow \text{RO}_2\text{R} \quad (2.30)$$
$$2\text{RO}_2^{\cdot} \longrightarrow \text{RO}_4\text{R} \longrightarrow \text{nonradical products} + \text{O}_2. \quad (2.31)$$

In principle a metal catalyst may intervene in any, or all, of these basic steps, thus influencing the course and rate of the autoxidation reaction. In practice, the key role of the metal complexes added to such liquid phase systems is to catalyse the decomposition of hydroperoxide species of the type RO_2H or, to a lesser extent, peroxy species of the type RO_2R. Such species may be initially present, in which case addition of a metal species can induce their decomposition and thus considerably reduce the induction period generally found in uncatalysed autoxidation reactions. Alternatively, if such species are formed during the course of the reaction, e.g. by the steps shown in Equation (2.28) or Equation (2.30), the presence of a metal catalyst can hasten their decomposition and thus increase the overall rate, or moderate the course of the autoxidation sequence.

The two basic steps in metal catalysed hydroperoxide decomposition, first proposed [291] to explain Fe(II) catalysed decomposition of hydrogen peroxide (as found in Fenton's reagent), are as follows:

$$RO_2H + M^{n+} \longrightarrow RO^{\cdot} + OH^- + M^{(n+1)+} \quad (2.32)$$

$$RO_2H + M^{(n+1)+} \longrightarrow RO_2^{\cdot} + H^+ + M^{n+}. \quad (2.33)$$

If a given metal ion is capable of effecting only one of these reactions, then stoichiometric, rather than catalytic, decomposition of hydroperoxide occurs unless a catalytic means is devised of returning the metal ion to its original oxidation state. Thus although copper(I) decomposes hydroperoxides, e.g. t-butyl hydroperoxide, via Equation (2.32), in the presence of certain organic substrates, RH, the reaction becomes catalytic via the sequence shown below [287]:

$$Bu^tO_2H + Cu^I \longrightarrow Bu^tO^{\cdot} + OH^- + Cu^{II}$$

$$Bu^tO^{\cdot} + RH \longrightarrow Bu^tOH + R^{\cdot}$$

$$R^{\cdot} + Cu^{II} \xrightarrow{HOAc} ROAc + H^+ + Cu^I$$

In general when the metal is a strong reducing agent, e.g. Cu^I, Cr^{II}, hydroperoxide decomposition occurs via Equation (2.32) while with strong oxidants, e.g. Pb^{IV}, Ce^{IV}, reaction (2.33) predominates. When the metal is equally happy in either the n or the $n + 1$ oxidation states, i.e. the two are of comparable stability, then both reactions (2.32) and (2.33) can occur concurrently and a catalytic cycle can be established with the metal cycling between the n and the $n + 1$ oxidation states. The overall reaction, i.e. that resulting from combining Equations (2.32) and (2.33) constitutes a catalytic decomposition of the hydroperoxide into alkoxy and alkylperoxyl radicals

$$2RO_2H \longrightarrow RO_2^{\cdot} + RO^{\cdot} + H_2O.$$

It is this latter reaction which lies at the heart of most metal catalysed autoxidation systems. Both cobalt (Co^{II}/Co^{III}) and manganese (Mn^{II}/Mn^{III}) compounds are capable of participating in such a cycle. It is, however, the cobalt system which is most widely used in catalysing homogeneous liquid phase homolytic oxidation reactions. Two such reactions, which are of particular relevance to the petrochemical industry, are the liquid phase oxidation of alkyl-aromatics, in particular *para*-xylene, and the liquid phase oxidation of alkanes, in particular cyclohexane.

(a) *Autoxidation of alkyl-aromatics – p-xylene*

Terephthalic acid, the product obtained by oxidizing both of the methyl groups of p-xylene, is used, generally in the form of its dimethyl ester, to make polyester fibres. Polyester fibre, an ethylene glycol/terephthalic acid co-polymer, is one of the most important of the synthetic fibres. In 1976 it held 41.5% of the total synthetic fibre market (cf. nylon 30%) [53]. The original oxidation process for *p*-xylene used nitric acid, under pressure at 150–250 °C, as oxidant. However, this process, although still in use, has now been widely replaced by processes using air or oxygen as oxidant in the presence of homogeneous cobalt or manganese catalysts. Generally the oxidations are carried out in acetic acid using either cobalt(III) acetate or manganese(III) acetate as catalyst, often in the presence of 'promotors' such as metal bromides (see below).

Typical operating conditions involve temperatures of between 100 °C and 200 °C and pressures in the range 15 atm to 30 atm, e.g. in the Amoco direct oxidation process the oxidation is carried out at 200 °C, 15–30 atm, using air as oxidant, with bromine promoted cobalt salt as catalyst in acetic acid; in the Mobil process, which uses pure oxygen, the temperature is *ca.* 130 °C at 15 atm pressure [53]. The first stage of the oxidation, i.e. *p*-xylene to *p*-toluic acid, occurs quite readily in the presence of small amounts of cobalt (or manganese) salts. To go through to the di-acid necessitates the presence of higher catalyst concentrations and/or promotors. An alternative process, the Witten process, involves esterifying the toluic acid initially formed to give, for example, methyl toluate. In the presence of excess *p*-xylene methyl toluate is readily oxidized to methyl terephthalate [52].

Alkyl aromatic oxidations involving cobalt(III) catalysts appear to proceed almost exclusively via an electron transfer mechanism [287]. Most of the experimental data are consistent with the initial formation of a radical cation of the type (*2.224*) via electron transfer oxidation of the methyl-aromatic by Co^{III}

$$ArCH_3 + Co^{III} \rightleftharpoons [ArCH_3]^{+\cdot} + Co^{II}. \qquad (2.34)$$

(*2.224*)

Loss of proton from (*2.224*) gives the benzyl radical (*2.225*)

$$[ArCH_3]^{+\cdot} \longrightarrow ArCH_2^{\cdot} + H^+. \qquad (2.35)$$

(*2.225*)

In the presence of oxygen (*2.225*) forms the peroxy species (*2.226*),

HOMOGENEOUS CATALYST SYSTEMS IN OPERATION

which can then go on to give the normal oxidation products

$$ArCH_2^{\cdot} + O_2 \longrightarrow ArCH_2O_2^{\cdot} \longrightarrow \text{products.}$$

(2.226)

The cobalt(III) species can be regenerated either by reaction with (2.226), Equation (2.36), or by reaction with the hydroperoxy species, (2.227), Equation (2.37), formed by the chain propagating step (2.38)

$$ArCH_2O_2^{\cdot} + Co^{II} \longrightarrow ArCHO + Co^{III} + OH^{-} \quad (2.36)$$

$$ArCH_2O_2H + Co^{II} \longrightarrow ArCH_2O^{\cdot} + Co^{III} + OH^{-} \quad (2.37)$$

(2.227) $\longrightarrow$ ArCHO

$$ArCH_2O_2^{\cdot} + ArCH_3 \longrightarrow ArCH_2O_2H + ArCH_2^{\cdot} \quad (2.38)$$

Both of these catalyst regeneration reactions lead to the formation of aromatic aldehydes, which are the primary products of the oxidation of methylbenzenes. The corresponding aromatic acids are produced by subsequent aldehyde oxidation via a peroxy acid intermediate. It has been suggest [289] that with very reactive radical cations, e.g. monoalkylbenzenes, rapid proton transfer (Equation (2.35)) may occur predominantly within the solvent cage such that free radical cations are never formed as discrete intermediates, e.g.

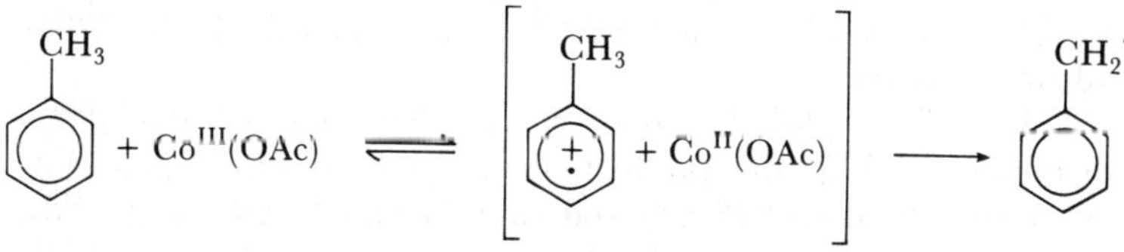

$$\text{PhCH}_3 + Co^{III}(OAc) \rightleftharpoons [\text{PhCH}_3^{+\cdot} + Co^{II}(OAc)] \longrightarrow \text{PhCH}_2^{\cdot}$$

$$+ Co^{II} + HOAc.$$

As mentioned above, metal bromides are frequently added as promotors to the cobalt catalysed oxidation of p-xylene. Thus, whereas in the absence of such promotors p-toluic acid is the principal product, with further oxidation being very slow, in the presence of sodium bromide further oxidation is rapid and high purity terephthalic acid is formed in near quantitative yield. Industrial processes, e.g. the Amoco process [292] have led to the commercial availability of high purity terephthalic acid which in turn has led to a trend to use the acid, rather than its dimethyl ester, in the production polyester fibre.

The promoting effect of bromide is ascribed to the formation of bromine atoms via electron transfer oxidation of bromide by cobalt(III)

$$Br^- + Co^{III} \longrightarrow Co^{II} + Br^{\cdot}$$

The bromine atom is an extremely efficient hydrogen abstractor which can rapidly form the benzyl radical (2.225) thus initiating the autoxidation sequence

$$ArCH_3 + Br^{\cdot} \longrightarrow ArCH_2^{\cdot} + HBr \quad (2.39)$$

$$(2.225)$$

$$ArCH_2^{\cdot} + O_2 \longrightarrow ArCH_2O_2^{\cdot} \text{ etc.}$$

The cobalt(III) species is regenerated via either Equation (2.36) or (2.37) and the hydroxide ion, or cobalt hydroxy species, concurrently formed can regenerate the bromide ion from the HBr of Equation (2.39)

$$Co^{III}(OH) + HBr \longrightarrow Co^{III}Br + H_2O.$$

In contrast to the situation found with cobalt-catalysed autoxidations carried out in the absence of bromide there is no evidence for the intermediacy of radical cations, and apparently no direct reaction between the Co^{III} complex and the hydrocarbon substrate occurs.

Although terephthalic acid is certainly one of the major chemicals used in the production of synthetic fibres two close rivals are adipic acid (2.228) and caprolactam (2.229) used in the production of nylon 6/6 and nylon 6, respectively. Metal catalysed alkane autoxidation, in this case of cyclohexane, plays a key role in the manufacture or both of these chemicals.

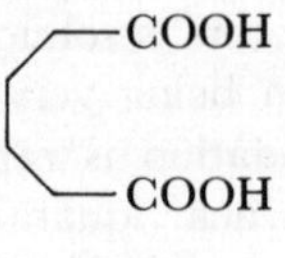

(2.228) 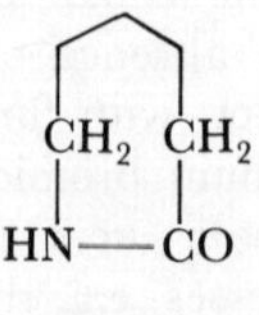   (2.229)

(b) *Autoxidation of alkanes – cyclohexane*

Hydrocarbon oxidation reactions are usually highly exothermic and, when carried out on a large scale, efficient heat removal constitutes a major practical problem. To aid heat removal such reactions are usually carried in the liquid phase. In the absence of suitable catalysts, alkane autoxidation reactions generally require high temperatures and since, at least in normal alkanes, there is little to choose, with regard to reactivity, between the various C–H bonds radical attack tends to be indiscriminate leading to a wide range of oxidation products. This said, there is one, essentially uncatalysed, alkane autoxidation reaction which has achieved considerable commercial success, namely the air oxidation of butane or paraffinic naphtha (an oil fraction containing C_4 to C_8 hydrocarbons) to acetic acid. This process accounts for nearly all the acetic acid produced in the United Kingdom and is widely used elsewhere in the world. Typically the reaction is run at between 160 °C and 200 °C at *ca.* 50 atm. Strict temperature control is necessary to prevent complete oxidation through to CO_2 and H_2O. Product distribution varies with operating conditions and feedstock; with naphtha the ratio of formic to acetic to propionic acid is *ca.* 1:6:0.8, with 1 tonne of alkane substrate yielding between 1 and 1.4 tonnes of acid products [193]. While this process does have the distinct advantage of going in one step from an oil product to a major chemical, it suffers from the disadvantage of producing a considerable number of co- and by-products.

Another direct alkane autoxidation reaction which is practised on a commercial scale, this time in the presence of a homogeneous cobalt catalyst, is the liquid phase air oxidation of cyclohexane to give cyclohexanone and cyclohexanol as major products. Typically a cyclohexane soluble cobalt salt, e.g. cobalt naphthenate, is used as catalyst and, in order to avoid excessive by-product formations, cyclohexane conversion is limited to *ca.* 10% per pass at which selectivity to cyclohexanol and cyclohexanone is *ca.* 70%. As in the alkylaromatic autoxidation reactions previously discussed the major task of the cobalt appears to be to catalyse the decomposition of the cyclohexylhydroperoxide formed during radical chain reaction. In Fig. 2.36 a suggested [52] set of catalysis cycles is illustrated. The Co(III)/Co(II) cycle indicated at the bottom of the figure is analogous to that previously described by Equations (2.32) and

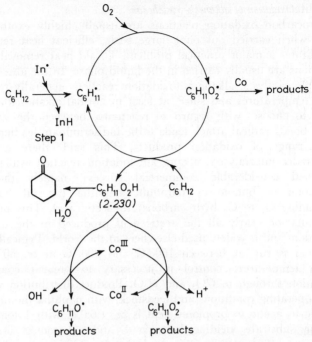

Fig. 2.36 *Catalysis cycles for the cobalt catalysed autoxidation of cyclohexane*

2.33). The ketone product can be formed directly from the hydroperoxide intermediate (2.230) via loss of water and, in the absence of metal catalysts, about twice as much ketone as alcohol is formed. However, in the presence of cobalt(III) salts the amount of cyclohexanol formed exceeds that of cyclohexanone. Experiments with iron and nickel stearates have indicated that, in the presence of these species, peroxy radicals can be converted directly into alcohol and ketone without necessarily passing through the hydroperoxide intermediate [294]. As with cobalt catalysed alkyl-aromatic autoxidation the rate of cyclohexane oxidation, catalysed by cobalt(III) acetate in acetic acid, is enhanced by adding bromide ions [295]. By analogy with the alkyl-aromatic systems this enhancement is probably due to the participation of bromine atoms in the radical chain reaction. With reference to the initiation process, step 1 in Fig. 2.36, there is some evidence [287] to suggest that alkyl radicals can be formed

HOMOGENEOUS CATALYST SYSTEMS IN OPERATION 181

by direct reaction between an alkane and a cobalt(III) complex

$$RH + Co^{III} \rightleftharpoons R^{\cdot} + Co^{II} + H^+.$$

Although the actual mode of interaction between the Co^{III} and the alkane is not yet clear it has been suggested [296] that such an interaction could involve electrophilic substitution by the metal at a saturated carbon centre via a three-centre transition state of the type (2.231).

$$-\overset{|}{\underset{|}{C}}-H + CoX_2^+ \longrightarrow \left[-\overset{|}{\underset{|}{C}} \cdots \overset{H}{\underset{CoX_2}{\diagdown}} \right]^{\ddagger +} \xrightarrow{-H^+}$$

$$(2.231)$$

$$-\overset{|}{\underset{|}{C}}-CoX_2 \longrightarrow -\overset{|}{\underset{|}{C}}{}^{\cdot} + CoX_2$$

Such an interaction, if proven, would have important implications in the field of transition-metal catalysed alkane activation.

An important outlet for the cyclohexanol/cyclohexanone mixture obtained from the autoxidation of cyclohexane is in the production of adipic acid. This latter conversion is generally carried in a separate reactor using 60% nitric acid as oxidant in the presence of a copper(II)/vanadium(V) catalyst at 60–80 °C and slightly elevated pressure, [52,53]. Clearly a one step oxidation process from cyclohexane to adipic acid would be of considerable commercial interest. Although it has been reported [297] that in the presence of relatively high concentrations of colbalt(III) acetate at 80–90 °C cyclohexane can be directly oxidized by air, or preferably oxygen, to adipic acid with selectivities of ca. 75% at 80% conversion, it appears that, under commercial operating conditions, the two step process is still preferred [53]. In the nitric acid oxidation step, adipic acid yields often exceed 90% whereas in the direct oxidation process there is considerable by-product formation necessitating the inclusion of costly product purification steps in the plant design.

By-product formation in the one-step process highlights one of the major problems encountered in homolytic oxidation reactions; that of selectivity. Since catalyst participation in such systems is essentially limited to the hydroperoxide decomposition step, and since there are many other steps in the autoxidation cycle in which the metal catalyst takes little or no part, selectivity improvements based on catalyst tailoring are somewhat restricted. More success in this direction might be anticipated with metal catalysed oxidation reactions in which the metal remains closely associated with either the substrate or the oxidant, or preferably with both, throughout as many, if not all, stages, of the oxidation. Such types of oxidation reactions are unlikely to involve free radical intermediates and will generally fall under the heading of heterolytic oxidation.

2.7.2 *Heterolytic oxidation*

(a) *Transition-metal catalysed epoxidation – propene*

In the metal catalysed homolytic oxidations discussed above the key interaction is that between the metal and the substrate hydroperoxide. Such an interaction is also of prime importance in homogeneous transition-metal catalysed epoxidation. However, in this latter case the interaction leads predominantly to hydroperoxide co-ordination rather than to hydroperoxide homolytic decomposition.

The net result of alkene epoxidation is the addition of an oxygen atom across the alkene double bond:

$$\text{>C=C<} + \tfrac{1}{2}O_2 \longrightarrow \text{>C} \overset{O}{-} \text{C<}$$

The most important epoxide, in terms of commercial volume, is ethylene oxide production of which is expected to exceed 4 million tonnes in western Europe and America in 1980, [53]. Almost all the ethylene oxide presently produced in the western world is manufactured by the direct air, or oxygen, oxidation of ethene using heterogeneous silver based catalysts. With temperatures of *ca.* 250 °C and pressures in the range of 10 to 30 atm, selectivities to ethylene oxide (i.e. moles of ethylene oxide produced for each mole of ethene reacted) of the order of 70% can be readily achieved. When propene is used in place of ethene the epoxidation selectivity drops markedly; only small amounts of propylene oxide are formed with the bulk of the propene being completely oxidized through to carbon dioxide and water. Most propylene oxide – in 1976 western European and American production was *ca.* 1.5 million tonnes – is produced using the chlorohydrin route originally developed for ethylene oxide

$$CH_3CH=CH_2 + Cl_2 + H_2O \longrightarrow CH_3CH(OH)CH_2Cl + HCl$$
$$\text{propylene chlorohydrin}$$

$$2CH_3CH(OH)CH_2Cl + Ca(OH)_2 \longrightarrow 2CH_3CH\underset{O}{\underset{\diagdown\!\diagup}{CH_2}} + CaCl_2 + H_2O.$$

This process has, however, several drawbacks, not least of which is the wasteful consumption, in the sense that it does not appear in the product, of expensive and corrosive chlorine. A relatively new process, first commercialized in 1969 [298] is based on the finding [287] that, in the presence of certain homogeneous metal catalysts, notably those based on molybdenum, organic hydroperoxides will react with alkenes to give epoxides in high yield

$$\text{>C=C< + RO}_2\text{H} \xrightarrow{\text{'cat'}} \text{>C}\overset{O}{-}\text{C< + ROH}. \qquad (2.40)$$

The essential step in this selective epoxidation reaction is thought to be the non-dissociative co-ordination of the hydroperoxide molecule to the metal centre via the R bonded oxygen to give a complex of the type (2.232). In such a complex

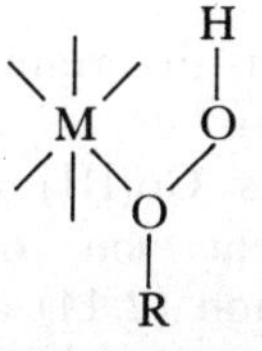

$$\begin{array}{c} \diagdown \;|\; \diagup \quad H \\ \quad M \quad\;\; | \\ \diagup\;|\;\diagdown \quad O \\ \quad\;\;\; | \\ \quad\;\;\; O \\ \quad\;\;\; | \\ \quad\;\;\; R \end{array}$$

(2.232)

the hydroperoxide is 'activated by co-ordination' in that the metal centre reduces the electron density at the peroxide oxygens thus rendering them more susceptible to nucleophilic attack by substrate alkene. Suitable catalysts involve early transition metals in high oxidation states, e.g. Mo(VI), W(VI) and Ti(IV), and the basic epoxidation process may be described by the following sequence of reactions

$$M^{n+} + RO_2H \rightleftharpoons [M^{n+}RO_2H] \qquad (2.41)$$

$$[M^{n+}RO_2H] + \text{>C=C<} \longrightarrow \text{>C}\overset{O}{-}\text{C<} + ROH + M^{n+}. \qquad (2.42)$$

When the metal species contains the M=O unit (e.g. molybdenyl, tungstenyl, vanadyl or titanyl) transfer of oxygen from the metal-hydroperoxide complex to the alkene is suggested [299,300] to occur via a cyclic transition state of the type (2.233), Equation (2.43)

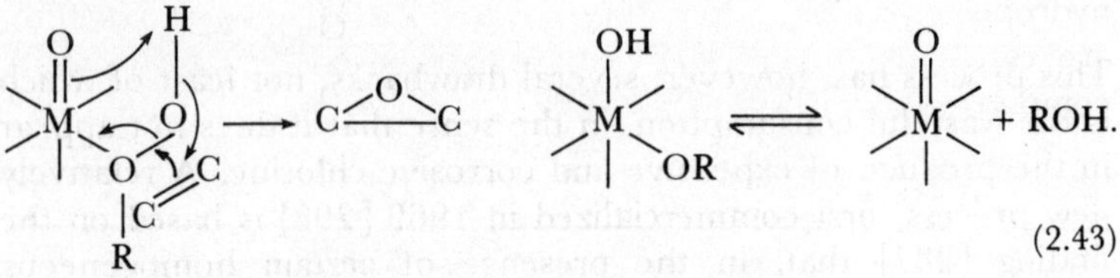

$$\underset{(2.233)}{\text{[structure]}} \longrightarrow C\overset{O}{\underset{}{\diagdown}}C \quad \underset{OR}{\overset{OH}{\text{M}}} \rightleftharpoons \overset{O}{\text{M}} + ROH. \tag{2.43}$$

In addition to the heterolytic process depicted in Equations (2.41) and (2.42), there is also the possibility of metal catalysed homolytic decomposition of the hydroperoxide via the sequence shown in Equations (2.44) and (2.45), i.e. analogous to that described in Section 2.7.1.

$$[M^{n+}RO_2H] \longrightarrow M^{(n-1)+} + RO_2^{\cdot} + H^+ \tag{2.44}$$

$$M^{(n-1)+} + RO_2H \xrightarrow{\text{fast}} M^{n+} + RO^{\cdot} + OH^-. \tag{2.45}$$

The selectivity to epoxide is determined by the relative rates of Equations (2.42) and (2.44).

Strong oxidants such as Co(III) and Mn(III), with redox potentials to the divalent ion of 1.82 eV and 1.51 eV respectively, favour Equation (2.44) and thus give rise to poor epoxidation selectivity. Very weak oxidants such as Mo(VI), W(VI) and Ti(IV), with one electron redox potentials of ~0.2 eV, −0.03 eV and −0.37 eV respectively, are poor catalysts for homolytic hydroperoxide decomposition and favour reaction via Equation (2.42).

Since a prime function of the catalyst is to increase the electrophilicity of the co-ordinated hydroperoxide the activity of the catalyst should increase with increasing Lewis acidity. This is found with Mo(VI) compared to W(VI). MoO_3 is a considerably stronger Lewis acid than WO_3 and epoxidation catalysts based on Mo(VI) are generally considerably more active than those based on W(VI). Similarly, increasing the electron withdrawing properties of the associated ligands would be expected to increase the activity of the catalyst. This is found to be the case in the initial stages of the epoxidation reaction. Thus $MoO_2(acac)_2$ is initially a more active epoxidation catalyst than the corresponding diol complex (2.234). However, under epoxidation conditions the ligands originally present are rapidly replaced or destroyed by substrate molecules, i.e. alkene and/or hydroperoxide, and, after an initial period, the epoxidation rates are essentially independent of the nature of the molybdenum complex originally added [300].

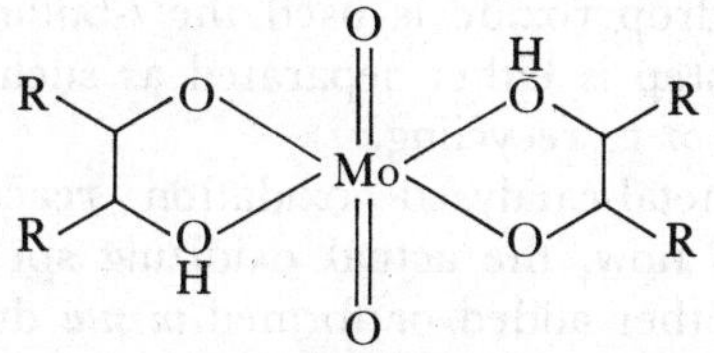

(2.234)

In Fig. 2.37 a generalized catalysis cycle for molybdenum catalysed epoxidation is illustrated. It should be emphasized that this cycle is somewhat speculative since, in spite of extensive study, there are still several uncertainties surrounding the precise mechanism of metal catalysed alkene epoxidation.

In commercial operations either ethylbenzene hydroperoxide, formed by the air oxidation of ethylbenzene, or *t*-butyl hydroperoxide, formed by the air oxidation of isobutane, are employed in the epoxidation of propene. When ethylbenzene

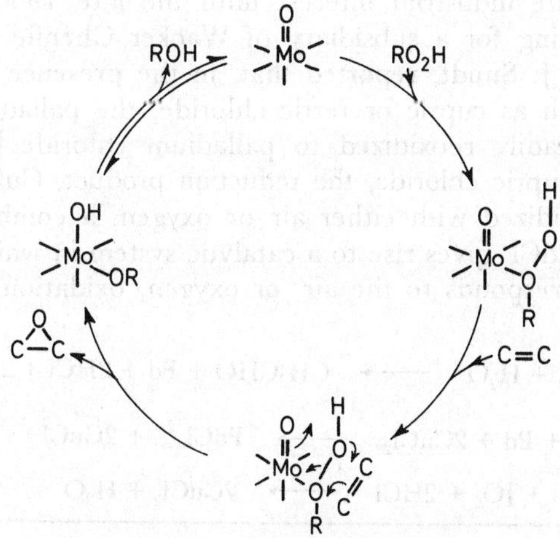

Fig. 2.37 *Catalysis cycle for molybdenum(VI) catalysed alkene epoxidation*

hydroperoxide is used, the product alcohol is dehydrated to yield styrene which is separated as co-product. The overall reaction is

$$PhCH_2CH_3 + CH_3CH=CH_2 + O_2 \longrightarrow PhCH=CH_2 + CH_3\overset{O}{\overset{\frown}{CH}-CH_2} + H_2O.$$

When t-butyl hydroperoxide is used the t-butanol produced in the epoxidation step is either separated as such, or dehydrated to isobutylene prior to recycling.

In all the metal-catalysed oxidation reactions we have considered up to now, the actual oxidizing species has been a hydroperoxide, either added or formed *in situ* during the course of the oxidation. In our final example of homogeneous transition-metal catalysed oxidation we turn our attention to a system in which the metal itself is the oxidant.

(b) *Palladium catalysed alkene oxidation – ethene*

In 1894, F. C. Phillips [301] reported that aqueous palladium chloride was reduced to palladium metal by ethene with concurrent formation of ethanal. This reaction which amounts to the stoichiometric palladium oxidation of ethene

$$CH_2=CH_2 + PdCl_2 + H_2O \longrightarrow CH_3CHO + Pd + HCl$$

attracted little industrial interest until the late 1950s when a group, working for a subsidiary of Wacker Chemie under the direction of J. Smidt, reported that, in the presence of certain oxidants such as cupric or ferric chloride, the palladium metal could be readily reoxidized to palladium chloride [302,303]. Since with cupric chloride, the reduction product, CuCl, can be readily reoxidized with either air or oxygen, a combination of $CuCl_2$ and $PdCl_2$ gives rise to a catalytic system in which the net reaction corresponds to the air, or oxygen, oxidation of ethene to ethanal

$$PdCl_4^{2-} + C_2H_4 + H_2O \longrightarrow CH_3CHO + Pd + 2HCl + 2Cl^- \quad (2.46)$$

$$2Cl^- + Pd + 2CuCl_2 \longrightarrow PdCl_4^{2-} + 2CuCl$$

$$2CuCl + \tfrac{1}{2}O_2 + 2HCl \longrightarrow 2CuCl_2 + H_2O$$

$$\overline{\tfrac{1}{2}O_2 + C_2H_4 \longrightarrow CH_3CHO}$$

This set of reactions form the chemical basis of the Wacker process which is presently used to produce some 1.5 million tonnes of ethanal per year.

Although there is still some doubt as to the details of precisely how the palladium effects the oxidation of ethene to ethanal, in broad outline the reaction can be represented by the following sequence [304]:

$$PdCl_4^{2-} + C_2H_4 \rightleftharpoons [PdCl_3(C_2H_4)]^- + Cl^- \quad (2.47)$$
$$(2.235)$$

$$[PdCl_3(C_2H_4)]^- + H_2O \rightleftharpoons PdCl_2(C_2H_4)(H_2O) + Cl^- \quad (2.48)$$
$$(2.236)$$

$$PdCl_2(C_2H_4)(H_2O) + H_2O \longrightarrow$$
$$[PdCl_2(CH_2CH_2OH)(H_2O)]^- + H^+ \quad (2.49)$$
$$(2.237)$$

$$[PdCl_2(CH_2CH_2OH)(H_2O)]^- \longrightarrow$$
$$CH_3CHO + Pd + 2Cl^- + H_3O^+ \quad (2.50)$$

which, at low Pd^{II} concentrations, corresponds to the rate expression

$$d[CH_3CHO]/dt = k[PdCl_4^{2-}][C_2H_4]/[Cl^-]^2[H^+].$$

The initial stage of the oxidation, represented by the Equilibria (2.47) and (2.48), involve π-co-ordination of the ethene substrate with concurrent, or subsequent, loss of chloride to form the 16 valence electron palladium(II) anion, (2.235), which in turn undergoes a ligand replacement reaction to yield the neutral palladium(II) aquo species (2.236). Experimentally it is found [302,303] that the overall reaction is inhibited by chloride ions which is consistent with presence of Equilibria (2.47) and (2.48). Co-ordination of the ethene to the palladium(II) centre will result in a decrease in the double bond electron density thus rendering it more susceptible to nucleophilic attack by either OH^- or H_2O as implicit in Equation (2.49).

It is not yet totally clear whether nucleophilic attack on the co-ordinated alkene occurs from within or from without the co-ordination sphere of the metal, i.e. whether or not prior co-ordination of OH^- or H_2O is required [304,305]. The original suggestion was that hydroxypalladation involved a ligand migration, co-ordinated OH^- to co-ordinated $CH_2{=}CH_2$, analogous to that previously discussed in hydrogenation, i.e. co-ordinated H^- to co-ordinated alkene (see Section 2.1). Such a migration requires that the hydroxy and the ethene groups be mutually *cis* and leads to syn addition via a four-centre transition state of the type (*2.238*).

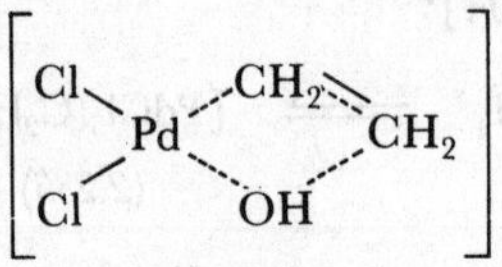

$$\left[\begin{array}{c} Cl \diagdown \diagup CH_2 \\ Pd \\ Cl \diagup \diagdown OH \end{array} \!\!\!\!\! \begin{array}{c} \\ CH_2 \\ \end{array} \right]^{\ddagger}$$

(*2.238*)

The alternative mechanism involving external attack of OH^-, i.e. without prior co-ordination, which would result in anti addition, has been rejected [306] on the basis that it would require a rate constant in excess of that possible for a diffusion controlled process. This argument does not, however, apply to external attack by water and, on the basis of stereochemical data obtained using partially deuterated alkene [307,308] which strongly indicated anti hydroxy addition, the following two sequences of reaction have been independently proposed [308] for the hydroxypalladation step.

Sequence 1

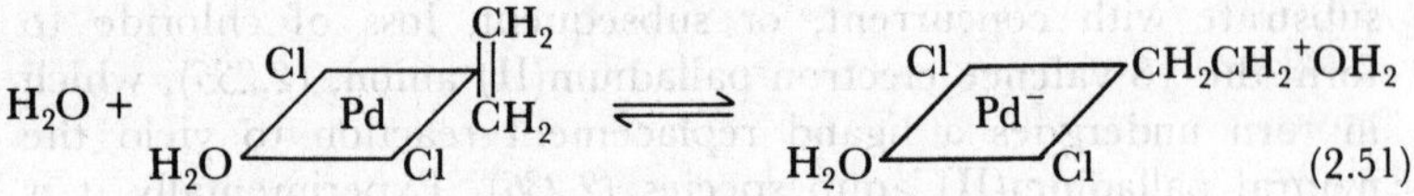

$$H_2O + \begin{array}{c} Cl \\ \diagup \diagdown \\ H_2O Pd CH_2 \\ \diagdown \diagup \\ Cl \end{array} \rightleftharpoons \begin{array}{c} Cl \\ \diagup \diagdown CH_2CH_2{}^+OH_2 \\ H_2O Pd^- \\ \diagdown \diagup \\ Cl \end{array} \quad (2.51)$$

$$H_2O + \begin{array}{c} Cl \\ \diagup \diagdown CH_2CH_2{}^+OH_2 \\ H_2O Pd^- \\ \diagdown \diagup \\ Cl \end{array} \rightleftharpoons$$

$$\left[\begin{array}{c} Cl \\ \diagup \diagdown CH_2CH_2OH \\ H_2O Pd \\ \diagdown \diagup \\ Cl \end{array} \right]^{-} + H_3O^+ \quad (2.52a)$$

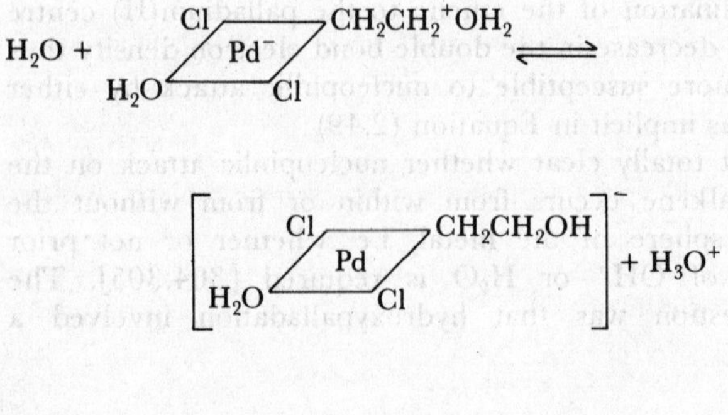

$$\left[\begin{array}{c} Cl\diagup Pd \diagdown CH_2CH_2OH \\ H_2O \diagdown \diagup Cl \end{array} \right]^{-} \xrightarrow{slow} \begin{array}{c} Cl\diagdown \\ H_2O \diagup \end{array} Pd-CH_2CH_2OH + Cl^{-}$$

(2.237a)   (2.53a)

Sequence 2

$$H_2O + \begin{array}{c} Cl\diagup Pd \diagdown \begin{array}{c} CH_2 \\ \| \\ CH_2 \end{array} \\ Cl \diagdown \diagup OH_2 \end{array} \rightleftharpoons \left[\begin{array}{c} Cl\diagup Pd \diagdown CH_2CH_2OH \\ Cl \diagdown \diagup OH_2 \end{array} \right]^{-} + H^{+}$$

(2.52b)

$$\left[\begin{array}{c} Cl\diagup Pd \diagdown CH_2CH_2OH \\ Cl \diagdown \diagup OH_2 \end{array} \right]^{-} \xrightarrow{slow} \begin{array}{c} Cl\diagdown \\ H_2O \diagup \end{array} Pd-CH_2CH_2OH + Cl^{-}$$

(2.237b)   (2.53b)

The only significant difference between the two schemes lies in the stereochemistry of the intermediate palladium species. The essential proposal underlying both is that the rate-determining step is not, as originally suggested, [306] the hydroxypalladation process but rather the loss of chloride from (2.237) to give the 3 co-ordinate, 14 valence electron species $PdCl(H_2O)(CH_2CH_2OH)$. The hydroxypalladation reaction is seen as a rapid equilibrium process, Equations 2.51 and 2.52a in sequence 1 and Equation 2.52b in Sequence 2. Although Smidt *et al* had previously [309] envisaged the hydroxypalladation as an equilibrium process, as shown in Equation 2.54, they had suggested prior coordination of the hydroxyl nucleophile.

$$PdCl_2(C_2H_4)(H_2O) \xrightleftharpoons{H_2O} [PdCl_2(CH_2CH_2OH)(H_2O)]^{-} \quad (2.54)$$

The stereochemical data obtained using partially deuterated ethene appears to be conclusive as regards *trans*(anti)-addition to the coordinated alkene, i.e. attack by non coordinated water. It is, however, important to stress that these stereochemical experiments were conducted under conditions of high chloride ion concentration (*ca.* 3 M) which could well mitigate against prior coordination of the OH^- or H_2O nucleophile. Under other experimental conditions, *e.g.* lower Cl^- concentrations, prior coordination could occur and it is perhaps premature to suggest that the mechanisms proposed by either Stille et al (Sequence 1) or Bäckvall et al (Sequence 2) are universally applicable under all experimental conditions.

Although there is some difference of opinion as to how it is formed, there is wide agreement that the β-hydroxyethylpalladium(II) complex (2.237) is formed as a result of the hydroxypalladation reaction. The next question is how does it break down to give product ethanal and palladium(0).

One of the more intriguing experiment results relating to the Wacker reaction is the finding that if ethene is oxidized in D_2O, non-deuterated ethanal is produced [310]

$$[PdCl_2(CH_2CH_2OD)(D_2O)]^- \longrightarrow CH_3CHO + Cl^- + Pd(0).$$

Thus any mechanistic proposal for the breakdown of the β-hydroxyethyl complex must, of necessity, involve a fast 1,2-hydrogen shift in the complex with loss of the hydroxy hydrogen.

Originally a concerted Pd(II)-assisted hydride shift via a dotted line transition state of the type (2.239) was proposed for this process [306]. Now, in the light of subsequent experimental data, a step-wise process involving hydrido-palladium intermediates seems more likely [306,308,309,311,312]. The type of process presently envisaged is shown in Fig. 2.38

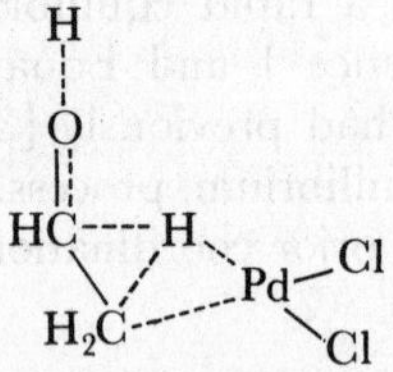

$$\text{(2.239)} \longrightarrow CH_3CHO + Pd(0) + 2Cl^- + H^+$$

Essentially what is involved is isomerization of the β-hydroxyalkyl complex (2.237) to the α-hydroxyalkyl species (2.242) via, a hydride elimination/addition sequence completely analogous to that described in Section 2.2.1a under metal catalysed alkene isomerization. To comply with the results of the deuteration studies [304] the vinyl alcohol in (2.240) must not dissociate before hydride addition to give (2.242) can occur. In other words hydride readdition in (2.240) must be considerably faster than

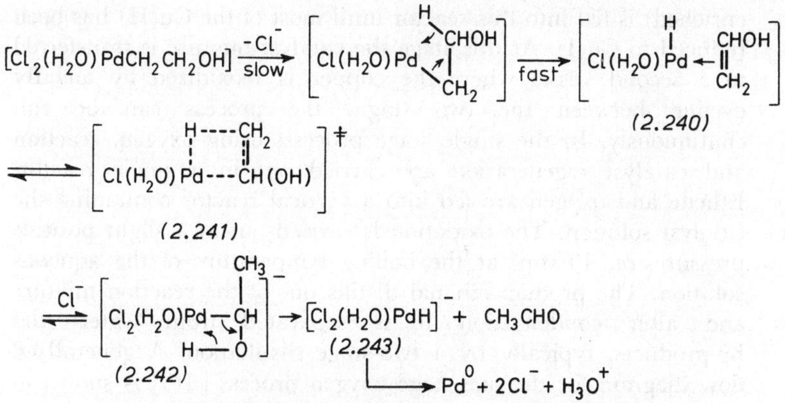

$$[Cl_2(H_2O)PdCH_2CH_2OH] \xrightarrow[slow]{-Cl^-} \left[Cl(H_2O)Pd \overset{H}{\underset{CH_2}{\overset{\diagdown CHOH}{\diagup|}}} \right] \xrightleftharpoons[]{fast} \left[Cl(H_2O)Pd \underset{CH_2}{\overset{H}{\underset{\|}{\overset{|}{\leftarrow}}} CHOH} \right]$$

(2.240)

$$\rightleftharpoons \left[Cl(H_2O)\overset{H----CH_2}{\underset{}{Pd---CH(OH)}} \right]^{\ddagger}$$

(2.241)

$$\underset{(2.242)}{\xrightleftharpoons[]{Cl^-} \left[Cl_2(H_2O)Pd \underset{H \longrightarrow O}{\overset{CH_3}{\underset{|}{\diagdown CH}}} \right]^{-}} \rightarrow \underset{(2.243)}{[Cl_2(H_2O)PdH]^-} + CH_3CHO$$

$$\longrightarrow Pd^0 + 2Cl^- + H_3O^+$$

Fig. 2.38 *Step-wise process involving hydrido-palladium intermediates*

either alkene dissociation from, or exchange with, (*2.240*). Such a requirement is in line with the known chemistry of hydrido-transition-metal-alkene species. Furthermore there is the possibility of orbital interaction between metal *d*-orbitals and those of the alkene-hydroxyl unit which would also mitigate against dissociation or exchange. Release of ethanal from (*2.242*) occurs by an irreversible β-hydroxy-hydrogen elimination. In the absence of any stabilizing ligands the resulting palladium(II)-hydrido species, e.g. (*2.243*), rapidly reductively eliminates HCl forming zero valent palladium metal.

The reaction sequences described above results in the stoichiometric oxidation of ethene or reduction of palladium(II). It is only the inclusion of a system capable of reoxidizing the palladium(0) which makes the process catalytic in palladium. In industrial Wacker processes, cupric chloride is almost universally used as the palladium oxidant since the resulting cuprous chloride can be readily reoxidized by either air or oxygen

$$Pd + 2Cl^- + 2CuCl_2 \longrightarrow PdCl_4^{2-} + 2CuCl \qquad (2.55)$$

$$2CuCl + \tfrac{1}{2}O_2 + 2HCl \longrightarrow 2CuCl_2 + H_2O. \qquad (2.56)$$

In the overall catalytic cycle the copper(I) reoxidation step, Equation (2.56), is rate determining. In commercial operation there are basically two variations of the process, one using oxygen and the other using air [313]. The air process is conducted in two separate reactors. Initially the first reactor contains a solution of $PdCl_2$ and $CuCl_2$; a mixture of ethene and air (possibly oxygen enriched) is fed into this reactor until most of the Cu(II) has been reduced to Cu(I). At this stage the catalyst mixture is transferred to a second vessel where the copper is reoxidized by air. By cycling between the two stages the process can be run continuously. In the single stage process, using oxygen, reaction and catalyst regeneration are carried out in a single reactor. Ethene and oxygen are fed into a vertical reactor containing the catalyst solution. The oxidation is carried out at a slight positive pressure, *ca.* 10 atm, at the boiling temperature of the aqueous solution. The product ethanal distills out of the reaction mixture and after condensation it is separated from water and by-products, typically by a two-stage distillation. A generalized flow diagram for the one stage oxygen process [313] is shown in Fig. 2.39 and in Fig. 2.40 a possible catalysis cycle, based on the above mechanistic discussion, is illustrated.

In commercial processes the concentration of copper far exceeds that of palladium and, although in the catalysis cycles of Fig. 2.40 the role of the copper is shown to be merely that of reoxidizing the palladium, it seems more likely that the copper chloride is involved in more than just the reoxidation step. Indeed, in the presence of acetic acid, which is frequently employed as a co-solvent, or produced by further oxidation of product ethanal, a variety of oxidants, including cupric chloride, will produce saturated esters from ethene in the presence of Pd(II) [314,315]

$$C_2H_4 + PdCl_2 + CuCl_2 \xrightarrow{HOAc} ClCH_2CH_2OAc, HOCH_2CH_2OAc,$$
$$AcOCH_2CH_2OAc.$$

Such products are always formed in small amounts during aqueous Wacker reactions and it appears that they result from

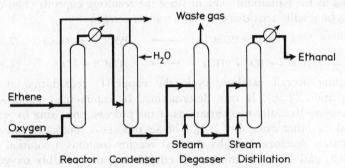

Fig. 2.39 *One stage process for the production of ethanal from ethene*

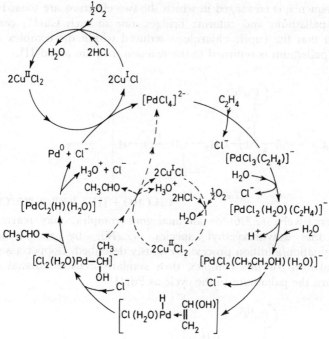

Fig. 2.40 *Catalysis cycles for the palladium(II)/copper(II) catalysed oxidation of ethene to ethanal*

interaction between the oxidant and an acetoxypalladium intermediate, (2.244).

$$\rightarrow PdCH_2CH_2OAc$$

(2.244)

Results of kinetic studies have suggested that the copper chloride is also involved in the main ethene oxidation process at high Cu^{2+} concentrations [316,317]. Intermediates of the type (2.245) have been proposed [318] and it has been suggested that the corresponding hydroxy ethyl species (2.246) may break down without the formation of Pd(O).

$$\begin{bmatrix} Cl_3Cu\text{—}Cl & & CH_2 \\ & \diagdown & \| \\ & Pd & CH_2 \\ & \diagup & \diagdown \\ Cl_3Cu\text{—}Cl & & OH_2 \end{bmatrix}^{2-}$$

(2.245)

A sequence is envisaged in which the two electrons are passed via the palladium and chlorine bridges, one to each Cu(II) centre such that the cupric chloride is reduced within the complex and the palladium is returned to the reaction medium as Pd(II).

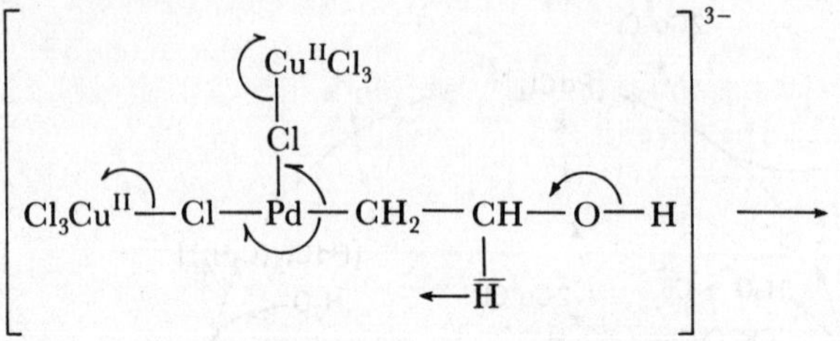

$$\left[Cl_3Cu^{II}-Cl-Pd-CH_2-CH(-H)-O-H \atop {Cl \atop Cu^{II}Cl_3}} \right]^{3-} \longrightarrow$$

(2.246)

$$CH_3CHO + H^+ + PdCl_2 + 2Cu^ICl_3^{2-}$$

Alternatively, (2.246) or an analogous complex, may rearrange to the α-hydroxyethyl species, (2.247), by the hydride elimination/addition process previously described. Copper assisted breakdown of this complex then similarly release ethanal and return the palladium to the cycle as. Pd(II)

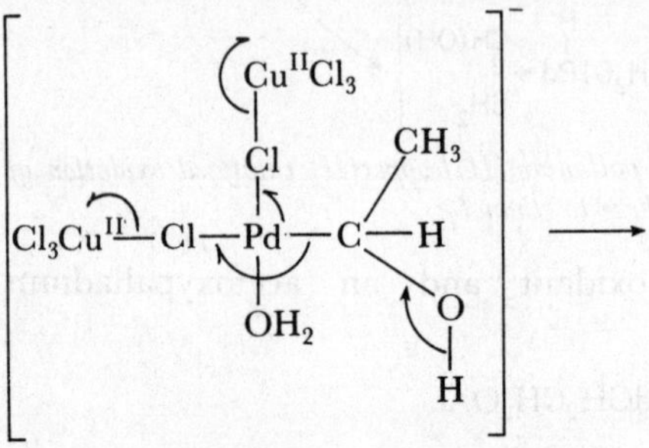

$$\left[Cl_3Cu^{II}-Cl-Pd(-OH_2)-C(CH_3)(-H)-O-H \atop {Cl \atop Cu^{II}Cl_3}} \right]^{-} \longrightarrow CH_3CHO + H_3O^+ + PdCl_2 + 2Cu^ICl_3^{2-}.$$

(2.247)

This latter sequence is indicated by the dashed cycle in Fig. 2.40. Although there seems little doubt that, under commercial Wacker conditions, i.e high [Cu^{2+}] relative to [Pd^{2+}], the copper participates in more than the oxidative regeneration step of the catalysis cycle, and although the synergistic participation implied in either (2.246) or (2.247), is aesthetically appealing, the precise nature of this, and indeed many of the intermediates in the Wacker reaction, must await more detailed mechanistic study. In the meantime, the Wacker process remains one of the most important applications of homogeneous catalysts in industrial chemistry.

HOMOGENEOUS CATALYST SYSTEMS IN OPERATION

Although all of the oxidation systems we have described involve molecular oxygen, in no case is this directly attached to the substrate via the metal centre. In all cases the molecular oxygen is introduced by a round about route, e.g. via the hydroperoxide in both the alkane oxidation and alkene epoxidation reactions and via water in the Wacker reaction. Many complexes of dioxygen are known [319] containing side-on, end-on and bridging oxygen units, e.g. (2.248), (2.249) and (2.250), respectively

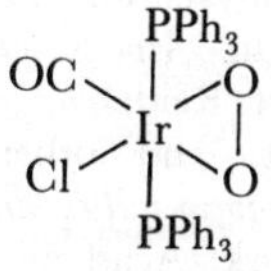

(2.248) 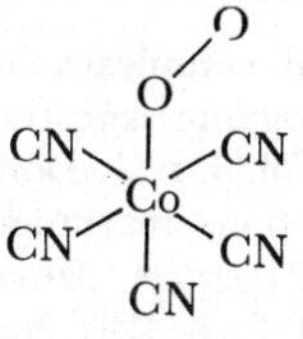 (2.249) 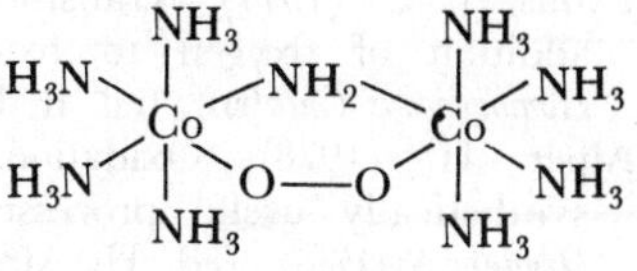 (2.250)

but up to now no catalytic reaction involving direct transfer of one oxygen atom of bound dioxygen to an organic substrate has been reported. At the time of writing the aim of many industrial oxidation chemists, not to mention quite a few academics – the direct oxidation of propene to propylene oxide

$$CH_3CH=CH_2 + \tfrac{1}{2}O_2 \xrightarrow{\text{'cat'}} CH_3\overset{O}{\overset{|}{CH}}{-}CH_2$$

still remains illusive.

Further reading

General texts and reviews

Waters, W. A. (1964), *Mechanisms of Oxidation of Organic Compounds*, London: Methuen.

Berezin, I. V., Denisov, E. T. and Emanuel, N. M. (1966), *The Oxidation of Cyclohexane*, Oxford: Pergamon.

Szonyi, G. (1968), Recent homogeneously catalysed commercial processes. *Advances in Chemistry Series*, **70**, Washington: American Chemical Society, p. 53.

Reich, L. and Stivala, S. S. (1969), *Autoxidation of Hydrocarbons and Polyolefins, Kinetics and Mechanisms*, New York: Marcel Dekker.

Stern, E. W. (1971), Homogenous metal catalysed oxidation of organic compounds, in *Transition Metals in Homogeneous Catalysis* (ed. G. N. Schrauzer), p. 93, New York: Marcel Dekker.

Maitlis, P. M. (1971), *The Organic Chemistry of Palladium. Vol. II, Catalytic Reactions*, Ch. 11, p. 77), Academic Press

Hucknall, D. J. (1974), *Selective Oxidation of Hydrocarbons*, New York, Academic Press.

Lyons, J. E. (1974), Oxidation of olefins in the presence of transition metal complexes. *Adv. Chem. Series*, **132**, p. 64, Washington: American Chemical Society.

Henry, P. M. (1975), Palladium-catalysed organic reactions. *Adv. Organometal. Chem.*, **13**, 363.

Benson, D. (1976), *Mechanisms of Oxidation by Metal Ions*, Amsterdam: Elsevier.

Sheldon, R. A. and Kochi, J. K. (1976), Metal-catalysed oxidations of organic compounds in the liquid phase: a mechanistic approach. *Adv. Catal.*, **25**, 272.

Lyons, J. E. (1977), Transition metal complexes as catalysts for the addition of oxygen to reactive organic substrates, in *Aspects of Homogeneous Catalysis*, (ed. R. Ugo), Vol. 3, p. 1, Dordrecht: Reidel

Alper, H. (1978), Oxidation, reduction rearrangement, and other synthetically useful processes, in *Transition Metal Organometallics in Organic Synthesis*. (ed. H. Alper) Vol. 2, p. 121, New York: Academic Press.

Davidson, J. M. (1978), Homogeneous catalytic oxidation, in *Catalysis*. *Vol. 2*, London: Chemical Society.

2.8 Metathesis

We started this chapter on homogeneous catalysts in operation by considering one of the oldest and probably most studied of transition-metal catalysed reactions, namely hydrogenation. We conclude by looking at one of the youngest and certainly one of the most intriguing of such reactions, namely alkene metathesis.

Following Rooney and Steward's definition [320], alkene metathesis is a process in which the alkene carbon–carbon double bonds are ruptured and reformed to give a statistical redistribution of the alkylidene entities such that, under steady state conditions, equilibrium concentrations of reactants and products are formed. Thus, alkene metathesis catalysts promote the establishment of the equilibria of the type shown in Equations (2.57), (2.58) and (2.59).

$$2R^1CH{=}CHR^2 \rightleftharpoons R^1CH{=}CHR^1 + R^2CH{=}CHR^2 \quad (2.57)$$

$$\begin{array}{c} R^2\ R^3 \\ |\ \ | \\ R^1{-}C{=}C{-}R^4 \\ + \\ R^5{-}C{=}C{-}R^8 \\ |\ \ | \\ R^6\ R^7 \end{array} \rightleftharpoons \begin{array}{c} R^1\diagdown\ \diagup R^2 \\ C \\ \| \\ C \\ R^5\diagup\ \diagdown R^6 \end{array} + \begin{array}{c} R^3\diagdown\ \diagup R^4 \\ C \\ \| \\ C \\ R^7\diagup\ \diagdown R^8 \end{array} \quad (2.58)$$

$$R^1-\underset{\underset{R^8}{|}}{\overset{\overset{R^2}{|}}{C}}=\underset{\underset{R^3}{|}}{\overset{\overset{R^5}{|}}{C}}-R^6 \quad \rightleftharpoons \quad \underset{\underset{R^8}{C}}{\overset{\overset{R^1}{C}}{\underset{\|}{C}}}\underset{R^7}{\overset{R^2}{}} \quad \underset{\underset{R^4}{C}}{\overset{\overset{R^5}{C}}{\underset{\|}{C}}}\underset{R^3}{\overset{R^6}{}}$$

(2.59)

$$R^7-\underset{\underset{R^8}{|}}{\overset{\overset{}{|}}{C}}=\underset{\underset{R^3}{|}}{\overset{\overset{}{|}}{C}}-R^4$$

The first recognized metal catalysed alkene metathesis reactions were reported in the late fifties and early sixties using heterogeneous catalyst systems [321, 322]. For example, molybdenum oxide on alumina 'promoted' with tri-isobutyl aluminium catalyses the transformation of propene into ethene and butene [321]

$$\begin{array}{c}CH_3CH=CH_2 \\ + \\ CH_3CH=CH_2\end{array} \rightleftharpoons \begin{array}{c}CH_3CH \\ \| \\ CH_3CH\end{array} + \begin{array}{c}CH_2 \\ \| \\ CH_2\end{array}$$

(2.60)

In 1967 the first homogeneous* alkene metathesis catalysts – combinations of WCl_6, C_2H_5OH and $C_2H_5AlCl_2$ – were reported [323], and since then there has been a virtual explosion of experimental effort dedicated to this topic. Metathesis catalysts incorporating most of the transition metals are now known with the most active systems being based on tungsten, molybdenum and, to a lesser extent, rhenium [48].

One of the most fascinating aspects of the reaction is the way in which the metal centre actually brings about the rupture and reformation of the alkene double bond. Early experiments [328,329], using ^{14}C-labelled alkenes, e.g. (2.251) in Equation (2.61), showed that product formation resulted from $C=C$ rupture rather than from alkyl group transfer. However, it is only

---

*There is still some debate [324–326] as to whether this particular catalyst system, and those based on WCl_6 in general, are truly homogeneous. This is frequently a problem with complex catalyst systems, i.e. those containing several catalyst components. Generally a good indication of catalyst homogenity is when the 'homogeneous' catalyst system gives rise to markedly different product selectivity compared to its heterogeneous counterpart. Although this is somewhat in doubt vis-à-vis the $WCl_6/EtOH/C_2H_5AlCl_2$ system, markedly different selectivities have been obtained [327] with other tungsten based metathesis catalysts, and there seems little doubt that truly homogeneous alkene metathesis catalysts do exist.

in the last four or five years that a consensus has been reached as to the most probable mechanism for the reaction

$$2CH_3-CH=\,^{14}CH_2 \rightleftharpoons \,^{14}CH_2=\,^{14}CH_2 + CH_3-CH=CH-CH_3. \quad (2.61)$$

(2.251)

2.8.1 Alkene metathesis mechanisms

Conceptually the most appealing mechanism is that suggested by the way in which we have shown the Equilibria (2.58) to (2.60), i.e. a concerted mechanism involving simultaneous co-ordination of a pair of substrate molecules, proceeding via a metal-cyclobutane transition state of the type (2.252).

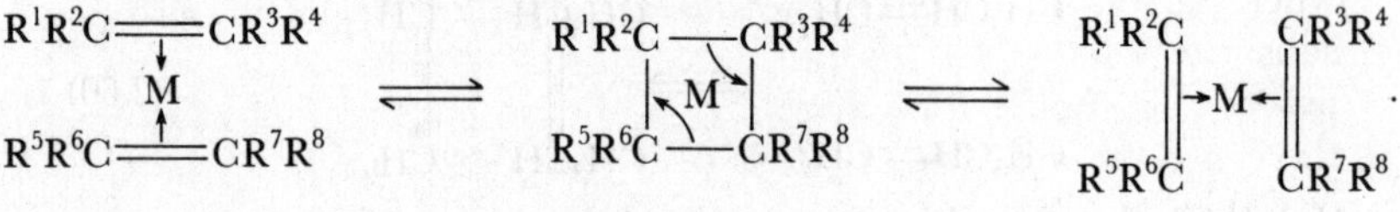

$$\begin{array}{c} R^1R^2C=\!\!=\!\!CR^3R^4 \\ \updownarrow M \updownarrow \\ R^5R^6C=\!\!=\!\!CR^7R^8 \end{array} \rightleftharpoons \begin{array}{c} R^1R^2C-\!\!-CR^3R^4 \\ | \searrow M \swarrow | \\ R^5R^6C-\!\!-CR^7R^8 \end{array} \rightleftharpoons \begin{array}{c} R^1R^2C \quad CR^3R^4 \\ \| \rightarrow M \leftarrow \| \\ R^5R^6C \quad CR^7R^8 \end{array}$$

(2.252)

(2.62)

Such a concerted, or pairwise mechanism, which is completely analogous to that described for skeletal isomerisation in Section 2.2.2a, was originally thought to apply to alkene metathesis [330–332]. However it is now widely recognised [48,320,333–335] that the reaction occurs via a chain process involving metal-carbenes as the key reaction intermediates

$$\begin{array}{c} R^1R^2C=\!\!=\!M \\ \\ R^5R^6C=\!\!=\!CR^7R^8 \end{array} \rightleftharpoons \begin{array}{c} R^1R^2C-\!\!-M \\ | \searrow \swarrow | \\ R^5R^6C-\!\!-CR^7R^8 \end{array} \rightleftharpoons \begin{array}{c} R^1R^2C \quad M \\ \| \quad \| \\ R^5R^6C \quad CR^7R^8 \end{array}$$

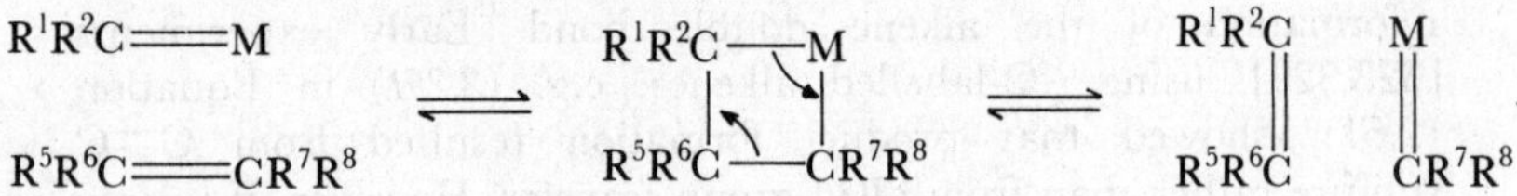

(2.63)

(a) Metal-carbenes as reaction intermediates

Much of the experimental evidence for the intermediacy of metal-carbene species in alkene metathesis reactions comes from detailed analyses of the products forming during the initial stages of the reaction [335]. Under metathesis conditions, a mixture of but-2-ene, oct-4-ene and cyclo-octene can give rise to C_6, C_{12}, C_{14} and C_{16} alkene products as follows:

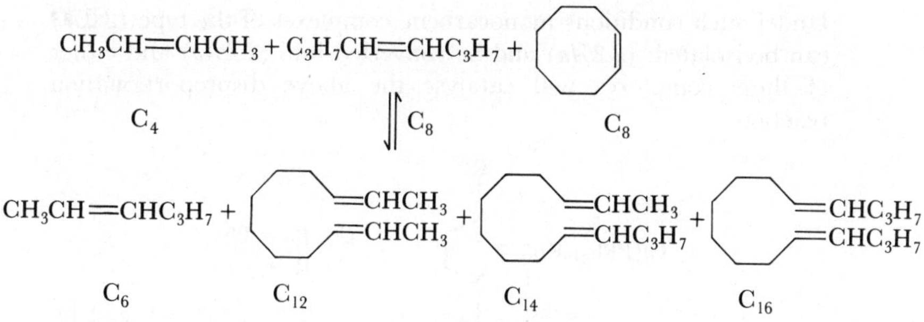

CH$_3$CH=CHCH$_3$ + C$_3$H$_7$CH=CHC$_3$H$_7$ + [cyclooctene]

C$_4$ C$_8$ C$_8$

CH$_3$CH=CHC$_3$H$_7$ + [ring]=CHCH$_3$/=CHCH$_3$ + [ring]=CHCH$_3$/=CHC$_3$H$_7$ + [ring]=CHC$_3$H$_7$/=CHC$_3$H$_7$

C$_6$ C$_{12}$ C$_{14}$ C$_{16}$

If metathesis occurs via the pairwise process, exemplified by Equation (2.62), then, in the initial stages of the reaction, the concentration of C$_{14}$ should be very much less than that of either C$_{12}$ or C$_{16}$ since C$_{14}$ would depend on the presence of product C$_6$ whereas both C$_{12}$ and C$_{16}$ could be formed directly from the substrate C$_4$ and C$_8$ alkenes. If, on the other hand, the chain process, exemplified by Equation (2.63), was taking place then the concentration of C$_{14}$ should be at least twice that of either C$_{12}$ or C$_{16}$.

With MoCl$_2$(NO)$_2$[P(C$_6$H$_5$)$_3$]$_3$/Al$_2$Cl$_3$(CH$_3$)$_3$ in chlorobenzene, as catalyst precursor, C$_{14}$ was always found to be formed in greater amounts than either C$_{12}$ or C$_{16}$ during the early stages of the metathesis [336]. If we accept that alkene exchange does not constitute the rate-determining step in the overall metathesis process [335] then the results of such 'cross' experiments effectively exclude the pairwise mechanism depicted in Equation (2.62).

Some of the first direct evidence for the intermediacy of metal-carbene species was that obtained using electron rich alkene of the type (2.253) [333,337]. In the presence of RhCl(PPh$_3$)$_3$, in xylene at 140 °C, a mixture of (2.253a) and (2.253b) gives rise to an approximately equilibrium amount of (2.253c) over 2 h.

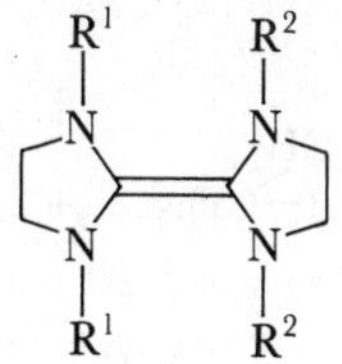

(a) R^1 = R^2 = Ph
(b) R^1 = R^2 = p-tol
(c) R^1 = Ph; R^2 = p-tol

(2.253)

Under such conditions monocarbene complexes of the type (2.254) can be isolated; (2.254a) and be converted into (2.254b) and either of these complexes will catalyse the above disproportionation reaction.

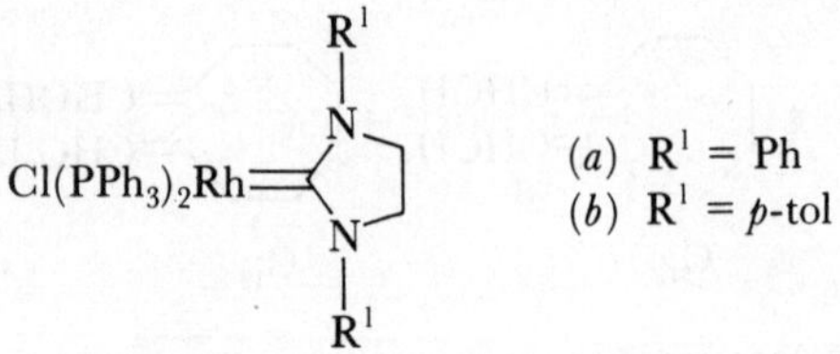

$$Cl(PPh_3)_2Rh=C\begin{pmatrix}N(R^1)-CH_2\\N(R^1)-CH_2\end{pmatrix}$$

(a) $R^1 = Ph$
(b) $R^1 = p\text{-tol}$

(2.254)

Mechanistic conclusions based on these later systems could be considered slightly dubious since such electron rich alkenes are somewhat removed from the more conventional alkenes used in the vast majority of metathesis reactions. This is frequently a problem when model, and thus generally more stable, complexes are used in an attempt to elucidate the mechanism of a catalytic reaction. However, in spite of such reservations the suggestions based on the particular systems have proved essentially correct in the light of more recent findings.

In 1973 Casey and Burkhardt reported [338] the synthesis and isolation of the metal-carbene complex (2.255) with R = Ph. This

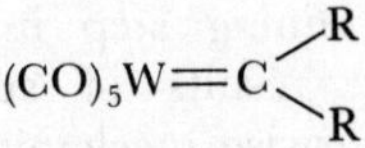

$$(CO)_5W=C\begin{pmatrix}R\\R\end{pmatrix}$$

(2.255)

complex is much closer to the type of carbene species envisaged as taking part in metathesis reactions in that it does not contain a hetero-atom bonded to carbene carbon. Its significance to the mechanism of alkene metathesis became clear the following year when the reaction shown in Equation (2.64) was demonstrated [339]

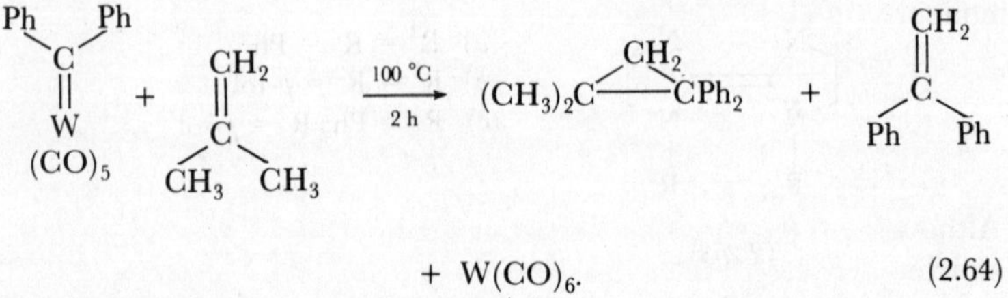

$$\underset{(CO)_5}{\overset{Ph}{\underset{W}{\overset{\|}{C}}}\overset{Ph}{}} + \underset{CH_3\ CH_3}{\overset{CH_2}{\underset{C}{\overset{\|}{C}}}} \xrightarrow[2\text{ h}]{100\ °C} (CH_3)_2C\text{------}CPh_2 + \underset{Ph\ Ph}{\overset{CH_2}{\underset{C}{\overset{\|}{C}}}}$$

$$+\ W(CO)_6. \qquad (2.64)$$

HOMOGENEOUS CATALYST SYSTEMS IN OPERATION 201

Using the sequence shown below, Casey rationalized these products by proposing the metallocyclobutane (2.256), R^1 = Ph, R^2 = Me, as the key intermediate. With this particular system only the diphenyl-dimethyl cyclopropyl compound,

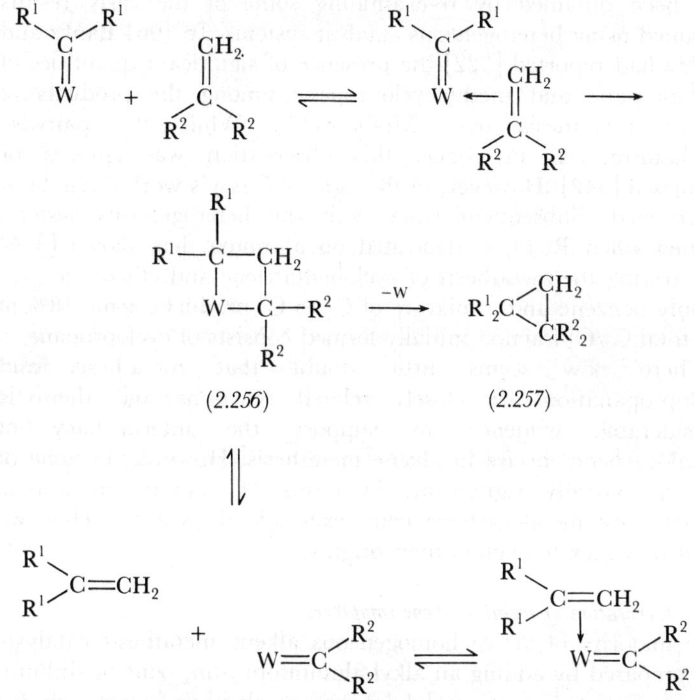

$$\underset{W}{\overset{R^1\diagdown\diagup R^1}{C}} + \underset{R^2\diagup\diagdown R^2}{\overset{CH_2}{\underset{C}{\|}}} \rightleftharpoons \underset{\underset{R^2\diagup\diagdown R^2}{C}}{\overset{R^1\diagdown\diagup R^1}{\underset{\|}{C}}} \longleftarrow \overset{CH_2}{\underset{}{\|}} \longrightarrow$$

$$\underset{(2.256)}{\begin{array}{c} R^1 \\ | \\ R^1-C-CH_2 \\ | \quad\quad | \\ W-C-R^2 \\ | \\ R^2 \end{array}} \xrightarrow{-W} \underset{(2.257)}{R^1{}_2C\overset{CH_2}{\underset{CR^2{}_2}{\diagup\diagdown}}}$$

$$\Updownarrow$$

$$\underset{R^1\diagup}{\overset{R^1\diagdown}{C}}=CH_2 \; + \; W=C\underset{R^2}{\overset{R^2}{\diagdown}} \rightleftharpoons \underset{R^1\diagup}{\overset{R^1\diagdown}{C}}=CH_2 \downarrow \; W=C\underset{R^2}{\overset{R^2}{\diagdown}}$$

(2.257) R^1 = Me; R^2 = Ph, was isolated, i.e. there is only circumstantial evidence for the presence of a dimethyl carbene species. However, when the di-*p*-tolyl-carbene complex, (2.255) R = *p*-tolyl, is used with styrene in place of isobutene both the expected cyclopropyl compounds, i.e. (2.258) and (2.259) are formed, indicating the presence of both di-*p*-tolyl- and the monophenyl-carbene intermediates [340].

$$(Tol)_2C\overset{CH_2}{\underset{}{\diagup\!\!\!\diagdown}}CHPh \quad\quad PhHC\overset{CH_2}{\underset{}{\diagup\!\!\!\diagdown}}CHPh$$
$$(2.258) \quad\quad\quad\quad (2.259)$$

Although the tungsten reactions described are essentially stoichiometric, (diphenylcarbene)pentacarbonyltungsten, (2.255)

R = Ph, will, albeit poorly, catalyse the metathesis of *cis*-pent-2-ene [341] and a number of *cis*-cycloalkenes [342] (see Section 2.8.2).

Further evidence for the intermediacy of metal–carbene species has been obtained by re-examining some of the early results obtained using heterogeneous catalyst systems. In 1964 Banks and Bailey had reported [322] the presence of significant quantities of cyclopropane and methylcyclopropane among the products of ethene metathesis over MoO_3/Al_2O_3. While the pairwise mechanism was in vogue, this observation was ignored or dismissed [342]. However, in the light of Casey's work it has been resurrected. Subsequent work with the heterogeneous system, formed when Re_2O_7 is deposited on aluminia, has shown [344] that during the metathesis of cycloheptatriene and ethene, to give mainly benzene and a mixture of C_1 to C_4 products, some 18% of the total C_3/C_4 fraction initially formed consists of cyclopropane.

There now seems little doubt that metathesis and cyclopropanation are closely related reactions and there is considerable evidence to support the intermediacy of metal–carbene species in alkene metathesis. However, in none of the catalytically significant, in terms of activity, metathesis systems are metal–carbene complexes added as such. Thus we must now seek to explain their origins.

(b)   Generation of metal–carbene complexes

The majority of active homogeneous alkene metathesis catalysts are prepared by adding an alkyl-aluminium, -tin, -zinc or -lithium species to a tungsten-, molybdenum-, or rhenium-halide complex in an inert solvent, such as chlorobenzene, frequently in the presence of small amounts of ethanol. Typical examples of active catalyst systems are WCl_6/R_xAlCl_y $(x + y = 3)$ WCl_6/R_4Sn, WCl_6/R_2Zn, WCl_6/RLi, $MoCl_5/R_3Al$, $ReCl_5/Et_3Al$ [48,320]. With such systems, transition-metal alkyl species could be readily formed by alkyl transfer from, for example, the aluminium alkyl to the tungsten chloride

$$WCl_6 + EtAlCl_2 \longrightarrow EtWCl_5 + AlCl_3.$$

α-Hydride elimination from such a tungsten-alkyl species would give the carbene species (*2.260*) which could then participate in the metathesis sequence. Although β-elimination, to give hydride alkene complexes, is the more well established reaction (see, for

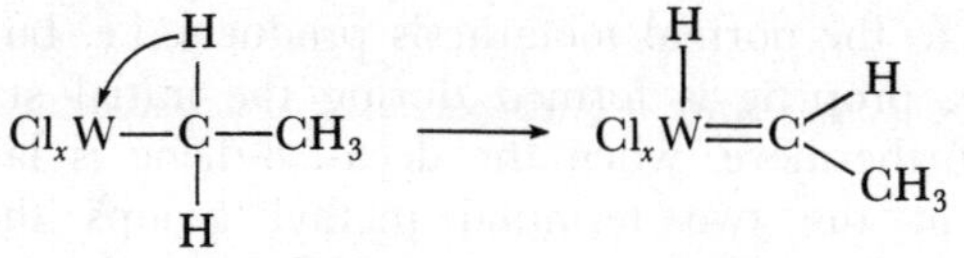

HOMOGENEOUS CATALYST SYSTEMS IN OPERATION

$$Cl_xW\text{—}\underset{H}{\overset{H}{C}}\text{—}CH_3 \longrightarrow Cl_xW\text{=}C\overset{H}{\underset{CH_3}{\diagdown}}^{H} \quad (2.65)$$

(*2.260*)

example, Section 2.2) α-elimination of the type shown in Equation (2.65) has been shown to occur in tungsten-methyl complexes formed on treating $[W(\eta^5\text{-}C_5H_5)_2CH_3(C_2H_4)]PF_6$ with PMe_2Ph [47]. Furthermore, when either WCl_6 is treated with tetramethyltin or $MoCl_2(NO)_2(PPh_3)_2$ is treated with $Me_3Al_2Cl_3$ in chlorobenzene, methane and ethene are produced. If $Sn(CD_3)_4$ is used then both the methane and ethene are perdeuterated [345]. The formation of methane can be rationalized by a sequence of reactions analogous to those first proposed by Muetterties [346] to explain the formation of methane in the $WCl_6/Zn(CH_3)_2$ system, i.e.

$$WCl_6 + Zn(CH_3)_2 \xrightarrow{-ZnCl_2} Cl_4W\overset{CH_3}{\underset{CH_3}{\diagdown}} \rightleftharpoons Cl_4\overset{H}{\underset{CH_3}{W}}\text{=}CH_2$$

$$\longrightarrow Cl_4W\text{=}CH_2 + CH_4$$

and/or

$$WCl_6 + Zn(CH_3)_2 \xrightarrow{-ZnClCH_3} Cl_5W\text{—}CH_3 \rightleftharpoons Cl_5\overset{H}{W}\text{=}CH_2$$

$$\xrightarrow[-ZnCl^+]{CH_3Zn^+} Cl_4W\text{=}CH_2 + CH_4.$$

In both cases the essential intermediate is a hydrido-carbene species formed by α-hydride elimination from a methyl-tungsten species. In the $WCl_6/SnMe_4$ system, it is suggested the ethene is formed via a carbene dimerisation reaction

$$2L_nW\text{=}CH_2 \longrightarrow 2L_nW + CH_2\text{=}CH_2.$$

When the catalyst system produced by adding $SnMe_4$ to WCl_6 is used for the metathesis of deca-2,8-diene (*2.261*),

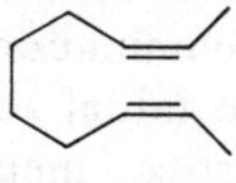

(*2.261*)

in addition to the normal metathesis products, i.e. but-2-ene and cyclohexene, propene is formed during the initial stages of the reaction. Furthermore, when the deca-2,8-diene is labelled with deuterium in the two terminal methyl groups the resulting propene is exclusively deuterated at C-3 and, when $Sn(CD_3)_4$ is used to generate the tungsten catalyst, metathesis of $[1,1,1,10,10,10-^2H_6]$ deca-2,8-diene yielded, in addition to the normal products, $[1,1,3,3,3-^2H_5]$-propene [345]. These extremely elegant labelling experiments, which are consistent with the sequence of reaction shown below, lend considerable credence to both the intermediacy of metal-carbene species and to their initial formation by α-elimination.

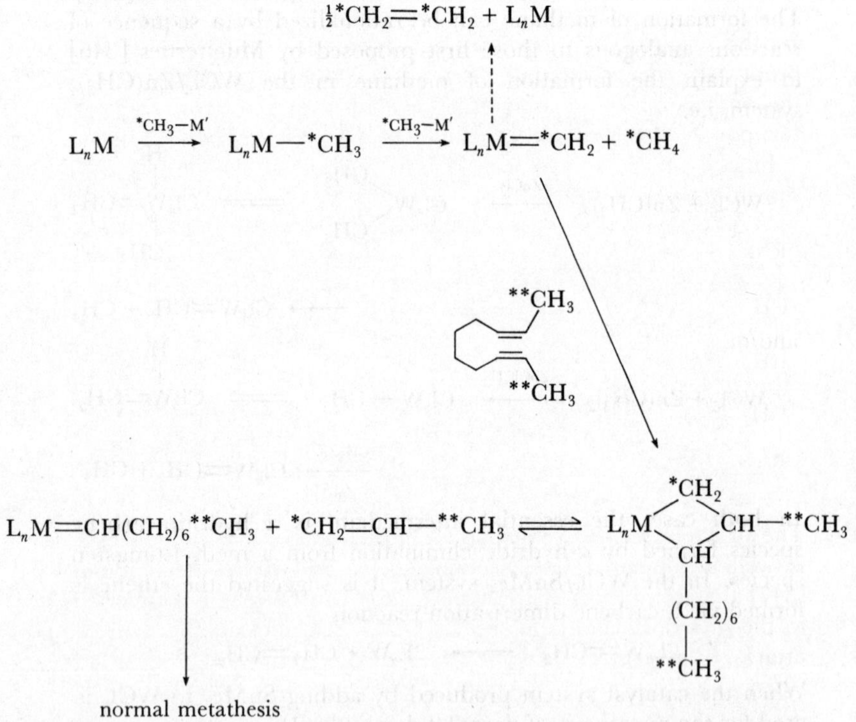

$$\tfrac{1}{2}{}^*CH_2={}^*CH_2 + L_nM$$

$$L_nM \xrightarrow{{}^*CH_3-M'} L_nM-{}^*CH_3 \xrightarrow{{}^*CH_3-M'} L_nM={}^*CH_2 + {}^*CH_4$$

$$L_nM=CH(CH_2)_6{}^{**}CH_3 + {}^*CH_2=CH-{}^{**}CH_3 \rightleftharpoons L_nM\underset{\underset{\underset{{}^{**}CH_3}{(CH_2)_6}}{CH}}{\overset{{}^*CH_2}{\diagup}}\overset{CH-{}^{**}CH_3}{\diagdown}$$

normal metathesis

When alkyl-metal species are used as co-catalysts then the transition-metal alkylation/α-elimination processes outlined above offer a plausible route to the initial carbene species. There are, however, a number of active metathesis catalyst systems, including nearly all the heterogeneous systems, which do not involve co-catalysts containing alkyl groups.

If a source of hydride is present then metal-alkyl formation can occur via the well established sequence

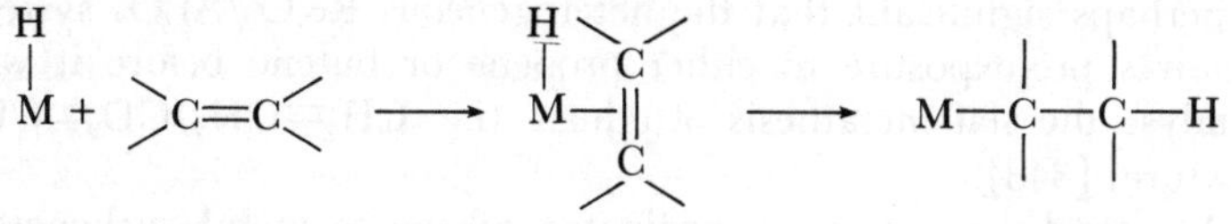

$$\underset{}{M}-H + \overset{}{\underset{}{>}}C=C\overset{}{\underset{}{<}} \longrightarrow M-\overset{H}{\underset{}{\|}}\overset{C}{\underset{C}{\|}} \longrightarrow M-\overset{|}{\underset{|}{C}}-\overset{|}{\underset{|}{C}}-H$$

(2.262)

with α-hydride elimination from (2.262) giving the initiating carbene species

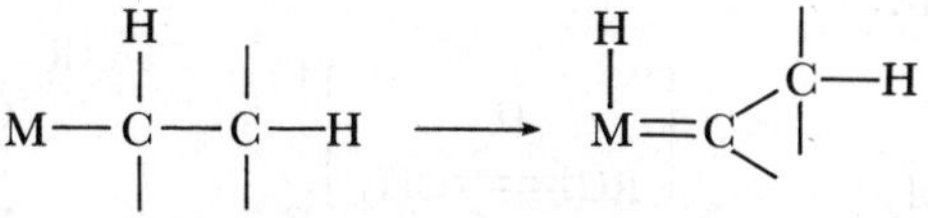

$$M-\overset{H}{\underset{|}{\overset{|}{C}}}-\overset{|}{\underset{|}{C}}-H \longrightarrow M=\overset{H}{\underset{}{C}}\overset{C-H}{\underset{}{<}}$$

In this context it is perhaps significant that trace amounts of protic solvents e.g. ethanol, often aid metathesis reaction [320]. In the absence of either a hydride or alkyl source we can envisage two possible routes, from co-ordinated alkene, to metal carbene.

Firstly, via a hydrido-π-allyl intermediate of the type (2.263) shown in Equation (2.66)

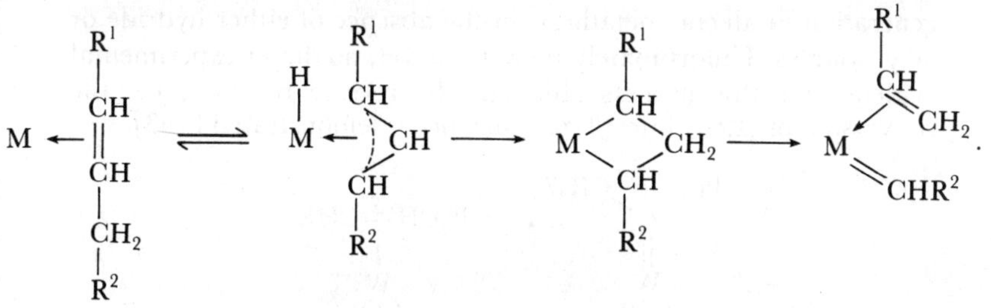

(2.263)   (2.66)

The first step in this sequence, i.e. formation of a π-allyl species, is well documented (see Section 2.2.1b) and it has been found [347] that complexes of the type (2.264) can be readily converted to the metallocyclobutane species (2.265) via nucleophilic hydride attack on the central carbon of the π-allylic group

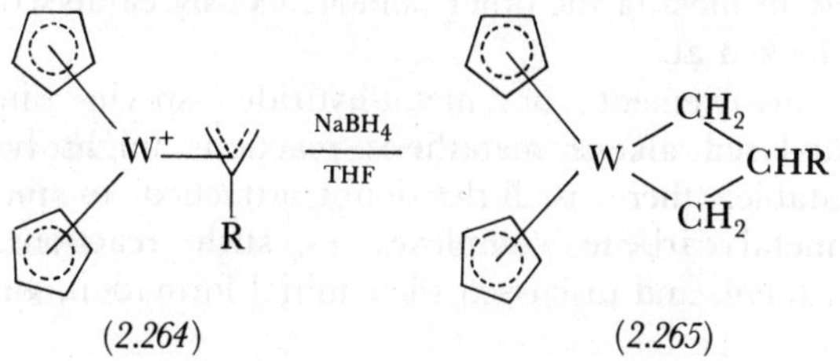

(2.264)   (2.265)   R = H, Me

This type of initiation sequence demands the presence of at least three carbon atoms in the initiating alkene and in this context it is perhaps significant that the heterogeneous Re_2O_7/Al_2O_3 system requires pre-exposure to either propene or butene before it will catalyse the self metathesis of ethene (i.e. $CH_2{=}CH_2/CD_2{=}CD_2$ mixture) [348].

A second route from co-ordinated alkene to metal–carbene is, as previously discussed in Section (2.2.1b), via a transition state of the type (2.266), i.e. via a process involving a 1,2-suprafacial hydrogen shift.

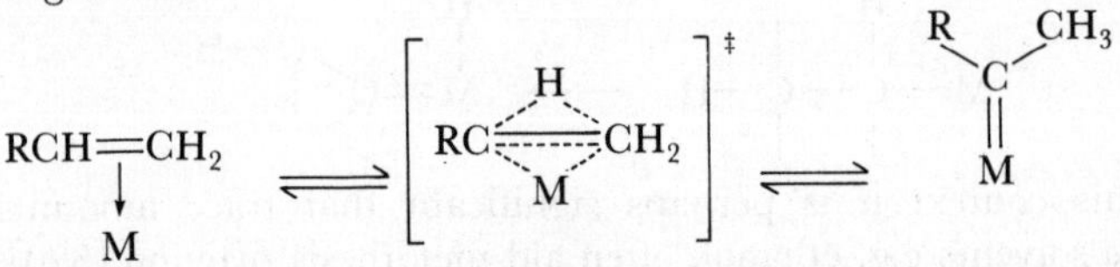

$$RCH{=}CH_2 \atop \phantom{RCH{=}}M \rightleftharpoons \left[RC{\cdots}{\cdots}CH_2 \atop M \right]^{\ddagger} \rightleftharpoons {R\diagdown C\diagup CH_3 \atop \|\atop M}$$

(2.266)

Such a sequence does not necessitate the presence of a three carbon fragment, i.e. R could equally well be hydrogen, and, as such, this sequence could constitute a universal process for initial carbene generation in alkene metathesis in the absence of either hydride or alkyl sources. Unfortunately there is, as yet, no direct experimental evidence for this process although the reverse reaction, i.e. the conversion of (2.63) into (2.267), has been demonstrated [163].

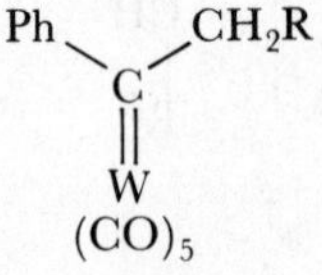

$$Ph\diagdown C \diagup CH_2R \atop \|\atop W \atop (CO)_5 \qquad\qquad PhCH{=}CHR \atop \|\atop W \atop (CO)_5$$

(2.63) 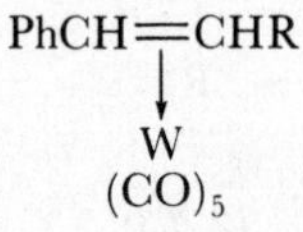   (2.267)

In all but this last process initial carbene generation involves, at some stage, the formation of a metal–hydride bond and there is an increasing amount of evidence [320] to suggest that metal-hydrides play as important a role in alkene metathesis reactions as they do in most of the other homogeneously catalysed reactions we have looked at.

Although the involvement of metal-hydride species in homogeneously catalysed alkene metathesis reactions might be regarded as debatable, there is little doubt attached to the involvement of metal–carbene complexes in such reactions. Having implicated them, and discussed their initial formation, we

HOMOGENEOUS CATALYST SYSTEMS IN OPERATION

now move on to consider how they can take part in a catalysis cycle.

(c) A possible catalysis cycle

Given the formation of an initial metal–carbene complex the next stage is to combine this species with substrate alkene to give the metallocyclobutane intermediate *(2.268)*

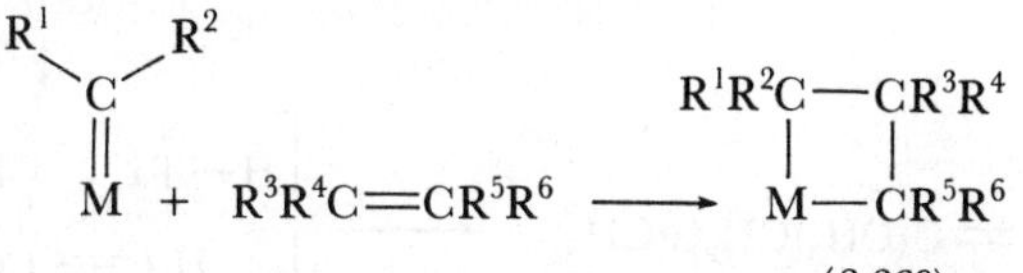

$$\underset{M}{\overset{R^1\diagdown\;\diagup R^2}{\underset{\|}{C}}} + R^3R^4C{=}CR^5R^6 \longrightarrow \begin{array}{c} R^1R^2C-CR^3R^4 \\ |\qquad\quad| \\ M-CR^5R^6 \end{array}$$

(2.268)

Electronically the course of this reaction will be determined to a large extent by the apparent charge separation in the metal–carbene linkage. Such a linkage may be represented by two extreme cannonical forms, namely *(2.269)* and *(2.270)*

$$L_n\overset{+}{M}-\overset{-}{C}R^1R^2 \longleftrightarrow L_nM{=}CR^1R^2 \longleftrightarrow L_n\overset{-}{M}-\overset{+}{C}R^1R^2$$

(2.269) $\qquad\qquad\qquad\qquad\qquad$ *(2.270)*

suggesting that, formally, the carbene moiety may function as either a nucleophile *(2.269)* or an electrophile *(2.270)*. Reactions characteristic of both types of behaviour are known and, as in the case of metal-hydrides, cf. Section 2.1, the relative charge separation appears to depend on the nature of M, and the associated ligands. Thus, in tantalum–carbene complexes of the type *(2.271)*, where R = $CH_2C(CH_3)_3$, the carbene moiety functions as a nucleophile, readily reacting with ketones to give alkenes of the type *(2.272)* with, for example, $R^1 = R^2 = CH_3$ [349].

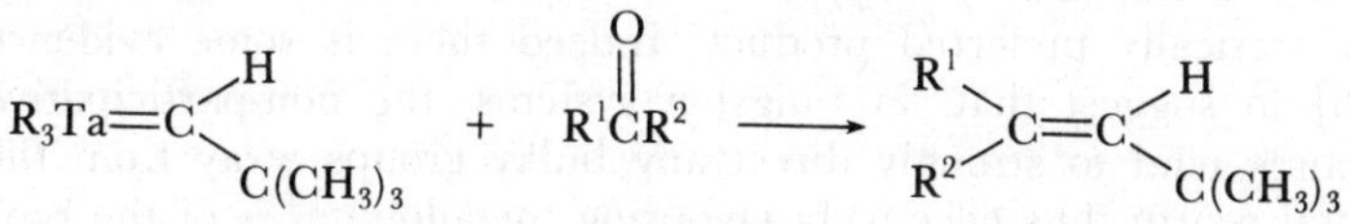

$$R_3Ta{=}C\diagup^{H}_{C(CH_3)_3} \;+\; R^1\overset{\overset{O}{\|}}{C}R^2 \longrightarrow \underset{R^2}{\overset{R^1}{\diagdown}}C{=}C\diagup^{H}_{C(CH_3)_3}$$

(2.271) $\qquad\qquad\qquad\qquad\qquad$ *(2.272)*

$+\; [R_3Ta(O)]_x.$

Similarly $(\eta^5C_5H_5)TaCl_2[{=}CHC(CH_3)_3]$ reacts with propene to give 2-methyl-4,4-dimethylpent-1-ene in 86% yield, by the sequence of reactions shown below, again implying nucleophilic carbene [350] (attack at the most substituted carbon of the alkene

unit)

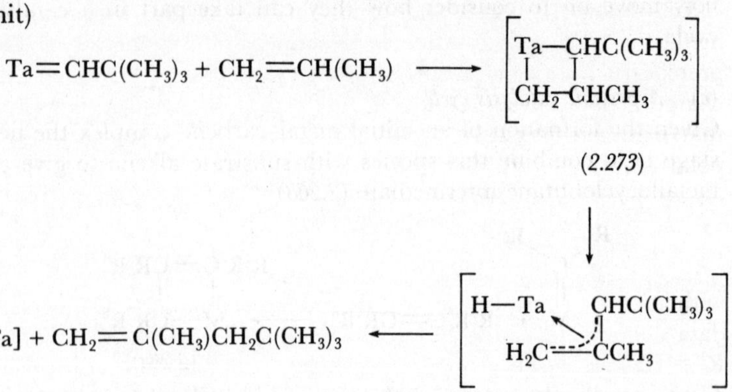

$$\text{Ta}=\text{CHC}(\text{CH}_3)_3 + \text{CH}_2=\text{CH}(\text{CH}_3) \longrightarrow \begin{bmatrix} \text{Ta}-\text{CHC}(\text{CH}_3)_3 \\ | \quad | \\ \text{CH}_2-\text{CHCH}_3 \end{bmatrix}$$

(2.273)

$$[\text{Ta}] + \text{CH}_2=\text{C}(\text{CH}_3)\text{CH}_2\text{C}(\text{CH}_3)_3 \longleftarrow \begin{bmatrix} \text{H}-\text{Ta} \quad \text{CHC}(\text{CH}_3)_3 \\ \text{H}_2\text{C}=\text{CCH}_3 \end{bmatrix}$$

On the other hand in complexes of the type $(CO)_5M=CR^1R^2$, where M = Cr, Mo, W; R^1 = aryl; R^2 = aryl or alkoxyl, the bulk of the experimental evidence [351,352], seems to favour the charge separation shown in (2.270).

A further complicating factor relating to the direction of initial addition is the steric requirement of the system. Viewed simply from the standpoint of the carbon ligand, the metallocyclobutane arrangement in (2.273) may appear sterically more constricted than that in the alternative (electrophilic) addition product (2.274).

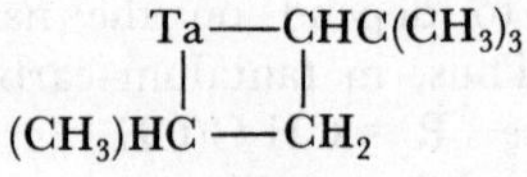

$$\begin{array}{c} \text{Ta}-\text{CHC}(\text{CH}_3)_3 \\ | \quad \quad | \\ (\text{CH}_3)\text{HC}-\text{CH}_2 \end{array}$$

(2.274)

However, given the presence of further bulky ligands co-ordinated to the metal, (2.273) may, in *total* complex terms, turn out to be the sterically preferred product. Indeed there is some evidence [48] to suggest that, in tungsten systems, the non-participative ligands tend to strongly direct any bulky groups away from the metal centre thus effectively favouring metallobutanes of the type (2.273) over those of the type (2.274).

The relative weight attached to electronic and steric factors will clearly depend on the detailed characteristics of the particular metathesis system under consideration and, although a number of detailed stereochemical rationales have been proposed [48,320,335], there is as yet no comprehensive theory to account for the overall stereochemistry of alkene metathesis reactions.

Yet another unresolved question associated with alkene metathesis reactions is whether π-complexation of the alkene is a prerequisite for metallocyclobutane formation. Assuming the charge separation shown in (2.270) it has been suggested [353] that metallocyclobutane formation involves a dipolar attack of a polarized metal-carbene on the alkene

$$L_n\bar{M}\!-\!\overset{+}{C}R^1R^2 \;+\; \!\!>\!\!C\!\!=\!\!C\!\!<\!\! \longrightarrow \;\; \begin{array}{c} >\!C\!\!-\!\!C\!\!< \\ | \quad\;\; | \\ L_nM\!-\!CR^1R^2 \end{array}$$

without the necessity of alkene pre-co-ordination. Conversely, data obtained with the $(CO)_5W\!\!=\!\!CR^1R^2$, $R^1 = R^2 = Ph$ or $R^1 = H$; $R^2 = Ph$, model systems indicate that alkene co-ordination precedes metallocycle formation

$$\begin{array}{c} R^1CR^2 \\ \| \\ W \end{array} \longleftarrow \begin{array}{c} \diagdown\!C\!\diagup \\ \| \\ \diagup\!C\!\diagdown \end{array} \longrightarrow \begin{array}{c} R^1R^2C\!-\!C\!- \\ | \quad\;\; | \\ W\!-\!C\!- \\ | \end{array}$$

In the light of the other transition-metal catalysed alkene reactions we have considered in previous sections, this latter route is more appealing. Metallocyclobutane formation can be envisaged as basically a ligand migration reaction occurring via a four-centre transition state analogous to those previously described for alkene hydrogenation (Section 2.1), alkene isomerization (Section 2.2) or alkene polymerization (Section 2.6):

$$\begin{array}{c} C \\ \| \\ M\!-\!\|\!C \\ C \end{array} \longrightarrow \left[\begin{array}{c} C\!\cdots\!C \\ \|\quad\| \\ M\!\cdots\!C \end{array} \right]^{\ddagger} \longrightarrow \begin{array}{c} C\!-\!C \\ | \quad\; | \\ M\!-\!C \end{array}$$

cf.

$$\begin{array}{c} \dot{H} \\ | \quad C \\ M\!-\!\| \\ \;\;\;\;C \end{array} \longrightarrow \left[\begin{array}{c} H\!\cdots\!C \\ \vdots\quad\| \\ M\!\cdots\!C \end{array} \right]^{\ddagger} \longrightarrow \begin{array}{c} H\!-\!C \\ | \quad\; | \\ M\!-\!C \end{array}$$

or

$$\begin{array}{c} R \\ | \quad C \\ M\!-\!\| \\ \;\;\;\;C \end{array} \longrightarrow \left[\begin{array}{c} R\!\cdots\!C \\ \vdots\quad\| \\ M\!\cdots\!C \end{array} \right]^{\ddagger} \longrightarrow \begin{array}{c} R\!-\!C \\ | \quad\; | \\ M\!-\!C \end{array}$$

As with both hydride and alkyl migration, the direction and rate of initial addition will depend on the nature of the catalyst; cf. the electrophilic nature of hydride in $CoH(CO)_4$ and the nucleophilic nature of the hydride in $CoH(CO)_3(PBu_3)$, the nature of the associated ligand system; cf. the use of asymmetric ligands in catalytic asymmetric hydrogenation or oligomerization, the nature of the substrate alkene; cf. the faster rate of hydrogenation of terminal compared to internal alkene, and, clearly intimately connected with that of the substrate alkene, the nature of the carbene ligand; cf. the nature of R in alkene polymerization.

The present state of knowledge does not allow a definitive statement on any of these aspects as far as the generality of alkene metathesis is concerned. However, in Fig. 2.41 a possible catalysis

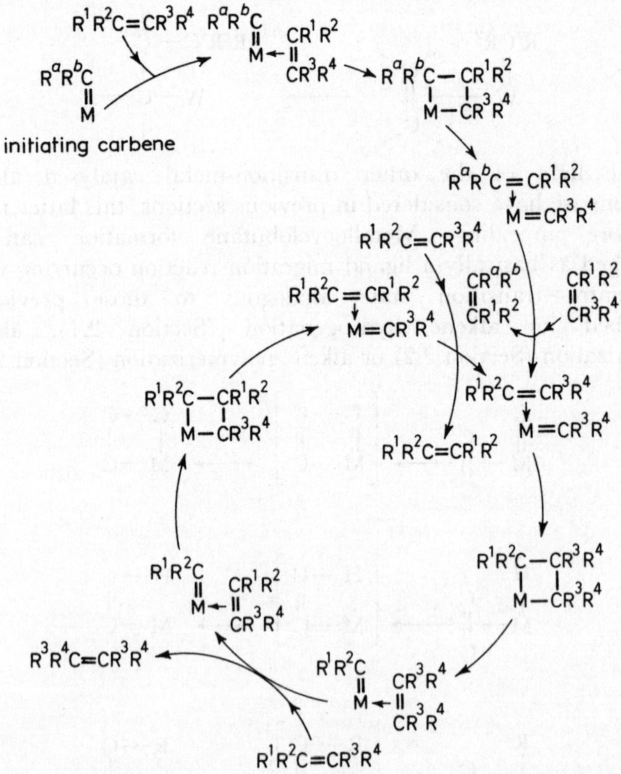

Fig. 2.41 *Generalized catalysis cycle for transition metal catalysed alkene metathesis*

cycle is shown which is consistent with the bulk of the presently available experimental data. Having glanced at the more esoteric, mechanistic side of alkene metathesis we can now turn our attention to some of the practical applications of this remarkable reaction.

2.8.2 Practical applications of alkene metathesis

Although most large-scale industrial applications of the alkene metathesis reaction involve heterogeneous catalyst systems [354], homogeneous systems are beginning to find some application in the production of smaller volume, high value, chemicals for use in the pharmaceutical, agrochemical and perfumery industries. In discussing practical applications basically two types of alkene substrate can be distinguished, namely acyclic and cyclic alkenes.

(a) *Acyclic alkenes as substrates*

The first commercial process incorporating alkene metathesis – the Phillips Triolefin Process – came on stream in 1966 [355]. This process was originally developed in the days of cheap, readily available propene, for the manufacture of ethene and butene

$$\begin{matrix} C \\ \diagdown \\ C \\ \| \\ C \end{matrix} + \begin{matrix} C \\ \diagup \\ C \\ \| \\ C \end{matrix} \rightleftharpoons \begin{matrix} C \diagdown \quad \diagup C \\ C{=}C \\ + \\ C{=}C \end{matrix}$$

However, with the present relative increased value of propene this particular process has largely become uneconomic. A more promising process which is expected to prove commercially attractive is that for producing isopentenes via the metathesis of a mixture of 2-methylpropene and but-2-ene (i.e. essentially the C_4 stream from a steam cracker)

$$\begin{matrix} C \\ \diagdown \\ C {\diagup} \| \\ \quad\; C \end{matrix} + \begin{matrix} C \\ \diagup \\ C \\ \| \\ C \diagdown \\ \quad C \end{matrix} \rightleftharpoons \begin{matrix} C \diagdown \quad \diagup C \\ \;\;\;C{=}C \\ C \diagup \;\;\; + \\ \quad C{=}C \\ \qquad\quad \diagdown C \end{matrix}$$

$$\begin{matrix} C \\ \diagdown \\ C {\diagup} \| \\ \quad\; C \end{matrix} + \begin{matrix} C \\ \diagup \\ C \\ \| \\ C \end{matrix} \rightleftharpoons \begin{matrix} C \diagdown \quad \diagup C \\ \;\;\;C{=}C \\ C \diagup \;\;\; + \\ \quad C{=}C \end{matrix}$$

$$\underset{C}{\overset{C}{>}}C\overset{C}{\underset{\|}{\|}} + \underset{C}{\overset{C}{\|}}C\overset{C}{\underset{}{<}} \rightleftharpoons \underset{C}{\overset{C}{>}}C=C\overset{C}{\underset{}{<}} + \underset{C=C}{\overset{}{}}$$

The product isopentenes find an important outlet as high octane fuel components, a market which is likely to grow with the move towards unleaded petroleum. The above equilibria can be shifted to the right, i.e. to the isopentene side, by continually removing co-product ethene.

The tactic of shifting metathesis equilibria by removing a volatile co-product can be used on preparative scale to prepare long chain internal alkenes, for example, tetradec-7-ene from oct-1-ene or hexacos-13-ene from tetradec-1-ene [354].

Co-metathesis of ethene and 4-methylpent-2-ene gives rise to 3-methylbut-1-ene and propene

$$\underset{C}{\overset{C}{>}}\underset{C}{\overset{C-C}{\underset{\|}{}}}\underset{}{} + \underset{C}{\overset{C}{\underset{\|}{}}}\underset{}{} \longrightarrow \underset{C}{\overset{C}{>}}\underset{C}{\overset{C-C=C}{}} + \underset{}{\overset{C=C}{}}$$

As described in Section 2.5.1*b*, with the aid of a tertiary-phosphine modified nickel catalyst propene can be dimerized to 4-methylpent-2-ene. Thus, the above metathesis sequence coupled with a propene dimerization reaction constitutes a catalytic route from propene plus ethene to 3-methylbut-1-ene

$$CH_3CH=CH_2 + CH_2=CH_2 \xrightarrow{\text{'cats'}} CH_3(CH_3)CHCH=CH_2 .$$

Polymerization of 3-methylbut-1-ene using, for example, a modified Ziegler catalyst (see Section 2.6) gives a high melting point speciality plastic material.

A recent application of alkene metathesis is in the synthesis of sex attractants. A problem in using conventional broad-spectrum insecticides is that not only do they kill harmful pests but they also tend to eliminate beneficial insects. Furthermore, to be effective the insecticide, while it is still active, and the insect must be in the same place at the same time. Since it is difficult to predict the exact whereabouts of a given insect, e.g. the harmful pest, in a given area (for instance a crop field) in a given time, such as the active life of the pesticide, it is usual to adopt a policy of 'over-kill' in applying conventional pesticides. If the pest in question could be persuaded to localize in a small area then much more effective and selective pest control would be possible. The problem is akin to co-reacting two substrates in catalysis. One of the functions of the catalyst is to bring the two substrates into range. Potential catalysts for pest control are pheromones, or sex attractants, which can be used to attract the pest to a given location. The pheromone appropriate to the common housefly is *cis*-tricos-9-ene, (*2.275*), and one of the more elegant synthetic routes to this product is via the cometathesis of dec-1-ene and tetradec-1-ene [356]:

$$\text{Me(CH}_2)_7\text{CH}=\text{CH}_2 + \text{Me(CH}_2)_{12}\text{CH}=\text{CH}_2 \longrightarrow \begin{array}{c}\text{Me(CH}_2)_7\text{CH} \\ \| \\ \text{Me(CH}_2)_7\text{CH}\end{array} + \begin{array}{c}\text{Me(CH}_2)_{12}\text{CH} \\ \| \\ \text{Me(CH}_2)_7\text{CH}\end{array} + \begin{array}{c}\text{Me(CH}_2)_{12}\text{CH} \\ \| \\ \text{Me(CH}_2)_{12}\text{CH}\end{array} + \begin{array}{c}\text{CH}_2 \\ \| \\ \text{CH}_2\end{array}$$

(*2.275*)

Metathesis gives a mixture of both the *cis* and *trans*-isomers, the *cis*-/*trans*-tricos-9-ene fraction can be isolated in 29% yield and the required *cis*-isomer separated with ca. 95% purity. Similarly, 2-methylheptadecane, the sex attractant of various species of tiger moths, can be prepared in 30% yield from 4-methylpent-1-ene and tetradec-1-ene by metathesis followed by hydrogenation.

As should be clear from the above brief discussion, the synthetic possibilities of catalytic metathesis of acyclic alkenes are extensive. The reaction is, however, not limited to acyclic alkenes. Cyclic alkenes can also participate.

(*b*) *Cyclic alkenes as substrates*

With the possible exception of cyclohexene, see below, cyclomonoalkenes in the range C_4 to C_{12} yield polymers, known as polyalkenamers, in the presence of alkene metathesis catalysts. For example, with the $WCl_6/C_2H_5OH/(C_2H_5)_2AlCl$ catalyst system cyclo-octene yields a polymeric material having a molecular weight in the range 200 000–300 000 [334]. As in the acyclic alkene metathesis reaction discussed above, a metal-carbene species is the essential chain carrier

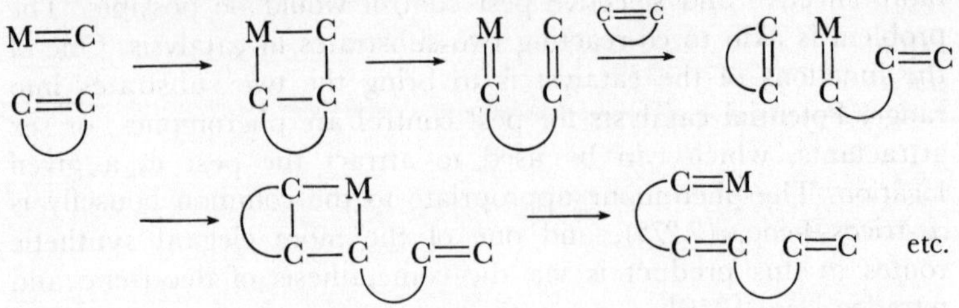

Only small quantities of the macrocyclic products, expected on the basis of a pairwise mechanism, i.e.

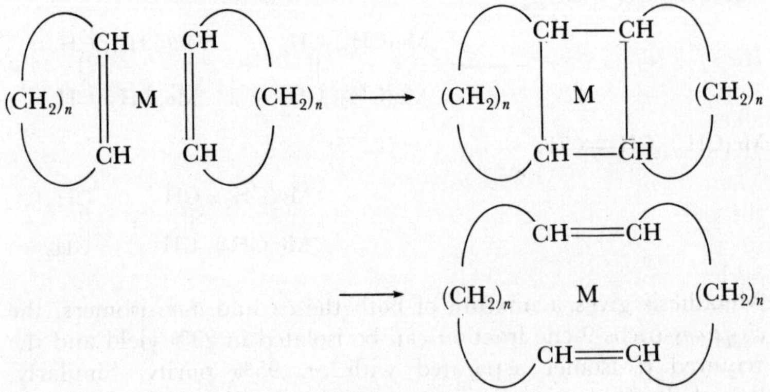

are formed and those which are, are thought to result from the propagating carbene species 'back biting' into growing polymer chain

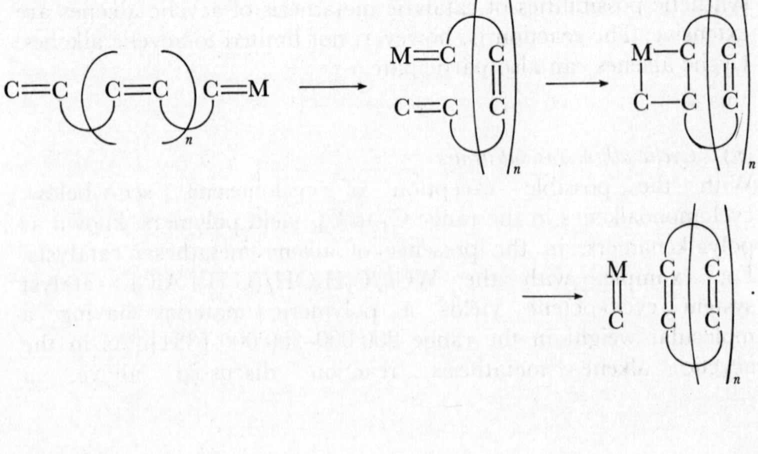

Introduction of acyclic alkenes into the polymerization system results in chain scission, via, for example, the sequence shown in Equation (2.67), and can thus be used to control the degree of polymerization

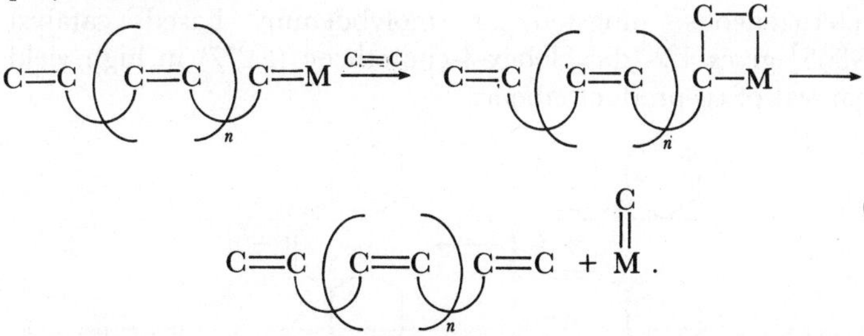

$$C=C(C=C)_n C=M \xrightarrow{C=C} C=C(C=C)_n \begin{matrix} C-C \\ | \quad | \\ C-M \end{matrix} \longrightarrow$$

$$C=C(C=C)_n C=C + \overset{C}{\underset{M}{\|}}. \qquad (2.67)$$

In acyclic alkene metathesis reactions the degree of stereospecificity, i.e. the selective production of either *cis* or *trans* alkene products, is generally, but not always, low. With cyclic alkenes as substrates high stereospecificity is quite common. Thus when diphenylcarbene(pentacarbonyl)tungsten is used to initiate the polymerization of cyclo-octene over 97% of the double bonds in the product polyoctenamer are *cis* [342]. Similarly, when cyclopentene is used as substrate it is possible, by careful choice of both catalyst and polymerization conditions, to obtain either essentially *cis*- [357] or essentially *trans*-polypentenamer [358]. The *cis*-polymer has good low temperature characteristics while the *trans*-polymer can be used as a substitute for natural rubber in the tyre industry. Commercial processes for the production of the latter polypentenamer have been developed by Bayer [359]. Polyalkenamerization of 2-methylocta-1,5-diene, (2.276), gives a polymeric material which is equivalent to that which would be obtained by the alternating copolymerization of isoprene and buta-1,3-diene, and as such can be used as a model system for this latter commercial copolymerization system [360].

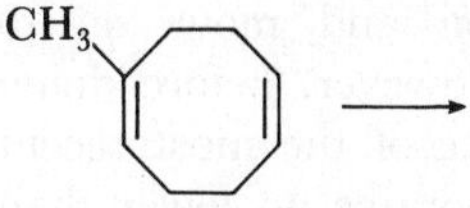

$$\underset{(2.276)}{\overset{CH_3}{\diagup}} \longrightarrow$$

$$(=CH-CH_2 \mathrel{\substack{|\\|}} CH_2 - \underset{\underset{\text{1 isoprene}}{|}}{\overset{CH_3}{\underset{|}{C}}} = CH-CH_2 \mathrel{\substack{|\\|}} CH_2 - CH=)_n$$

$$\tfrac{1}{2}BD \qquad\qquad\qquad \tfrac{1}{2}BD$$

As mentioned at the beginning of this sub-section cyclohexene does not participate in polyalkenamerization reactions. Its derivatives do, however, take part in metathesis reactions. One such derivative, namely 4-vinylcyclohex-1-ene, on metathesis over a heterogeneous tungsten or molybdenum based catalyst [361–363] gives 1,2-(dicyclohex-3-ene)ethene (*2.277*) in high yield on removal of co-product ethene.

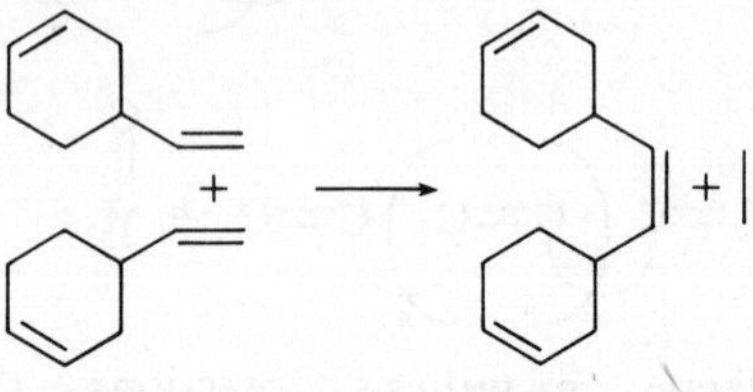

(*2.277*)

Since 4-vinylcyclohex-1-ene can be readily prepared from butadiene by nickel catalysed dimerization, see Section 2.5.2*a*, the above reaction constitutes half of a catalytic process for the preparation of (*2.277*) from butadiene. (*2.277*) has a number of specialist applications including, on bromination, as a flame retardent additive, on hydroxylation, use as a UV stabilizer for cellulose ester, and, after conversion to the ketone, as a perfume constituent.

The above examples have been chosen in an attempt to illustrate the present and potential synthetic scope of the alkene metathesis reaction. The range of possible products is almost as wide as the range of available substrates. Although we have concentrated on simple, i.e. non-hetero atom substituted alkenes, catalysts systems, albeit of low activity, are also known which will promote the metathesis of alkenes containing halogen, nitrogen, oxygen and even silicon substituents [354,367,368]. Metathesis is presently one of the most rapidly developing areas of catalysis and provides a suitable topic on which to leave our discussion of homogeneous catalyst systems in operation and move on to consider some presently in development. However, before doing so let us take a few minutes to describe one of the most recent large-scale industrial-processes which incorporates no fewer than three of the catalytic reactions we have discussed in this second section, namely oligomerization, isomerization and metathesis.

2.8.3 *The catalytic production of higher olefins*

The Shell Higher Olefins Process (SHOP) has been developed by the Shell Group of Companies for converting ethene to detergent-range (C_{11} to C_{14}) alkenes [369]. The process involves three distinct catalyst technologies. In the first stage ethene is oligomerized to give a mixture of α-olefins ranging from C_4 to C_{20+}. After separation of the C_{10} to C_{20} alk-1-enes, the remaining alkene fraction is fed over a heterogeneous isomerization catalyst. Extensive isomerization results in the formation of an almost statistical (thermodynamic) distribution of n-alkene isomers (cf. Section 2.2.1). This mixture, which is rich in lower alkenes, is now fed to a reactor containing a heterogeneous metathesis catalyst. Because of the preponderance of lower alkenes the resulting metathesis product tends to contain an excess of α-olefins which are formed by reactions of the type

$$\begin{matrix} C_{10}-C \\ \| \\ C_9-C \end{matrix} + \begin{matrix} C \\ \| \\ C \end{matrix} \longrightarrow \begin{matrix} C_{10}-C=C \\ + \\ C_9-C=C \end{matrix}$$

The desired detergent-range products (C_{11}-C_{14}) are separated by normal distillation. The remaining alkenes are recycled to the metathesis reactor; the C_{10-} products directly and the C_{15+} via the isomerization stage. By careful control of operating conditions excellent yields of high purity, typically 99% mono and 94–97% normal, α-olefins are obtained. The number of reaction variables available allow for the selective manufacture of a wide range of products. One of the main uses of C_{10} to C_{14} α-alkenes is in the production of detergent alcohols. This latter conversion is accomplished using homogeneous ligand modified cobalt hydroformylation/hydrogenation catalysts (cf. Section 2.4.1). Thus in going from ethene to detergent alcohols, half of the catalytic reactions we have described up to now are concerned. Furthermore, in the various process steps the split between the use of homogeneous and heterogeneous catalyst systems is 50/50. On that note it really is time to go on to look at some homogeneous catalyst systems in development.

*Further reading on alkene metathesis**

General texts and reviews
Bailey, G. C. (1969), Olefin disproportionation. *Catal. Rev.*, **3**, 37.
Khidekel, M. L., Shebaldova, A. D. and Kalechits, I. V. (1971),

---

*As will be apparent from the titles of some of the earlier review articles alkene metathesis was originally widely referred to as alkene disproportionation

Catalytic disporptionation of unsaturated hydrocarbons. *Russ. Chem. Rev.*, **40**, 669.

Banks, R. L. (1972), Catalytic olefin disproportionation. *Topics in Current Chem.*, **25**, 39.

Calderon, N. (1972), The olefin metathesis reaction. *Acc. Chem. Res.*, **5**, 127.

Cardin, D. J., Cetinkaya, B., Doyle, M. J. and Lappert, M. F. (1973), The chemistry of transition-metal carbene complexes and their role as reaction intermediates. *Chem. Soc. Rev.*, **2**, 99.

Evans, D. A. and Green, C. L. (1973), Insect attractants of natural origin. *Chem. Soc. Rev.*, **2**, 75.

Haines, R. J. and Leigh, G. J. (1975), Olefin metathesis and its catalysis. *Chem. Soc. Rev.*, **4**, 155.

Mol, J. C. and Moulijn, J. A. (1975), The metathesis of unsaturated hydrocarbons catalysed by transition metal complexes. *Adv. Catal.*, **24**, 131.

Streck, R. (1975), Olefin metathesis, a versatile tool for petro- and polymer-chemistry. *Chem. Ztg.*, **99**, 397 (German text).

Calderon, N., Ofstead, E. A. and Judy, W. A. (1976), Mechanistic aspects of olefin metathesis. *Angew. Chem. Int. Ed. Engl.*, **15**, 401.

Katz, T. J. (1977), The olefin metathesis reaction. *Adv. Organometal. Chem.*, **16**, 283.

Rooney, J. J. and Steward A. (1977), Olefin metathesis, in *Catalysis*, Vol. 1, Specialist Periodical Reports, Chem. Soc., London, p. 277

Calderon, N., Lawrence, J. P. and Ofstead, E. A. (1979), Olefin metathesis. *Adv. Organometal. Chem.*, **17**, 449.

3 Homogeneous catalyst systems in development

The distinction between homogeneous catalyst systems in operation and those in development is, of necessity, artificial, since most catalytic processes, and certainly those we have discussed in Chapter 2, are in a continual state of development. However, the stage of development of all the homogeneous catalysts we have considered up to now is such that these systems are presently capable of practical application. In this Chapter we will be concentrating on three areas of catalysis in which, although the problem is well defined, the solution in the form of a practical homogeneous catalyst is yet to be found. All three problems can be grouped under the general heading of activation of small molecules. The small molecules in question are nitrogen, carbon monoxide and methane. The problem areas are nitrogen fixation, reductive oligomerization/polymerization of carbon monoxide and alkane activation.

3.1 Nitrogen fixation

Nitrogen is one of the essential elements for all plant and animal life. It constitutes some 80% of the air we breathe. However, before it can be effectively used it must be 'fixed', i.e. converted into a form which plants and animals can easily assimilate. In biological systems this fixing process is carried out by certain bacteria using the enzyme nitrogenase and, in the natural nitrogen cycle, the compounds produced are stored by plants which animals, in turn, use as food. The animal breaks down these compounds, some of the stored nitrogen is converted into materials necessary for nutrition, while some is returned to the soil in the form of animal waste which, after bacterial conversion, is again utilized by the plants. With the development of intensive farming techniques around the turn of the century, necessary to meet the food demands of an ever increasing population, the need to augment the natural supply of nitrogen compounds became acute. This led to the development of the Haber–Bosch process in which ammonia is produced from nitrogen and hydrogen over a heterogeneous iron based catalyst at between 400 °C and 540 °C with synthesis pressures in the range 80 atm to 350 atm [370]. Since its inception in 1913 this process has brought world output of ammonia up to over 70 million tonnes per year. There are hundreds of Haber–Bosch plants throughout the world and in 1979, a further 30 plants, representing a capital investment of some $2500 million and a combined capacity of 10 million tonnes, came on stream [371]. About eighty per cent of the ammonia so produced is used directly or indirectly as fertilizer, the remaining twenty or so per cent being used in the manufacture of synthetic fibres, thermosetting resins and explosives.

Although over the years the Haber–Bosch process has been made more efficient through improved catalysts, better engineering etc., it remains essentially a high temperature, high pressure process requiring high inputs in terms of both energy and capital. With the increasing scarcity of both of these resources the search for a lower cost nitrogen fixation process, preferably operating at atmospheric pressure and ambient temperature, has intensified in recent years. The natural system, based on the nitrogenase, represents one such process and, in fact, still constitutes the major world source of agricultural nitrogen 'fixing' some 90 million tonnes of nitrogen per year [372].

Basically two main approaches have been adopted in the study of nitrogen fixation, i.e. the biological and the chemical approach. The former has concentrated on the natural nitrogen-fixing systems found in bacteria and has been concerned with elucidating the nature and *modus operandi* of nitrogenase proteins. The latter has concentrated on attempting to model the natural systems and on gaining a greater understanding of the way in which nitrogen bonds to, and reacts at, a metal centre. Clearly the two approaches are closely related and although in this section we shall restrict the discussion to the chemical approach, a number of references to the biological approach are included under 'further reading'.

3.1.1 Nitrogen bonded to a metal centre

Historically, molecular nitrogen, or dinitrogen as it is generally referred to in discussing nitrogen chemistry to avoid any possible confusion between N_2 and N, has been regarded as an 'inert' gas. Indeed, it is still widely used in preparative chemistry to provide an 'inert' atmosphere. Thus one of the first problems in transition-metal dinitrogen activation was persuading it to interact with the metal centre. This was first consciously accomplished in 1965 when Allen and Senoff isolated $[Ru(NH_3)_5(N_2)]^{2+}$ from the reaction between $RuCl_3.3H_2O$ and hydrazine in water [373]. Since then dinitrogen complexes involving most of the transition metals have been isolated [374].

A cursory examination of the electronic structure of the dinitrogen molecule suggests essentially two ways in which it can bond to a metal centre, i.e. via the lone pairs located on the nitrogen atoms to give end-on complexes of the types (*3.1*) and (*3.2*) or via the π-electrons of the NN triple bond to give side-on complexes of the types (*3.3*) or (*3.4*).

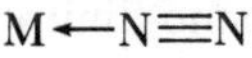

$$M \leftarrow N \equiv N \qquad M \leftarrow N \equiv N \rightarrow M$$
end-on end-on bridging
(*3.1*) (*3.2*)

$$M \leftarrow \overset{N}{\underset{N}{|||}} \qquad M \leftarrow \overset{N}{\underset{N}{|||}} \rightarrow M$$
side-on side-on bridging
(*3.3*) (*3.4*)

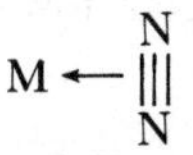

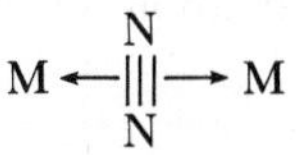

All four types of complex are known [374] although those containing end-on bonded dinitrogen are by far the most common.

The electronic structure of dinitrogen is similar to that of carbon monoxide and, as is the case with carbon monoxide, the bonding to the metal centre can be viewed as consisting of two components: a sigma component formed by transfer of electron density from the filled lone pair σ-orbital on the dinitrogen into vacant σ-orbital of the metal, and a π-component formed by transfer of electron density from a filled $d\pi$-, or hybrid $dp\pi$-, metal orbital into an empty $p\pi$-antibonding orbital of the dinitrogen.

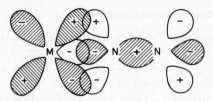

Fig. 3.1 σ and π-components of the dinitrogen–transition metal bond

Both bonding components, which are illustrated schematically in Fig. 3.1, result in a weakening of the nitrogen–nitrogen bond since they remove electron density from the N–N bonding system and inject electron density into the N–N antibonding system. However, compared to carbon monoxide, and indeed most other analogous σ-donor/π-acceptor ligands, dinitrogen is an extremely poor σ-donor. Its ionization potential is some 1.6 eV above that of carbon monoxide and at 15.6 eV is almost equal to that of argon (15.7 eV). As far as π-acceptor ability is concerned, spectroscopic data indicate that that of dinitrogen is moderate, lying between CO and RCN [374]. Theoretical calculations, using an extended Hückel treatment [375], are consistent with the spectroscopic findings and suggest that, at least for mono-dinitrogen complexes, the bonding is predominantly π-acceptor in character. Thus co-ordination of dinitrogen to a metal centre results in a net transfer of electron density from the metal to the dinitrogen ligand and thus to activation of the dinitrogen towards electrophilic attack.

3.1.2 Hydrogenation of bound nitrogen

Having activated the dinitrogen towards electrophilic attack, then, given N_2 hydrogenation is the goal, the obvious electrophile of choice is the proton. Several systems are now known in which protonation of a dinitrogen complex results in the formation of a reduced nitrogen molecule, e.g. N_2H_4 or NH_3. Thus treatment of cis-$W(N_2)_2(PMe_2Ph)_4$ with a fifteen-fold excess of sulphuric acid in methanol at 20 °C results in the almost quantitative (98%) formation of ammonia according to Equation (3.1) [376].

$$\text{cis-}W(N_2)_2(PMe_2Ph)_4 \xrightarrow{H_2SO_4/MeOH} 2NH_3 + N_2 + 4[PMe_2PhH]HSO_4$$
$$(3.5) \qquad\qquad\qquad + \text{ oxidized metal products}$$

(3.1)

With the analogous molybdenum complexes essentially the same behaviour is observed although the yield of ammonia is only about half that found with tungsten. In both cases small quantities of hydrazine are also produced.

As already mentioned, co-ordination of dinitrogen to a metal centre increases the electron density in the dinitrogen unit. However, in a mono-hapto complex such as (*3.5*), in which only one end of the dinitrogen ligand is attached to the metal, the resulting increased electron density is, as might be expected, not evenly distributed over the dinitrogen molecule but preferentially located at one of the two nitrogen centres. The clear resolution of the two nitrogen atoms in the X-ray photoelectron spectra of such complexes [374] together with the high intensity of $\nu(N_2)$ stretch in the infrared spectra confirm the presence of charge asymmetry. The experimental evidence suggests that the nitrogen atom, not co-ordinated to the metal centre, bears the greater negative charge [377]. Thus, in the sequence summarized by Equation (3.1) the first step involves protonation of the terminal nitrogen atom to give the N_2H species (*3.6*) which on further terminal nitrogen protonation gives the hydrazide (2^-), $(=N-NH_2)$, species (*3.7*). At this stage it appears that the reaction can follow one of two routes, depending on whether M = Mo or W. With M = Mo ammonia could be formed by the disproportional reaction shown in Equation (3.2), whereas in the case of M = W, in order to account for the formation of almost two moles of ammonia per mole of complex, some at least of the ammonia must be formed by further protonation of (*3.7*), i.e. via (*3.8*) [376,378]

$$M(N_2)_2L_4 \xrightarrow[-N_2]{+HX} MX(N_2H)L_4 \xrightarrow[-(LH)X]{+2HX} MX_2(NNH_2)L_3$$

$$(3.6) \qquad\qquad (3.7)$$

$$\xrightarrow[\text{(M = W or Mo)}]{HX} \qquad \Big|\begin{array}{c}+2HX\\-(LH)X\\(M=W)\end{array}$$

$$2NH_3 + \text{oxidized metal products} \xleftarrow{HX} MX_3(NHNH_2)L_2$$

$$(3.8)$$

$$Mo\equiv N=NH_2 \longrightarrow \text{'Mo'} + \tfrac{2}{3}N_2 + \tfrac{2}{3}NH_3. \qquad (3.2)$$

Complexes containing N_2H and NNH_2 groups have been isolated from this system and a number of NNH_2 and related, e.g.

NNH(Me), compounds have been fully characterized using single crystal X-ray crystallography.

When the dinitrogen ligand is symmetrically bridged between two metal atoms in a complex such as (*3.9*) [379,380], there is nothing to choose between the two ends of the bridging dinitrogen as far as the incoming proton is concerned. However, the reduction still occurs in a stepwise manner giving the monomeric species (*3.10*) in the first stage of the reaction [381]

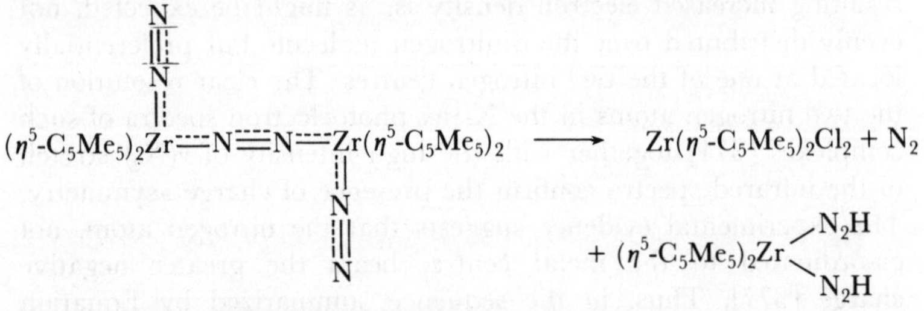

$$(\eta^5\text{-}C_5Me_5)_2Zr\text{=}N\text{≡}N\text{=}Zr(\eta^5\text{-}C_5Me_5)_2 \longrightarrow Zr(\eta^5\text{-}C_5Me_5)_2Cl_2 + N_2$$

$$+ (\eta^5\text{-}C_5Me_5)_2Zr\begin{smallmatrix}N_2H\\N_2H\end{smallmatrix}$$

(*3.9*)   (*3.10*)

Further protonation occurs in a manner analogous to that described for WX(N$_2$H)(PMe$_2$Ph)$_4$, (*3.6*) M = W, L = PMe$_2$Ph, although hydrazine, rather than ammonia is the predominant hydrogenated nitrogen containing product of the reaction

$$(3.10) \xrightarrow{\text{HCl}} Zr(\eta^5\text{-}C_5Me_5)_2Cl_2 + N_2H_4 + N_2.$$

^{15}N labelling experiments [381] have shown that essentially equal amounts of hydrazine are formed from both the bridging and non-bridging nitrogen ligands. So, at least in this system, as far as reactivity towards reduction is concerned, di co-ordinated dinitrogen appears to have little advantage over the mono co-ordinated type. Protonation of the dinitrogen bridged species (Ph$_3$P)$_2$H(Pr^i)Fe(N$_2$)FePr^i(PPh$_3$)$_2$ with the HCl at −50 °C gives small quantities, *ca.* 0.1 mol per N$_2$, of hydrazine and in this case symmetric protonation to give a diazene intermediate, L$_n$Fe=NH−NH=FeL$_n$, has been suggested [382] although as yet this system has not been fully characterized.

The idea of dual co-ordination being a requisite of dinitrogen activation is aesthetically appealing − almost the idea of tearing the dinitrogen molecule into two halves. However, few of the model systems described to data explicitly support this idea and, as suggested by Chatt *et al.* [374] the mono-metallic catalysis

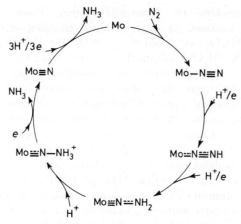

Fig. 3.2 *Hypothetical catalysis cycle for the reduction of molecular dinitrogen*

cycle shown in Fig. 3.2 could accommodate the catalytic reduction of molecular dinitrogen. There seems little doubt that more than one metal atom is involved in both biological nitrogen fixation [372] – nitrogenase contains six metal atoms, two molybdenum and four iron – as well as in most of the presently known homogeneous catalyse, see below. The function of such a multitude of metal centres could, however, be to provide the electrons necessary in the reduction rather than specifically to bind the dinitrogen. As indicated in Fig. 3.2, a total of six electrons are required to effect the N_2 to NH_3 transformation.

3.1.2 *Hydrogenation of atmospheric nitrogen*

In all of the systems described above, with the exception of the biological system to which we alluded, prior to hydrogenation the dinitrogen is present in a well-defined complex. None of the reactions is catalytic in that reduction ceases once the metal atom has reached its highest available oxidation state, i.e. M(VI) for M = Mo or W and Zr(IV) for the zirconium system. In order to produce a catalytic system it is clearly necessary to incorporate a reducing agent (the problem is the reverse of that encountered in the Wacker process, see Section 2.7.2b, where it is necessary to incorporate an oxidizing agent to return the inactive palladium(0) to catalytically active palladium(II)). Chemical reducing agents have been incorporated with molybdenum species with the aim of mimicking the biological system. Thus, in aqueous methanolic solution, a molybdenum complex derived from either $Na_2[Mo_2O_4(cys)_2]$, where cys = $[SCH_2CH(NH_2)CO_2]^{2-}$, or $K_2[MoO(CN)_4(H_2O)]$ will, in the presence of an excess of sodium borohydride, effect the reduction of dinitrogen gas to ammonia [383,384]. Although the yields are extremely low, seldom exceeding a third of a mole of hydrogenated nitrogen product per molybdenum atom, these systems do show several of the characteristics of natural nitrogenase, e.g. they will reduce ethyne to ethene and nitriles to hydrocarbons and ammonia [383]. Analogous molybdenum systems have also been developed incorporating either titanium(III) [385,386] which is readily oxidized to titanium(IV), or sodium amalgam as the reducing agent. Again under ambient conditions only low conversions are observed. However, under more forcing conditions, 90–110 °C and 100–150 atm N_2, conversions (moles of NH_3/N_2H_4 produced per mole of Mo) of the order of 100 have been reported [386].

These systems are often, perhaps somewhat unkindly, referred to as 'soups' since they generally incorporate several components in addition to the molybdenum complex and the reducing agent. To avoid undesired side reactions it is frequently necessary to work at very low concentrations, typically of 10^{-5} M Mo, and consequently the systems are sensitive to trace impurities with experimental reproducibility being a serious problem. Mechanistic conclusions are at best tenuous. Nevertheless, results obtained with such 'soups', taken together with data obtained with the discrete transition-metal complexes previously described, have unambiguously demonstrated that the reduction of dinitrogen under ambient conditions is a chemically feasible process which, on the right transition-metal site, can be carried out reasonably efficiently. The problem still remains of developing a practical nitrogen fixation system. Whether or not such a system could, within the next 20 or so years, successfully compete with the Haber–Bosch process in the large-scale production of ammonia for sophisticated markets is a moot point. Where such a process could make a significant contribution is in the small scale manufacture of fixed nitrogen in areas where the potential market together with the presently available infrastructure does not justify the construction of a 1000 tonnes per day – the generally accepted minimum viable size [371] – Haber–Bosch plant or in areas where transportation costs are prohibitive.

Further reading on nitrogen fixation

General texts and reviews

Chatt, J. and Leigh, G. H. (1972), Nitrogen fixation. *Chem. Soc. Rev.*, **1**, 121.
Allen, A. D., Harris, R. O., Loescher, B. R., Stevens, J. R. and Whiteley, R. N. (1973), Dinitrogen complexes of the transition metals. *Chem. Rev.*, **73**, 11.
Shilov, A. E. (1974), Fixation of nitrogen in solution in the presence of transition metal complexes. *Russ. Chem. Rev.*, **43**, 378.
Sellmann, D. (1974), Dinitrogen-transition metal complexes: synthesis properties and significance. *Angew. Chem. Internat. Ed. Engl.*, **13**, 639.
Schrauzer, G. N. (1975), Nonenzymatic simulation of nitrogenase reactions and the mechanism of biological nitrogen fixation. *Angew. Chem. Internat. Ed. Engl.*, **14**, 514.
Wentworth, R. A. D. (1976), Mechanisms for the reactions of molybdenum in enzymes. *Co-ord. Chem. Rev.*, **18**, 1.
Stiefel, E. I. (1977), The co-ordination and bioinorganic chemistry of molybdenum. *Prog. Inorg. Chem.*, **22**, 1.
Chatt, J., Dilworth, J. R. and Richards, R. L. (1978), Recent advances in the chemistry of nitrogen fixation. *Chem. Rev.*, **78**, 589.
Richards, R. L. (1979), Nitrogen fixation. *Educ. in Chem.*, **16**, 66.

3.2 Reductive oligomerization/polymerization of carbon monoxide

The rapid, and at times drastic, increase in the price of crude oil which has occurred over the last six years coupled with the realization that crude oil supplies, whether limited by technical or political factors, are likely to fall short of potential demand between 1985 and 1995 [387] has rekindled interest in coal both as a source of energy and as a feedstock for the petrochemical industry. At the current rate of energy consumption, known coal reserves are in excess of 250 years [388].

The most direct method of using coal as a primary energy source is combustion. However if coal is to augment and eventually replace crude oil in, for example, the production of transportation fuels and organic chemicals, it will be necessary to develop efficient ways of converting coal into gas, liquid hydrocarbon products, and preferably discrete organic chemicals. One method of at least partially accomplishing these goals is via the Fischer–Tropsch synthesis, in which synthesis gas – a mixture of carbon monoxide and hydrogen produced by burning coal in the presence of oxygen and steam – is converted into a wide range of hydrocarbon products.

The Fischer–Tropsch synthesis, which may be broadly defined as the reductive polymerization of carbon monoxide, can be schematically represented as shown in Equation (3.3)

$$CO + H_2 \xrightarrow{\text{'catalyst'}} \text{'CHO' products} \quad (3.3)$$

where the 'CHO' products are any organic molecules containing carbon, hydrogen and oxygen whose formation and presence is thermodynamically feasible under the synthesis conditions. With most heterogeneous catalysts the primary products are straight-chain alkanes, while the secondary products include branched-chain hydrocarbons, alkenes, alcohols, aldehydes and carboxylic acids [389]. Since its discovery in the early 1920s considerable progress has been made in developing heterogeneous catalyst systems for the Fischer–Tropsch synthesis and in 1976 over a quarter of a million tonnes of primary product, ranging from methane to paraffin wax, were produced in South Africa using this process [390].

As far as the conventional Fischer–Tropsch reaction is concerned there is little practical commercial incentive to develop a homogeneous catalyst system. However, from the standpoint of the chemical industry, the complex mixture of products obtained with present catalysts is generally unattractive owing to the economic constraints imposed by costly separation/purification processes. Thus there is considerable interest in developing selective catalysts for the selective reductive polymerization or, probably more realistically, oligomerization of carbon monoxide in the presence of hydrogen. It is in this area that homogeneous systems might be expected to make a significant contribution.

3.2.1 Model systems and possible mechanisms

In terms of the two basic molecular activation processes we discussed in Section 1.4, catalytic process involving carbon monoxide and hydrogen may be envisaged as involving both activation by co-ordination and activation by addition with the carbon monoxide undergoing the former and the hydrogen the latter. Adopting the view that in the transition-metal hydrido complex, resulting from hydrogen activation by addition, the hydrido moiety generally reacts as though it were present as an anionic hydride ligand (:H⁻) then, given the change separation

$$\overset{\delta+}{:}C\overset{\delta-}{\equiv}O$$

in carbon monoxide, initial addition could be pictured as involving hydride attack at the carbonyl carbon. Thus, with a transition-metal hydrido carbonyl complex having mutually *cis* CO and H, we might expect the proximity interaction, i.e. hydride migration, shown in Equation (3.4)

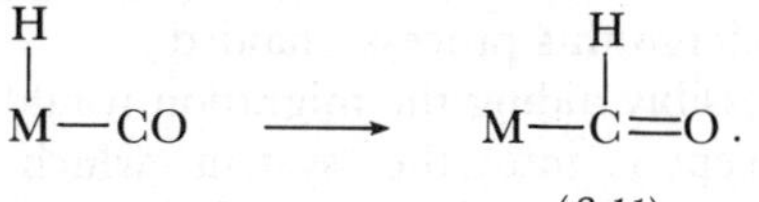

$$\underset{M-CO}{\overset{H}{|}} \longrightarrow \underset{M-C=O}{\overset{H}{|}}. \quad (3.4)$$

(3.11)

The feasibility of hydride attack at co-ordinated carbonyl has now been demonstrated in a number of model systems. Thus a wide range of anionic metal-formyl complexes have been prepared by treating metal carbonyl compounds with trialkoxyborohydrides, Equation (3.5) [391]. Neutral species can similarly be prepared starting from cationic metal carbonyls, e.g. Equation (3.6) [392,393]

$$[HB(OR)_3] + L_xM(CO) \longrightarrow B(OR)_3 + [L_xM(CHO)]^- \quad (3.5)$$

L = PPh$_3$, P(OPh)$_3$, CO; R = CH$_3$, CH(CH$_3$)$_2$; M = Fe, Cr, W

$$[(\eta^5\text{-}C_5H_5)Re(CO)(PPh_3)(NO)]^+ + [HB(C_2H_5)_3]^- \longrightarrow$$

$$(\eta^5\text{-}C_5H_5)Re(CHO)(PPh_3)(NO) + B(C_2H_5)_3. \quad (3.6)$$

Although inter-molecular hydride transfer leading to the formation of a metal-formyl complex is well established there are, as yet, no reports of the 'simple' intra-molecular hydride to carbonyl migration implied in Equation (3.4). Indeed there is some evidence to suggest that such a process would, for most transition metals, be highly, *ca.* 125 KJ mol⁻¹, endothermic. This is based on the dual observation that the superficially analogous process,

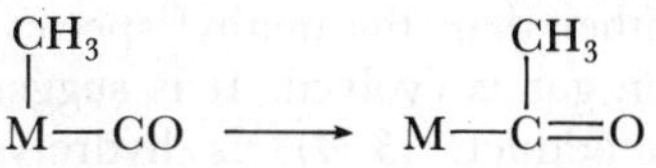

$$\underset{M-CO}{\overset{CH_3}{|}} \longrightarrow \underset{M-C=O}{\overset{CH_3}{|}}$$

which we discussed in Section 1.5, is, in many cases, approximately thermoneutral (± 20 KJ mol^{-1}) and that, for a given metal, the metal–hydride bond is some 125 KJ mol^{-1} stronger than the analogous metal–methyl bond [394]. Assuming that the metal–formyl and metal–acyl bonds are comparable in energy would imply that the intramolecular hydrido migration process of Equation (3.4) has an endothermicicity of *ca.* 125 ± 20 KJ mol^{-1}. Although it is premature to regard the reaction as universally endothermic none of the myriad of mononuclear hydrido-carbonyl species known has yet been found to show much inclination to undergo this process unaided.

One way of possibly aiding the migration would be to introduce an electron acceptor into the system which is capable of co-ordinating to the carbonyl oxygen. This could (*a*) reduce the overall electron density in the carbonyl unit and thus make the carbonyl carbon more susceptible to hydride attack and, (*b*) reduce the endothermicity of the process by the energy content of the carbonyl oxygen-electron acceptor bond. Thus the overall process now becomes

$$\begin{array}{c} H \\ | \\ M \leftarrow :C \equiv O + A \end{array} \longrightarrow \begin{array}{c} H \\ | \\ M - C = O: \rightarrow A. \end{array}$$

Whether such a sequence is actually operative in the reactions outlined in Equations (3.5) and (3.6) – in both cases the reaction co-product is an electron acceptor/Lewis acid, i.e. $B(OR)_3$ and $B(C_2H_5)_3$ – is not clear. What is clear is that incorporating an electron acceptor containing hydridic hydrogen leads to rapid reduction of co-ordinated formyl to co-ordinated methyl. Thus addition of BH_3 in THF to $(\eta^5\text{-}C_5H_5)Re(CHO)(CO)(NO)$, (*3.12*), results in the rapid, a few minutes at room temperature, formation of $(\eta^5\text{-}C_5H_5)ReCH_3(CO)(NO)$, (*3.13*) [392,395]. The overall reaction, i.e. reduction of co-ordinated carbonyl to co-ordinated methyl, can be accomplished using $NaBH_4$. Thus treatment with $[(\eta^5\text{-}C_5H_5)Re(CO)_2(NO)]^+$ with $NaBH_4$ in THF at 25 °C results in the essentially quantitative, 95%, formation of (*3.13*) [393]. The reaction is thought to proceed via a BH_3 adduct of the type (*3.14*) since, in the presence of water, the formyl complex (*3.12*), rather than the methyl species, is formed in 93% yield and hydrogen gas is evolved. It is suggested [393] that the intermediate BH_3 adduct, (*3.14*), is hydrolysed before further hydride transfer to give the methyl complex can occur.

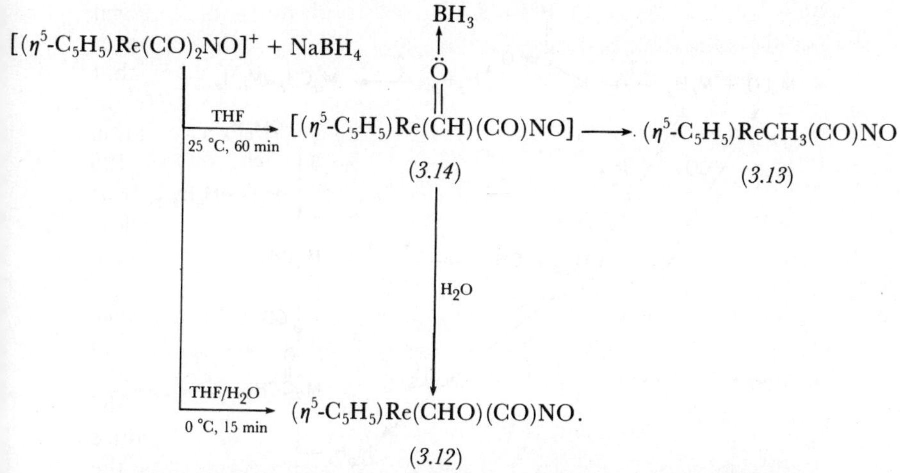

$[(\eta^5\text{-}C_5H_5)Re(CO)_2NO]^+ + NaBH_4$

$$[(\eta^5\text{-}C_5H_5)Re(CO)_2NO]^+ \xrightarrow[25\,°C,\,60\,min]{THF} [(\eta^5\text{-}C_5H_5)Re(\overset{\overset{BH_3}{\uparrow}}{\underset{\|}{\overset{O}{C}H}})(CO)NO] \longrightarrow (\eta^5\text{-}C_5H_5)ReCH_3(CO)NO$$

$$(3.14) \qquad\qquad (3.13)$$

↓ H_2O

$\xrightarrow[0\,°C,\,15\,min]{THF/H_2O}$ $(\eta^5\text{-}C_5H_5)Re(CHO)(CO)NO.$

$$(3.12)$$

Treatment of (3.12) with a further equivalent of $NaBH_4$, or alternatively treatment of $[(\eta^5\text{-}C_5H_5)Re(CO)_2NO]^+$ with two mole equivalents of $NaBH_4$ in aqueous THF, gives the hydroxymethyl species $(\eta^5\text{-}C_5H_5)Re(CH_2OH)(CO)NO$, (3.15) [393]. Schematically the sequence of reaction may be represented as

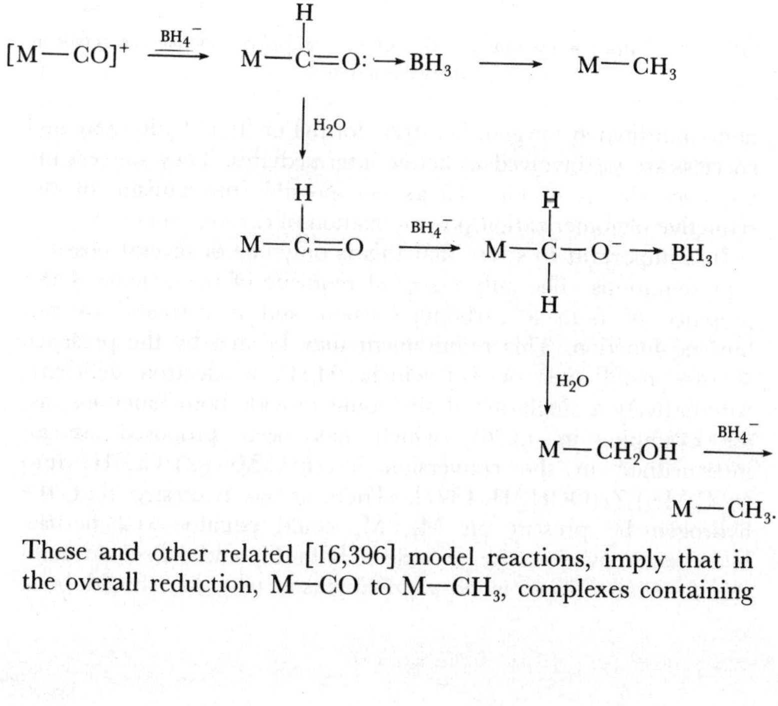

$$[M\text{—}CO]^+ \xrightarrow{BH_4^-} M\text{—}\overset{H}{\underset{}{C}}\!\!=\!\!O\!:\!\rightarrow BH_3 \longrightarrow M\text{—}CH_3$$

↓ H_2O

$$M\text{—}\overset{H}{\underset{}{C}}\!\!=\!\!O \xrightarrow{BH_4^-} M\text{—}\overset{H}{\underset{H}{C}}\text{—}O^- \rightarrow BH_3$$

↓ H_2O

$$M\text{—}CH_2OH \xrightarrow{BH_4^-} M\text{—}CH_3.$$

These and other related [16,396] model reactions, imply that in the overall reduction, M—CO to M—CH$_3$, complexes containing

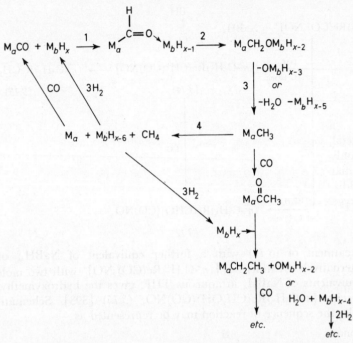

Fig. 3.3 *Possible mechanism for the reductive oligomerization/polymerization of carbon monoxide*

non co-ordinated oxygen, i.e. 'free' formyl or 'free' hydroxymethyl species, are *not* involved as active intermediates. They suggest the sequence shown in Fig. 3.3 as one possible mechanism for the reductive oligomerization/polymerization of carbon monoxide.

It is important to stress that this is only one of several possible representations. The only essential requisite of the scheme is the presence of both a carbonyl carbon and a carbonyl oxygen binding function. This requirement may be met by the presence of two metal sites one of which, M_bH_x, is electron deficient. Alternatively a single metal site could provide both functions, as, for example, in (3.16), which has been proposed as an intermediate in the conversion of $(\eta^5\text{-}C_5Me_5)_2Zr(CO)H_2$ into $(\eta^5\text{-}C_5Me_5)_2Zr(OCH_3)H$ [397]. There is no necessity that the hydrogen be present on M_b, M_a could equally well be the hydrogen activation site or indeed both M_a and M_b could fulfil this function. There is no *a priori* reason why M_b should be a

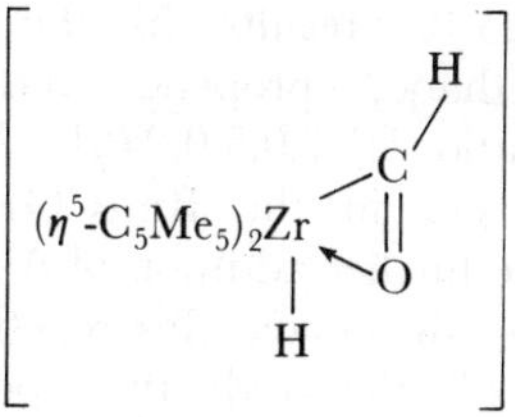

$$\left[(\eta^5\text{-}C_5Me_5)_2Zr \underset{H}{\overset{}{\diagdown}} \underset{O}{\overset{H}{\underset{\|}{C}}} \right]$$

(3.16)

metal, indeed rather an attractive possibility, as far as designing a catalyst system is concerned, is with $M_b = H^+$. In this case the initial reductive step could be envisaged as involving heterolytic addition of hydrogen to the metal centre with concurrent carbonyl oxygen protonation as shown in Equation (3.7)

$$M^n\text{---}CO + H_2 \longrightarrow \overset{H}{\underset{|}{M^{n+1}}}\text{---}COH^+ \longrightarrow M^{n+1}\overset{H}{\underset{|}{C}}\!=\!OH^+$$

(3.7)

With the possible exception of that involving $(\eta^5\text{-}C_5H_5)_2NbH_3$, all the model systems reported to date have involved hydride reducing agents involving elements which readily form very strong bonds with oxygen, e.g. B, Zr and Al (we will discuss the aluminium system shortly). Thus, an oxidized co-product of the type $O_xM_yH_z$ is formed which cannot readily be reduced with molecular hydrogen and the CO reduction is, at best, stoichiometric in the M_bH_x reductant. Clearly, in designing a catalyst based on the concept of dual co-ordination it will be necessary to choose M_b such that, while co-ordinating the oxygen function, it is not readily oxidized, i.e. M_b should be a noble metal possibly in a high oxidation state, or of course a proton when there is no problem attached to the reaction being stoichiometric in the reductant! The mechanism outlined in Scheme 3.1 may be described as step-wise reduction/insertion growth sequence, i.e. with growth termination occurring by hydrogenation of the metal-alkyl species or, in the case of MCH_2CH_3 and higher alkyl species, by β-elimination. An alternative growth sequence, which could be significant in the design of selective CO/H_2 catalyst systems, is that suggested by the products obtained from the reaction between certain transition-metal carbonyl complexes and alane [398].

Treatment of $Ru_3(CO)_{12}$ with a 15-fold excess of AlH_3 in tetrahydrofuran at 25 °C results in the rapid evolution of methane, ethene, ethane, propene and propane in the approximate molar ratio 1:1.7:0.5:0.2:0.1. Although only about 10% of the CO present in the $Ru_3(CO)_{12}$ is converted into hydrocarbons, with the further addition of AlH_3 or CO gas having no visible effect on the conversion, this reaction constitutes one of the first examples of the reductive polymerization, albeit stoichiometric, or co-ordinated carbon monoxide under ambient condition; a sort of room temperature Fischer–Tropsch reaction. However, possibly of greater significance is the hydrocarbon product distribution. Ethene is the single major product, making up some 52% of the total product volume. The concentrations of the other hydrocarbon products decreases on going from C_1 to C_3, following roughly the type of geometric progression expected on the basis of the step-wise sequence shown in Equation (3.8) [16,399].

$$MCO \xrightarrow[-H_2O]{H_2 \text{ reduction}} MCH_3 \xrightarrow{CO \text{ insertion}} M\overset{O}{\overset{\|}{C}}CH_3 \xrightarrow[-H_2O]{H_2 \text{ reduction}} MCH_2CH_3 \longrightarrow \text{etc.}$$

$$\downarrow \qquad\qquad\qquad\qquad\qquad\qquad\qquad\qquad \downarrow$$

$$CH_4 \qquad\qquad\qquad\qquad\qquad\qquad\qquad C_2H_6 \qquad (3.8)$$

With Group 6B metal carbonyls the selectivity to ethene is significantly increased. Thus treatment of $M(CO)_6$, where M = Cr, Mo, or W with a six-fold excess of AlH_3 in THF solution at 22 °C results in the selective (>95%) formation of ethene. With $Cr(CO)_6$ conversion of carbonyl ligands into ethene increases with increasing AlH_3 concentration reaching a maximum of 17 ± 3% when the $Cr(CO)_6$:AlH_3 molar ratio is 1:6. With a six-fold excess of alane ethene evolution is essentially complete within 15 min at 22 °C. With AlD_3 in place of AlH_3, C_2D_4 is formed. Similar results are obtained with $Mo(CO)_6$ and $W(CO)_6$. However, although the selectivity to ethene is equally high the maximum conversion, CO to ethene, is considerably lower (4% for molybdenum and 6% for tungsten).

These results strongly suggest the presence of an alternative growth sequence, or rather dimerization process, which leads exclusively to ethene. The experimental data is consistent with the intermediacy of carbenoid metal species which may be schematically represented as (*3.17*). Once formed such a species can be envisaged as either undergoing a reduction-carbonyl insertion-reduction sequence (path A) to give C_1, C_2 and C_3 products or undergoing dimerization (path B) to give ethene

$$(CO)_xM-CH_3 \xrightarrow{} (CO)_{x-1}M-\overset{\overset{O}{\|}}{C}-CH_3$$
$$(3.18) \qquad\qquad (3.19)$$

$$(CO)_xM=CH_2 \nearrow^A$$
$$(3.17) \quad {}_B\searrow$$

$$(CO)_xM\underset{CH_2}{\overset{CH_2}{\cdots\cdots}}M(CO)_x \longrightarrow 2M(CO)_x + C_2H_4$$

with CH$_4$ above (3.18) and C_2H_6 above $(CO)_{x-1}M-C_2H_5$

Hydrogenation of a carbene fragment and reduction of co-ordinated acetyl to ethyl, as implicit in path A, have been demonstrated using model systems [400,401] as has the carbene dimerization implicit in path B [402–404]. The high selectivity to ethene clearly results when path B predominates and that this is found with Group 6B metal carbonyls is, superficially at least, not too surprising since it is these metals which are most closely associated with metal carbene formation and reactions involving such intermediates (cf. alkene metathesis, Section 2.8).

The possible intermediacy of primary carbene species in CO oligomerization/polymerization reactions suggests yet another possible chain growth sequence. In the step-wise growth in path A of the above sequence the first stage, following primary carbene formation, involves reduction to the metal methyl intermediate (*3.18*) which can either undergo methyl to carbonyl migration to give (*3.19*) – chain growth, or further reduction to give methane – chain termination. There is some evidence to suggest that primary carbene units co-ordinated to transition metals can act as nucleophiles [349] (see Section 2.8.1*c*) and as such attack the carbonyl carbon of co-ordinated carbon monoxide [405]. This being the case the sequence shown in Fig. 3.4 may be proposed as a possible growth mechanism. Although illustrated as a bi-metallic mechanism there is, as previously pointed out, no *a priori* reason to include two discrete metals, merely the requisite to include at least two different types of co-ordination site in the catalyst system. In this sequence the first propagation step, i.e. formation of a C_2 unit, involves carbene attack on carbonyl

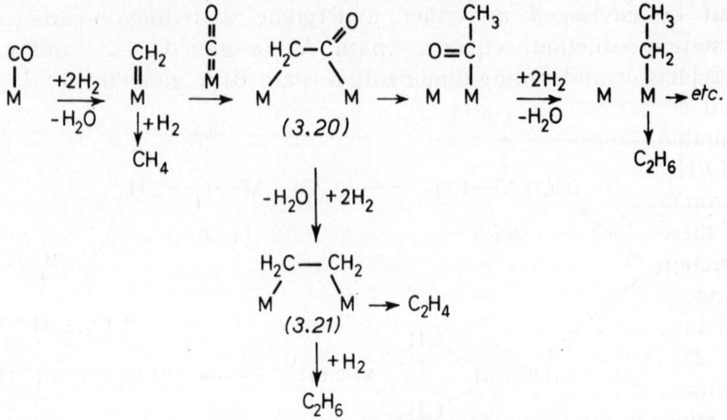

236   HOMOGENEOUS TRANSITION-METAL CATALYSIS

$$\begin{array}{c} CO \\ | \\ M \end{array} \xrightarrow[-H_2O]{+2H_2} \begin{array}{c} CH_2 \\ || \\ M \end{array} \xrightarrow{|+H_2} \begin{array}{c} CH_4 \end{array} \quad \begin{array}{c} O \\ || \\ C \\ || \\ M \end{array} \longrightarrow \begin{array}{c} H_2C-C{\diagup}^O \\ \diagup \quad \diagdown \\ M \quad\quad M \\ (3.20) \end{array} \longrightarrow \begin{array}{c} O=C \\ | \\ M \end{array} \begin{array}{c} CH_3 \\ | \\ M \end{array} \xrightarrow[-H_2O]{+2H_2} \begin{array}{c} \\ M \end{array} \begin{array}{c} CH_3 \\ | \\ CH_2 \\ | \\ M \end{array} \to etc.$$

$$\downarrow$$
$$C_2H_6$$

$$\downarrow {-H_2O \; |+2H_2}$$

$$\begin{array}{c} H_2C - CH_2 \\ \diagup \quad\quad \diagdown \\ M \quad\quad\quad M \\ (3.21) \end{array} \longrightarrow C_2H_4$$

$$\downarrow +H_2$$
$$C_2H_6$$

Fig. 3.4 *Alternative mechanism for the reductive oligomerization polymerization of carbon monoxide*

carbon rather than the methyl to carbonyl carbon migration suggested in Fig. 3.3. Also the methane formation in Fig. 3.4 differs from that of the other hydrocarbons in that it involves hydrogenation of a metal–carbene rather than a metal–alkyl bond. Since the rate of metal–carbene hydrogenation might be expected to differ from that of metal–alkyl hydrogenation the mechanism shown in Fig. 3.4 would imply that methane formation will strongly depend on the nature of M and will not necessarily conform to the statistical progression [16,399] expected in hydrocarbon (C_1 to C_n) formation. With most heterogeneous catalyst systems this is indeed the case and the methane make almost always exceeds that statistically expected [14]. The sequence illustrated in Fig. 3.4 would suggest that high methane production is not a necessary accompaniment and would further predict that, by correct choice of catalyst system, i.e. possibly one including the earlier transition metals, methane production could be radically reduced if not completely eliminated. This sequence would also suggest that catalyst systems capable of promoting the selective reductive dimerization of carbon monoxide are a realizable goal.

3.2.2 *Homogeneous catalysts*

In economic terms ethylene glycol is a very desirable product from the reaction between carbon monoxide and hydrogen. The first report of its production directly from syngas using a homogeneous catalyst appeared in 1951 when W. F. Gresham [406,407] of E. I. Du Pont de Nemours and Company claimed that certain cobalt complexes, e.g. cobalt acetate, dissolved in a suitable solvent, e.g. acetic acid, catalysed the conversion of CO/H_2 mixtures to polyfunctional oxygen-contained organic compounds. Ethylene glycol, higher polyhydric alcohols and esters of these alcohols were the principal products recovered. For this reaction pressures in excess of 1000 atm and temperatures in the range 180 °C to 300 °C were required. In a typical experiment a 2:1 mixture of $CO:H_2$, when heated under a pressure of 3000 atm at 225 °C to 246 °C for 1 h in the presence of acetic acid containing cobalt acetate (0.035 gcm^{-3}) gave 13.7 g of a mixture of ethylene glycol diacetate and glycerol triacetate. In 1974 R. L. Pruett and W. E. Walker, of Union Carbide Corporation, disclosed that under similar high pressure conditions, $ca.$ 3000 atm CO/H_2, certain soluble rhodium complexes were capable of catalysing the reaction between carbon monoxide and hydrogen to give as major product ($ca.$ 70%) ethylene glycol [408].

The catalyst precursor generally used for the reaction is rhodium dicarbonyl acetylacetonate. However, detailed infrared studies under the reaction conditions ($ca.$ 1000 atm CO/H_2 and 200 °C) have shown both the $[Rh(CO)_4]^-$ and the $[Rh_{12}(CO)_{34-36}]^{2-}$ anions to be present in various concentrations at different stages of the reaction [409,410]. It is suggested that rhodium carbonyl clusters, characterized as having three intense infrared absorptions at 1868 ± 10 cm^{-1}, 1838 ± 10 cm^{-1} and 1758 ± 10 cm^{-1}, are responsible for the catalysis [409] and it is believed that the reaction is dependent upon the existence of the following equilibria:

$$2[Rh_6(CO)_{15}H]^- \underset{H_2}{\rightleftharpoons} [Rh_{12}(CO)_{30}]^{2-} \underset{}{\overset{CO}{\rightleftharpoons}} [Rh_{12}(CO)_{34}]^{2-} \quad (3.9)$$

The addition of organic nitrogen ligands, such as 2-hydroxypyridine or piperidine, aids the establishment of the equilibria shown in Equation (3.9) possibly by acting as counter-ions (on protonation) for the anionic metal clusters [409].

A wide range of transition metal complexes have been screened as possible catalysts for the reaction but so far only systems containing rhodium have been reported to show any appreciable activity for the formation of polyhydridic alcohols. With respect to the organic products of the rhodium catalysed reaction, methanol and ethylene glycol are the major components, the ratio of the two appearing to depend on the nature of the organic ligand and on the CO/H_2 pressure. At a CO/H_2 (mole ratio 65/35) pressure in the range 1260–1400 atm with 2-hydroxypyridine as ligand or, after protonation, as counter-ion (reaction temperature 230 °C for 2 h) the ratio of ethylene glycol to methanol was found to be about 7:1 whereas with 3-hydroxypyridine as ligand/counter-ion, under the same conditions of temperature and pressure, the ratio decreases to about 2:1. Similarly, decreasing the CO/H_2 pressure from *ca.* 1300 atm to *ca.* 650 atm caused a decrease in the ethylene glycol:methanol ratio. The other minor products of the reaction include methyl formate, ethanol, glycerine and in some cases propylene glycol. It is claimed [408] that the process proceeds in the virtual absence of the methanation reaction, Equation (3.10)

$$CO + H_2 \rightleftharpoons CH_4 + H_2O. \tag{3.10}$$

Although, as yet, no mechanistic details of the process have been published it is tempting, in view of the high selectivity to ethylene glycol, to speculate on the intermediacy of hydroxyl carbene intermediates of the type (*3.22*) which, after dimerization followed by hydrogenation could yield the C_2 glycol

$$2M{=}C{\genfrac{}{}{0pt}{}{H}{OH}} \longrightarrow \begin{array}{c} CH(OH) \\ \| \\ CH(OH) \end{array} \xrightarrow{+H_2} \begin{array}{c} CH_2OH \\ | \\ CH_2OH \end{array}$$

(*3.22*)

As far as commercialization is concerned this homogeneous system suffers from the dual disadvantage of a high pressure requirement and a high initial catalyst investment and there are, as yet, no announcements of plans to produce ethylene glycol using such homogeneous rhodium cluster catalysts. These systems do, however, unequivocally demonstrate the possibility of selectively producing chemicals by the reductive oligomerization of carbon monoxide.

Other transition-metal cluster components, e.g. $Ir_4(CO)_{12}$, $Ru_3(CO)_{12}$ and $Os_3(CO)_{12}$, have been reported [16,411] to catalyse the hydrogen reduction of CO. However, under mild conditions (140 °C, CO/H_2 2 atm) the rates are extremely low (~1% conversion in 3–5 days with 3–5 catalyst turnovers) and under more forcing conditions (250 °C, CO/H_2 100 atm) there is considerable doubt attached to the homogeneity of these catalyst systems [412]. As must be clear from several of the preceding sections, carbon monoxide and hydrogen are two of the most widely studied molecules in homogeneous catalysis, cf. hydrogenation, hydroformylation, carbonylation, and the design of systems capable of persuading them to interact in a selective manner remains an exciting challenge. What we have tried to do in this section is highlight some of the current research, mechanistic proposals and speculative suggestions presently (1979) prevalent in this area.

Further reading on both heterogeneous and homogeneous CO/H_2 reactions

General texts and reviews

Storch, H. H., Golumbic, N. and Anderson, R. B. (1951), *The Fischer Tropsch and Related Synthesis*, New York, Wiley.

Nefedov, B. K. and Eidus, Ya. T. (1965), The development of catalytic synthesis of organic compounds from carbon monoxide and hydrogen. *Russ. Chem. Rev.*, **34**, 272.

Vannice, M. A. (1976), The catalytic synthesis of hydrocarbons from carbon monoxide and hydrogen. *Catal. Rev. -Sci. Eng.*, **14**, 153.

Henrici-Olivé, G. and Olivé, S. (1976), The Fischer–Tropsch synthesis: molecular weight distribution of primary products and reaction mechanism. *Angew. Chem. Internat. Ed. Engl.*, **15**, 136.

Denny, P. J. and Whan, D. A. (1978), The heterogeneously catalysed hydrogenation of carbon monoxide, in *Catalysis*, Vol. 2, p. 46, London: Chemical Society.

Masters, C. (1979), The Fischer–Tropsch reaction. *Adv. Organometal. Chem.*, **17**, 61.

Of the above, only the last reference deals specifically with homogeneous systems, the rest concentrate on heterogeneous CO/H_2 catalysts.

3.3 Alkane activation

Alkanes are among the most abundant of naturally occurring hydrocarbons. They make up an important part of most crude oils, the actual percentage depending on the particular crude type. Natural gas, world consumption of which corresponded to about 44% of total crude oil demand, consists almost exclusively of C_1 to C_4 alkanes. The natural abundance of alkanes bears witness to their chemical inertness and most chemical conversions involving such molecules tend to involve high temperature processes which often suffer from disappointingly low selectivities.

The development of active catalysts capable of selectively activating alkanes would make a significant contribution towards making the most efficient use of the world's hydrocarbon resources.

Although some progress has been made in developing homogeneous transition-metal catalyst systems capable of improving the selectivity of alkane oxidation reactions (see Section 2.7.1*b*), and although several homogeneous complexes are known which will effectively catalyse skeletal rearrangements in strained hydrocarbons (see Section 2.2.2), knowledge of the factors important in persuading simple, non-strained, alkanes to interact with metal centres is in its infancy. Basically two approaches have been adopted in studying sp^3C–H/metal interactions; one involves using hydrocarbon molecules containing 'activated' sp^3C–H bonds and the other, puristically more appealing, involves using simple alkanes, e.g. methane, ethane.

3.3.1 *Metal interaction with activated aliphatic C—H bonds*

(a) Steric activation

Steric activation of an aliphatic C–H bond involves bringing the sp^3C–H bond into frequent and close proximity to the metal centre by attaching it to the metal centre, via a co-ordinating unit, A. A can, in principle, be any moiety capable of bonding to the metal centre, e.g. a phosphorus atom, a double bond, an oxygen atom, or an aromatic ring. One of the first examples of this type of activation was reported [413] in 1965 when it was shown that the ruthenium(0) in Ru(dmpe)$_2$ (*3.23*), where dmpe = $(CH_3)_2PCH_2CH_2P(CH_3)_2$, can interact with a sp^3C–H bond in the dmpe ligand. The product was subsequently shown to be the dimeric alkyl-hydrido ruthenium(II) species (*3.24*) [414].

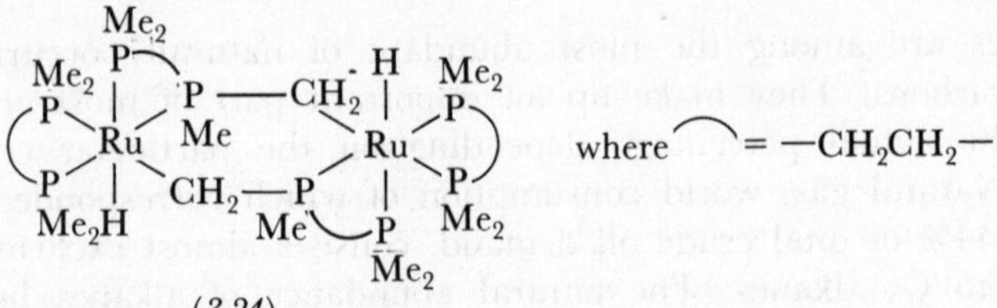

(*3.24*)

Formally this reaction may be seen as the oxidative addition of a sp^3C–H bond to the ruthenium (0) centre

$$\text{C—H} + \text{Ru}(0) \longrightarrow \overset{\text{C}}{\underset{|}{\text{Ru(II)—H}}} \qquad (3.11)$$

and as such may be likened to the oxidative addition of dihydrogen

$$\text{H—H} + \text{M}(n) \longrightarrow \overset{\text{H}}{\underset{|}{\text{M}(n+2)\text{—H}}}. \qquad (3.12)$$

As far as dissociation energy is concerned that of an sp^3C–H bond is the same (CH$_4$, D(C–H) = 435 KJ mol^{-1}) if not less (C$_2$H$_6$, D(C–H) = 410 KJ mol^{-1}) than that of H–H (D(H–H) = 435 KJ mol^{-1}). However, care must be taken in carrying this analogy too far, since, as pointed out in Section 3.2, the energy content of a M–C bond does not necessarily approximate to that of an M–H bond. Also entropy factors may be more significant in reaction of the type shown in Equation (3.11) than in those of the type illustrated in Equation (3.12).

Over the last few years many complexes containing alkyl-phosphine ligands have been found to undergo C–H insertion reactions analogous to that of the Ru(dmpe)$_2$ [415,416]. Increasing the steric bulk of the tertiary-phosphine ligand aids the insertion reaction. Thus, in refluxing 2-methoxyethanol, *trans*-PtCl$_2$(PBu$^t_2$Pr^n)$_2$ gives rise to the insertion product (*3.25*) in a matter of hours [417] whereas with *trans*-(PtCl$_2$(PPr$^n_3$)$_2$ no C–H insertion product is formed even after several weeks of refluxing.

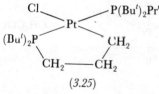

(*3.25*)

It is thought that both kinetic and thermodynamic effects are involved in the promoting influence of bulky substituents; not only do they force the alkyl C–H closer to the metal but they tend to favour a given conformation within the associated alkyl chain, i.e. the propyl unit in the case of the above example, such that the loss of freedom, i.e. decrease in internal rotational entropy, associated with forming the ring closed product is less than that which would be involved with a less restricted system, e.g. PtCl$_2$(PPr$^n_3$)$_2$ [418]. In the above examples, proof of metal/C–H interaction is provided by isolation of the insertion products, e.g. (3.24) and (3.25). An alternative, less direct but potentially experimentally more versatile method, is hydrogen for deuterium exchange. When dimeric complexes of the type (3.26), where L = a range of alkyl tertiary phosphine ligands, are heated in D_2O/CH_3CO_2D hydrogens of the alkyl groups are exchanged for deuterium [419].

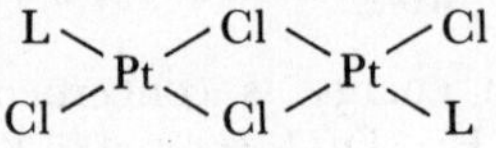

$$\text{L}\diagdown_{\text{Pt}}\diagup^{\text{Cl}}\diagdown_{\text{Pt}}\diagup^{\text{Cl}}$$
$$\text{Cl}\diagup\phantom{_{\text{Pt}}}\diagdown^{\text{Cl}}\diagup\phantom{_{\text{Pt}}}\diagdown_{\text{L}}$$
(3.26)

The exchange is regiospecific. Thus, with L = PPr_3 exchange occurs exclusively at the terminal carbon of the propyl chain; with L = PBu_3, although exchange occurs at both the terminal (C-4) and penultimate (C-3) carbon atoms, exchange at C-3 is some 11 times faster than at C-4; with L = PEt_3 no deuterium incorporation is observed even after prolonged reaction times. It is thought that exchange occurs via oxidative addition of an sp^3C–H bond at the metal centre accompanied by partial rupture of the chloro-bridge to give an intermediate of the type (3.27), for L = PPr_3. Reductive elimination of HCl from (3.27) followed by oxidative addition of DCl exchanges H for D and subsequent reductive elimination of DC would incorporate the deuterium into the alkyl chain.

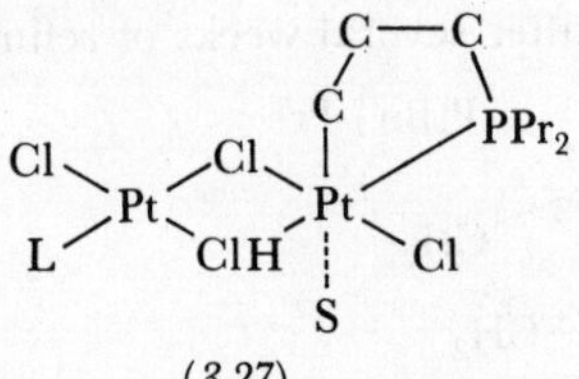

(3.27)   S = CH_3COOD or D_2O

The predominance of exchange at C-3 is taken to indicate the preference of the system to form 5-membered rings, i.e.

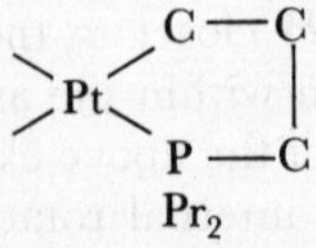

cf. (3.25). Although with this particular system the presence of bulky groups in the tertiary phosphine is not a prerequisite for C–H activation, i.e. H/D exchange, the presence of such groups has a marked influence on the rate of the exchange reaction. Thus the first order rate constant for H/D exchange at C-3 of the propyl group increases in the ratio 1:2.4:26.7 as L is changed from PPr_3 to PBu^tPr_2 to PBu^t_2Pr. With $L = PBu^t_2Pr$ deuterium incorporation also occurs into the tertiary butyl units indicating the formation of 4-membered ring intermediates.

Steric activation is not limited to the use of phosphorus as a co-ordinating centre. Indeed one of the first examples of H/D exchange at an sp^3 carbon utilized an aromatic ring as the co-ordinating unit [420]. Thus, in the presence of a homogeneous platinum(II) catalyst (e.g. $PtCl_4^{2-}$) in aqueous (D_2O) acetic acid (CH_3COOD), 1,1-dimethylpropylbenzene undergoes side chain deuteration exclusively at C-3. Similar results, i.e. preferential deuteration at C-5, were obtained with alkenes of the type (3.28), with $R = H$ or CH_3 [421].

$$R \overset{5}{-} CH_2 \overset{4}{-} CH_2 \overset{3}{-} \underset{\underset{CH_3}{|}}{\overset{\overset{CH_3}{|}}{C}} \overset{2}{-} CH \overset{1}{=} CH_2$$

(3.28)

In this particular system the geminal methyl groups at C-3 were originally included to prevent π-allyl formation, via β-hydrogen elimination, leading to alkene isomerization (see Section 2.2.1b). However, in the light of more recent work involving large ring metal chelate systems [418] is now appears likely that these groups actually promote activation at C-5 by favouring a particular conformation of the alkyl chain, much in the same way as the t-butyl groups do in the tertiary-phosphine systems previously discussed. Again the mode of activation is thought to involve oxidative addition of an sp^3C–H bond to the metal centre and, as with the $Pt_2Cl_4L_2$ systems, dimeric platinum complexes of the type (3.29) appear, from kinetic data, to play a significant role in the exchange reaction [421]. Unlike the tertiary-phosphine system, that which involves alkene is catalytic in platinum since, in the presence of excess alkene, there is a rapid exchange between free and co-ordinated alkene.

Another reaction sequence, which could be significant in developing homogeneous alkane activation catalysts, is that observed with the iridium analogue of (3.23), i.e. $[Ir(dmpe)_2]^+$. In

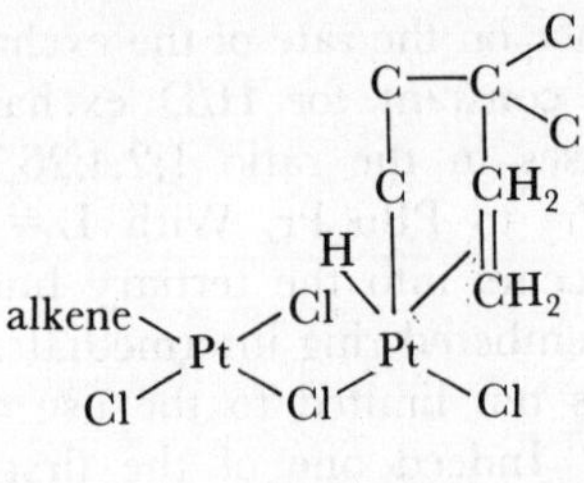

$$\text{alkene} \cdots \text{Pt}(Cl)(H)(Cl)\text{-Pt}(Cl)(Cl)\cdots \text{C-C-C-C-CH}_2\text{=CH}_2$$

(3.29)

solution this complex is in equilibrium with the C–H bond addition product (*3.30*) which, in the presence of carbon dioxide, gives rise to the carboxylate complex (*3.31*) [422]. Overall this sequence represents the conversion of a saturated C—H unit into a carboxylate anion and, at first sight, would seem to indicate the feasibility of Reaction (3.13)

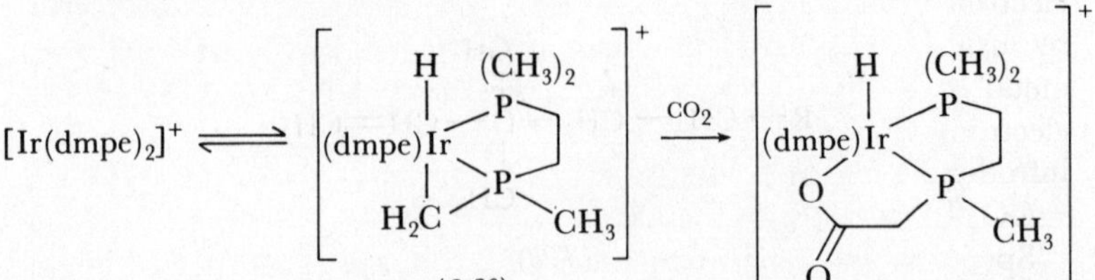

$$[Ir(dmpe)_2]^+ \rightleftharpoons \begin{bmatrix} (dmpe)Ir(H)(CH_2C)\cdots P(CH_3)_2/P\text{-}CH_3 \end{bmatrix}^+ \xrightarrow{CO_2} \begin{bmatrix} (dmpe)Ir(H)(O\text{-}C(=O)\text{-})\cdots P(CH_3)_2/P\text{-}CH_3 \end{bmatrix}^+$$

(*3.30*) (*3.31*)

$$RH + CO_2 \xrightarrow{\text{'catalyst'}} RCO_2H. \qquad (3.13)$$

There is, however, one slight problem – the change in Gibbs energy (ΔG) for Reaction (3.13), where RH represents most readily available alkane molecules, is strongly positive under most experimentally realistic reaction conditions, e.g. for RH = CH_4 $\Delta G°$ increases from 80.5 KJ mol^{-1} at 400 K to 150.2 KJ mol^{-1} at 1000 K. Thermodynamic feasibility is perhaps the major restriction in designing commercially useful alkane activation catalysts. Many sequences which appear attractive on the basis of known metal-alkyl or metal-hydrido-alkyl chemistry, e.g. Equations (3.14) or (3.15), fall down when the thermodynamic feasibility of the reaction is examined ($\Delta G°_{400} = 67.2$ KJ mol^{-1} for RH = CH_4 in Equation (3.14); $\Delta G°_{298} = 101$ KJ mol^{-1} for R = H in Equation (3.15))

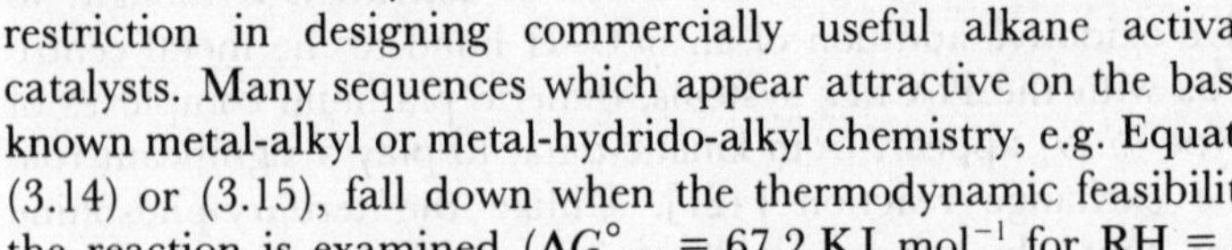

$$RH + M \rightarrow M\text{-}H \xrightarrow{CO} M(R)(H)\text{-}CO \rightarrow M(H)\text{-}C(=O)\text{-}R \rightarrow M + RC(=O)H \qquad (3.14)$$

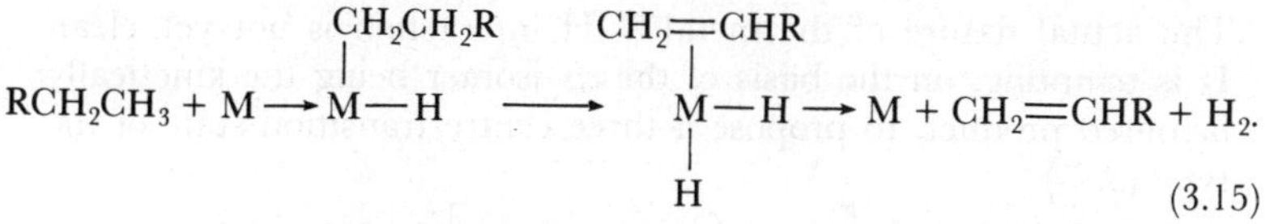

$$\text{RCH}_2\text{CH}_3 + \text{M} \longrightarrow \underset{\underset{\text{H}}{|}}{\overset{\overset{\text{CH}_2\text{CH}_2\text{R}}{|}}{\text{M}}}-\text{H} \longrightarrow \overset{\overset{\text{CH}_2=\text{CHR}}{|}}{\text{M}}-\text{H} \longrightarrow \text{M} + \text{CH}_2=\text{CHR} + \text{H}_2.$$

(3.15)

It is extremely important in alkane activation catalysis, as indeed in all catalysis, to ascertain the thermodynamic viability of a reaction before attempting to design a catalyst system. Chemical reactions, like water, do not take kindly to being made to move uphill!

(b) *Electronic activation*

On the supposition that oxidative addition provides one route to alkane activation such activation should be favoured by increasing the electron density at the metal centre and by decreasing the electron density in the sp^3C–H unit. The former can be achieved by using co-ordinatively unsaturated metal species having the metal in a low oxidation state, preferably zero, surrounded by electron donating, i.e. basic, ligands. The latter implies introducing electron withdrawing substituents, e.g. —CN, —NO$_2$, —CO$_2$R, —O—, into the alkane substrate.

Spectroscopic measurements [423] confirm the expected high electron density in the M(dmpe)$_2$ fragment of complexes of the type HMNp(dmpe)$_2$ (3.32), where M = Fe, Ru or Os; Np = 2-naphthyl and dmpe = bis-1,2-dimethylphosphinoethane. Fe(dmpe)$_2$, the iron analogue of (3.23) formed by reductive elimination of HNp from (3.32), readily reacts with activated CH$_3$X compounds (X = CN, —COR, —CO$_2$R, —SOR, —SO$_2$R) to form HFe(CH$_2$X)(dmpe)$_2$ complexes [424]. Generally the complex having mutually *trans* H and CH$_2$X groups is the final addition product. However, kinetic studies using ^{1}H n.m.r. spectroscopy have shown that the reaction proceeds through *cis*-HFe(CH$_2$CN)(dmpe)$_2$, (3.33), with subsequent isomerization giving the thermodynamically favoured *trans* isomer (3.34)

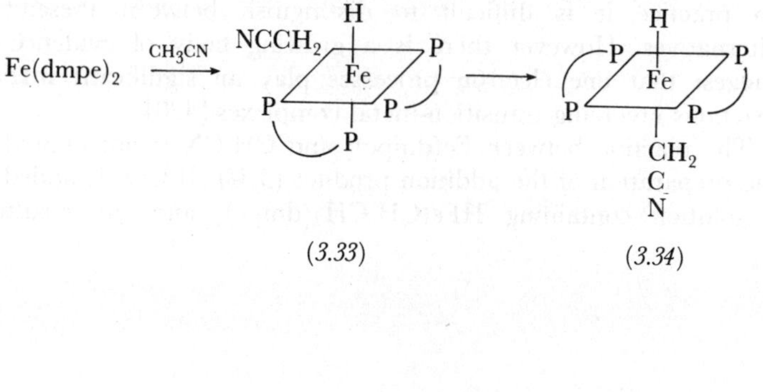

Fe(dmpe)$_2$ $\xrightarrow{\text{CH}_3\text{CN}}$

(3.33)   (3.34)

The actual nature of the metal/C–H interaction is not yet clear. It is tempting, on the basis of the *cis* isomer being the kinetically favoured product, to propose a three centre transition state of the type (*3.35*)

$$\left[(dmpe)_2Fe \begin{array}{c} \cdots H \\ \vdots \\ \cdots CH_2CN \end{array} \right]^{\ddagger}$$

(*3.35*)

However, in several complexes in which nonproductive interaction, i.e. metal/C–H interaction which does not result in addition, has been established the interaction appears to occur via a linear 3-centre 'bond', i.e. (*3.36*) rather than via the triangular equivalent (*3.37*) [425–427].

$$\begin{array}{c} \rangle C-H\cdots M \\ (3.36) \end{array} \qquad \begin{array}{c} \rangle C-H \\ \diagdown \diagup \\ M \\ (3.37) \end{array}$$

Furthermore, there is some evidence to suggest that the transition state of the metal plus C–H reaction is similar to that of the insertion reaction of a singlet carbene into a C–H bond [51]. For this latter reaction theoretical calculations [428,429] have indicated that insertion is a concerted reaction with CH_2 being attacked almost along the C–H axis. If the conclusions of these calculations may be extended to the M/C–H system a transition state of the type (*3.38*) would be preferred over (*3.35*) with the oxidative addition being envisaged as a stepwise homolytic process, i.e. initial hydrogen atom transfer to give (*3.39*) and ˙CH_2CH which rapidly undergo a radical recombination reaction giving (*3.33*).

$$[(dmpe)_2Fe\cdots H\cdots CH_2CN]^{\ddagger} \longrightarrow (dmpe)_2Fe^I\!\!-\!\!H + \, \dot{}CH_2CN \longrightarrow (3.33)$$

(*3.38*)   (*3.39*)

In practice it is difficult to distinguish between these two alternatives. However there is a growing body of evidence to suggest that one electron processes play an significant role in reactions involving transition-metal complexes [430].

The reaction between $Fe(dmpe)_2$ and CH_3CN is not limited to the preparation of the addition product (*3.34*). If CO_2 is added to a solution containing $HFe(CH_2CH)(dmpe)_2$ and the resulting mixture either heated or treated with Br_2 or I_2 then free cyanoacetic acid is liberated [424]:

$$CH_3CN + CO_2 \xrightarrow{Fe(dmpe)_2} CNCH_2CO_2H.$$

Another alkane activation reaction, which can lead to potentially interesting organic products and which is also catalytic, is transfer hydrogenation (see Section 2.1.4) in which an activated alkane, e.g. dioxane, is converted into the corresponding alkene. The thermodynamic feasibility of this later process is ensured by coupling the dehydrogenation with a suitable hydrogenation reaction [146–148]

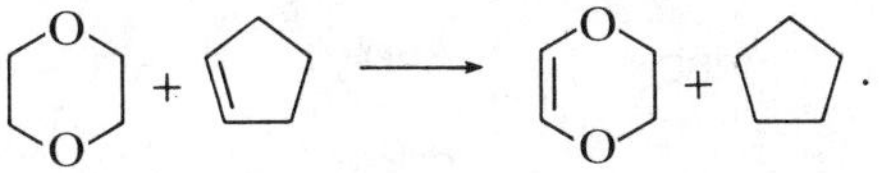

3.3.2 Metal interaction with unactivated aliphatic C–H bonds

Although all of the above systems demonstrate that a suitably activated saturated C–H unit can productively interact with a metal centre under mild conditions, the crucial experiment in this field is undoubtedly that which unequivocally demonstrates metal activation of a simple, unactivated alkane. Such an experiment was described in 1969 when it was reported that in the presence of $PtCl_4^{2-}$ simple alkanes undergo hydrogen for deuterium exchange in a $D_2O/CH_3CO_2D/DCl$ solvent at between 80 °C and 100 °C [431]. Originally there was some doubt surrounding the homogeneity of this system owing to the frequent presence of platinum metal under the reaction conditions, resulting from the disproportionation Reaction (3.16). However, subsequent work [51,415] has removed this ambiguity and there is now no doubt that Pt(II) in solution, rather than either platinum metal or Pt(IV), is the active catalyst

$$2Pt(II) \longrightarrow Pt(0)\downarrow + Pt(IV). \qquad (3.16)$$

The catalysis cycle shown in Fig. 3.5 is proposed for the reaction with $PtCl_2S_2$ where S = solvent, formed via the solvolysis equilibria Equations (3.17) and (3.18), being the major catalytically active species. The essential step in the alkane activation is oxidative addition of RH to (*3.40*) to give the platinum(IV) hydrido-alkyl species (*3.41*). As previously discussed it now appears likely that this 'oxidative addition' occurs via two

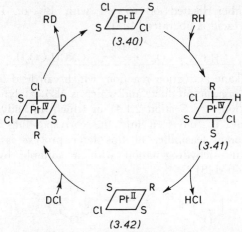

Fig. 3.5 *Catalysis cycle for platinum catalysed H/D alkane exchange (the configurations shown are illustrative rather than absolute)*

$$[PtCl_4]^{2-} + S \rightleftharpoons [PtCl_3S]^- + Cl^- \qquad (3.17)$$

$$[PtCl_3S]^- + S \rightleftharpoons PtCl_2S_2 + Cl^- \qquad (3.18)$$

one-electron steps rather than via a concerted, two-electron process. Subsequent reductive elimination of HCl followed by oxidative addition of DCl serves to exchange H for D in the alkyl fragment.

Examination of the deuterated products of these reactions show that the alkanes undergo multiple exchange, i.e. during a single contact with the metal centre the alkane can exchange more than one hydrogen atom. In the case of ethane and higher alkanes this multiple exchange can be rationalized by assuming that *(3.42)* R = CH_2CHR^1, undergoes a β-H elimination reaction to give an alkene intermediate of the type *(3.43)* which can then further exchange H for D. Multiple exchange when R = CH_3 is seen as occurring via an intermediate of the type *(3.44)*, i.e. a primary carbene species, resulting from an α-hydrogen elimination. An alternative to the π-alkene intermediate *(3.43)* is *(3.45)*, i.e. the direct analogue of *(3.44)*. Indeed the bulk of the available evidence [51], including the observation that introducing free alkene into the exchange medium effectively kills and H/D alkane exchange activity, favours the intermediacy of *(3.45)* over that of *(3.43)*.

$$\begin{array}{ccc}
\text{H} \quad \text{CHR}^1 & \text{H} & \text{H} \\
\diagdown \mid \diagup\!\!\!\!\diagup & \diagdown \mid \diagup\!\!\text{CH}_2 & \diagdown \mid \diagup\!\!\text{CHCH}_2\text{R}^1 \\
\text{Pt}\!\diagdown\!\text{CH}_2 & \text{Pt} & \text{Pt} \\
\diagup \mid \diagdown & \diagup \mid \diagdown & \diagup \mid \diagdown \\
(3.43) & (3.44) & (3.45)
\end{array}$$

An alternative suggestion [415] for the mechanism of multiple exchange involves the intermediacy of dimeric platinum species of the type (3.46) in which C–H addition has occurred at both platinum atoms.

$$\begin{array}{c}
C_nH_{2n} \\
\text{H}\diagdown\;\diagup\;\;\text{Cl}\diagdown\;\;\;\diagup\text{H} \\
\text{Pt}\quad\quad\text{Pt} \\
\text{S}\diagup\;\mid\;\diagdown\text{Cl}\diagup\;\mid\;\diagdown\text{S} \\
\text{Cl}\quad\quad\text{Cl} \\
(3.46)
\end{array}$$

Although there is presently little direct experimental data to support this suggestion, the involvement of dimeric platinum species in alkyl-tertiary-phosphine and alkyl-alkene H/D exchange systems (*vide supra*) lends some circumstantial evidence to the idea.

Alkane activation has come a long way since the first demonstration of metal/sp^3C–H interaction some ten years ago. However, as with most of the homogeneous catalyst systems in development which we have discussed in this chapter, the challenge still remains of extending what are up to now essentially academic observations and developing practical catalyst systems for the productive activation of, not only alkanes, but also nitrogen and synthesis gas.

Further reading on alkane activation

General texts and reviews

Parshall, G. W. (1970), The homogeneous catalytic activation of C–H bonds. *Accounts Chem. Res.*, **3**, 139.

Parshall, G. W. (1977), in *Catalysis*, Vol. 1, p. 335, London: Chemical Society.

Webster, D. E. (1977), Activation of alkanes by transition metal compounds. *Adv. Organometal. Chem.*, **15**, 147.

Shilov, A. E. and Shteinman, A. A. (1977), Activation of saturated hydrocarbons by metal complexes in solution. *Co-ord. Chem. Rev.*, **24**, 97.

Shilov, A. E. (1978), Activation of alkanes by transition metal complexes. *Pure & Appl. Chem.*, **50**, 725.

4 Where to now?

A better question is perhaps, where not to now? Although the number, variety and our understanding of homogeneous catalysts has increased markedly over the last few decades – most of the systems we have described have been developed in the last thirty years – there are still many areas where our knowledge is, at best, scanty with many exciting challenges still left to be taken up.

With the possible exception the rhodium, CO/H_2 to ethylene glycol, catalyst (Section 3.2.2) the homogeneous catalysts we have considered have involved essentially one transition-metal centre. As we have seen, most catalysis cycles involve the metal participating in a variety of processes. For example, in a hydroformylation sequence the metal centre must be able to break up the hydrogen molecule, co-ordinate the alkene, facilitate the migration of H to alkene, co-ordinate carbon monoxide, assist in the alkyl to carbonyl carbon transfer and finally help to combine the acyl and H moieties to give product aldehyde. Not content with requiring the metal to star in most, if not all, of these acts we then expect it, albeit with the assistance of the co-starring ligands, to carry out some of the steps in a highly specific way. Although there are frequently similarities between the individual processes, e.g. H to alkene and alkyl to CO migrations can both be thought of as nucleophilic attacks, it is stretching credibility to suggest that a given metal, say rhodium in the case of hydroformylation, is the optimum choice for all the individual steps in catalysis cycle. It may well be the best choice as far as the alkyl to CO migration step is concerned but perhaps another system could improve on the hydrogen activation step. Considerations such as these suggest the possibility of further refining catalyst systems by introducing two different metal

WHERE TO NOW?

centres into the catalysis cycle. It is now well established that ligands such as hydride [432], alkyl [433,434] and carbonyl [435] – all important intermediates in catalytic reactions – can freely move between different metal centres. Thus in chloroform at 0 °C, $PtMe_2(COD)$, where COD = cyclo-octadiene and $PdCl_2(PhCN)_2$ rapidly exchange chloro and methyl groups

$$PtMe_2(COD) + PdCl_2(PhCN)_2 \xrightarrow[CDCl_3]{0\ °C} PtCl(Me)(COD) + PdCl(Me)(PhCN)_2.$$

Not only can this exchange be detected spectroscopically (^{1}H n.m.r.) but also by observing reactions specific to one of the products of the exchange reactions. In the above case, addition of styrene to the reaction mixture results in the quantitative formation of β-methylstyrene [433]. In the absence of $PdCl_2(PhCN)_2$, neither $PtMe_2(COD)$ nor PtCl(Me)(COD) will methanate styrene. Similarly, in the presence of MeOH and $[Rh(CO)_2Cl]_2$, $PtMe_2(COD)$ gives rise to methyl acetate via transfer of methyl from platinum to rhodium followed by methyl to carbonyl migration on rhodium and methanolysis [436].

Combining the properties of two different transition metals in a catalysis cycle opens up some exciting possibilities. For example hydridochlorobis(cyclopentadienyl)zirconium, (*4.1*), reacts with an isomeric mixture of linear alkenes to selectively form the α-metallated species (*4.2*) [437].

$$(\eta^5\text{-}C_5H_5)_2Zr\begin{matrix}H\\Cl\end{matrix} + \text{\Large /\!\!\!\!\!\diagup\!\!\!\!\!\diagdown\!\!/} \xrightarrow[C_6H_6]{RT} (\eta^5\text{-}C_5H_5)_2Zr\begin{matrix}\text{\small /\!\!\!\!\!\diagup\!\!\!\!\!\diagdown\!\!/}\\Cl\end{matrix}$$

(*4.1*)   (*4.2*)

If this reaction could be combined with a rhodium hydroformylation system, as illustrated in the *hypothetical* catalysis cycle shown in Fig. 4.1, selective conversion of internal alkenes to terminal aldehydes might be possible.

Combining two different metals in one complex also suggest the possibility of being able to selectively activate a given site in a bi-functional substrate. For example, given a dialkene substrate of the type (*4.3*) and being posed the problem of catalytically hydrogenating only the substituted double bond we might conceive

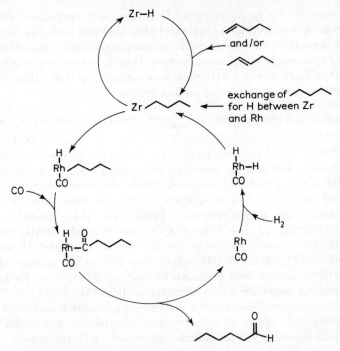

Fig. 4.1 *Hypothetical zirconium/rhodium hydroformylation catalysis cycles*

of a catalyst system of the type (*4.4*) in which M_a forms strong bonds with alkenes but is a poor hydrogenation catalyst, e.g. platinum(II) with tertiary phosphine ligands, and M_b not only co-ordinates alkenes but is also an effective hydrogenation catalyst, e.g. rhodium(I) with tertiary phosphine ligands. By tailoring the electronic and steric properties of the respective ligands systems, (*4.5*) could be favoured and, hypothetically at least, selective hydrogenation achievable.

$$R_1R_2C{=}C(CH_2)_xCH{=}CH_2 \qquad L_aM_a\text{\textemdash}M_bL_b$$

$$(4.3) \qquad\qquad (4.4)$$

$$\underset{L_aM_a\text{\textemdash}M_bL_b}{CH_2{=}CH(CH_2)_xC{=}CR_1R_2}$$

$$(4.5)$$

WHERE TO NOW?

Of course there is nothing easier than paper chemistry; nevertheless the above bi-heterometallic systems are only slight and indeed fairly obvious extensions of the present state of the art.

The participation of more than one metal centre in a catalytic reaction is central to the concept of cluster catalysis [435,438]. Transition-metal cluster compounds, i.e. discrete molecules or molecular ions in which two or more metal atoms are in a bonding interaction, exhibit several of the properties normally associated with metal surfaces [435,439]. For example, in metal carbonyl cluster compounds; the carbonyl ligands frequently undergo rapid exchange between the different mental atoms of the cluster, a phenomenon paralleling the mobility of chemisorbed species on metal surfaces; average metal–carbonyl bond energies are of the same order of magnitude as the heats of CO chemisorption on a metal surface; average metal–metal bond energies in dinuclear metal complexes closely resemble average metal–metal bond energies for the crystalline metal state. However, potentially more significant that these physical similarities are the different types of molecular activation possible in bi- or multimetal systems. Some of the various bonding interactions possible for carbon monoxide are shown in (4.5) to (4.8).

$$
\begin{array}{cccc}
\underset{M \quad M}{\overset{\overset{\displaystyle O}{\|}}{\underset{}{C}}} & \underset{M \mid M}{\overset{\overset{\displaystyle O}{\|}}{\underset{M}{C}}} & \underset{M \quad M}{\overset{C=O}{\underset{M}{}}} & \underset{M \quad M}{\overset{C=O}{}} \\
(4.5) & (4.6) & (4.7) & (4.8)
\end{array}
$$

As already mentioned (Section 3.2) interaction of the type exemplified by (4.7) and (4.8) may well be important in activating CO towards hydrogen reduction. A further possibility in complexes containing multiple metal–metal bonds is substrate activation by addition across the multiple bond with or without substrate bond rupture [38], e.g.

$$M{=}M + AB \longrightarrow \underset{M-M}{\overset{A \quad B}{| \quad |}} \qquad (4.1)$$

$$M{=}M + A: \longrightarrow M\overset{A}{\underset{}{\triangle}}M \qquad (4.2)$$

These processes are thought to closely resemble the activation process occurring on heterogeneous catalyst surfaces, and, since reactions occurring on clusters should be more amenable to detailed study than those taking place on a surface, it is hoped that a detailed study of transition-metal clusters will lead to a better understanding of heterogeneous systems. Furthermore there are several areas where heterogeneous catalysts are distinctly superior to their homogeneous counterparts, e.g. in activating C–H and C–C bonds in saturated hydrocarbons, and it has been suggested [435] that suitably designed cluster systems could serve to close this superiority gap.

Although metal clusters may serve as soluble analogues of heterogeneous catalysts, the ability to immobilize a homogeneous catalyst on a surface in such a way that the selectivity and activity advantages of the homogeneous system are preserved is of considerable commercial significance. Several approaches to this problem have been adopted [440,441] and, although in many of the cases reported up to now the results have been somewhat disappointing in terms of retention of homogeneous selectivity and activity, considerable progress has been made over the last ten years. The potential of such systems to combine the best of both the homogeneous and heterogeneous world is clear. Indeed, it is at this, and at other disciplinary interfaces that we might expect the most significant future advances to be made.

As chemical knowledge has expanded we have seen the emergence of a multitude of different chemical disciplines; electrochemistry, bio-inorganic chemistry, photochemistry, theoretical chemistry, homogeneous catalysis, heterogeneous catalysis. Each has evolved its own literature and jargon which, while possibly assisting intra-discipline communication, does little to aid inter-discipline transfer of ideas and approaches. Over the last decade considerable cross-fertilisation has occurred between heterogeneous and homogeneous catalysis and, as briefly described above, many of the existing new challenges in catalysis appear to lie almost midway between the two disciplines. Electrochemistry offers the transition-metal chemist the possibility of preparing and stabilising the metal in a particular oxidation state which may not be readily available by more conventional redox techniques [442,443]. Metals in such oxidation states could well exhibit novel catalytic activity. Alternatively, electrode reactions could be utilized to return a reduced or oxidized metal

WHERE TO NOW?

to a catalytically active state e.g. Pd(0) in the Wacker process or Mo(x) in a nitrogen fixation system [372]. The fruitfulness of close collaboration between bio-chemists and transition-metal chemists in designing metal systems capable of catalysing the reduction of dinitrogen has already been demonstrated [374].

Extension of such collaboration into the field of selective alkane oxidation is self-evident. Several recent publications have begun to explore the potential of photoexcited transition-metal complexes in catalysis, [444–446]. The incredibly rapid advances in computer technonology which have occurred in the last few years, coupled with the increasing sophistication of theoretical chemistry suggest the possibility of mathematically modelling complex catalytic reactions and gaining a greater understanding of processes occurring at an atomic level. We have concentrated on catalytic reactions involving the main block transition metals but even a cursory glance at the periodic tables reveals another, almost equally extensive, block, i.e. the lanthanides and actinides, of which our knowledge, in terms of catalytic properties, is minimal. What little is known of these elements suggest they have an exciting future in catalysis [447,448].

Advances in homogeneous catalysis over the last thirty years have been impressive. However, in many ways we are just beginning to scratch the surface of what is potentially achievable. We have come a long way but there is still no shortage of road left to cover. Like all arts, catalysis is continually evolving with only the physical laws of thermodynamics and the imagination of the chemist acting as constraints.

References

1. Pruett, R. L. (1979) *Adv. organometal. Chem.*, **17**, 1.
2. Ostwald, W. (1902) *Phys. Z.*, **3**, 313.
3. Emmett, P. H., Sabatier, P. and Reid, E. E. (1965) *Catalysis Then and Now*, Englewood Cliffs, New Jersey: Franklin.
4. Stull, D. R., Westrum, E. F. and Sinke, G. C. (1969) *The Chemical Thermodynamics of Organic Compounds*, p. 158, New York: Wiley.
5. Zeise, W. C. (1827) *Progg. Ann.*, **9**, 632.
6. Zeise, W. C. (1831) *Progg. Ann.*, **21**, 497.
7. Dewar, M. J. S. (1951) *Bull. Soc. Chim. France*, **18**, C79.
8. Chatt, J. and Duncanson, L. A. (1953) *J. chem. Soc.*, 2939.
9. Alderman, P. R. H., Owston, P. G. and Rowe, J. M. (1960) *Acta. Cryst.*, **13**, 149.
10. Fisher, E. O. and Werner, H. (1966) *Metal π Complexes*, Vol. 1 Amsterdam: Elsevier.
11. Cotton, F. A. (1964) *Inorg. Chem.*, **3**, 702.
12. Albano, V. G., Bellon, P. L. and Ciani, G. (1969) *J. chem. Soc., Chem. Commun.*, 1024.
13. Commons, C. J. and Hoskins, B. F. (1975) *Austral. J. Chem.*, **28**, 1663.
14. Colton, R. and Commons, C. J. (1975) *Austral. J. Chem.*, **28**, 1673.
15. Manassero, M., Sansoni, M. and Longoni, G. (1976) *J. chem. Soc., Chem. Commun.*, 919.
16. Masters, C. (1979) *Adv. organometal. Chem.*, **17**, 61 (and references therein).
17. Mason, R. and Meek, D. W. (1978) *Angew. Chem. Int. Ed. Engl.*, **17**, 183.
18. Halpern, J. (1968) *Advances in Chemistry Series No. 70*, p. 1, Washington: American Chemical Society.
19. James, B. R. (1973) *Homogeneous Hydrogenation*, New York: John Wiley (and references therein).
20. Collman, J. P. and Roper, W. R. (1968) *Adv. organometal Chem.*, **7**, 53 (and references therein).
21. Chatt, J. and Leigh, G. J. (1978) *Angew. Chem. Int. Ed. Engl.*, **17**, 400.

REFERENCES

22. Knowles, W. S., Sabacky, M. J. and Vineyard, B. D. (1970) *Ann. N.Y. Acad. Sci.*, *172*(9), 232.
23. Bogdanovic, B. (1979) *Adv. organometal Chem.*, **17**, 105 (and references therein).
24. Basolo, F., Chatt, J., Gray, H. B., Pearson, R. G. and Shaw, B. L. (1961) *J. chem. Soc.*, 2207.
25. Cramer, R. (1965) *Inorg. Chem.*, **4**, 445.
26. Tolman, C. A. (1977) *Chem. Revs.*, **77**, 313.
27. Tolman, C. A. (1970) *J. Amer. chem. Soc.*, **92**, 2953.
28. Immirzi, A., Musco, A. and Mann, B. E. (1977) *Inorg. Chim. Acta.*, **21**, L37.
29. Tolman, C. A. (1970) *J. Amer. chem. Soc.*, **92**, 2956.
30. Tolman, C. A., Seidel, W. C. and Gosser, L. W. (1974) *J. Amer. chem. Soc.*, **96**, 53.
31. Ugo, R. (1975) *Catal. Rev — Sci Eng.*, **11**, 225.
32. Birch, A. J. and Jenkins, I. D. (1976) In *Transition Metal Organometallics in Organic Synthesis*, ed. Alper, H. Vol. 1, p. 1, New York: Academic Press.
33. Cramer, R. (1968) *Accounts chem. Res.*,**1**, 186.
34. Deeming, A. J. and Shaw, B. L. (1969) *J. chem. Soc. (A)*, 1.
35. Shaw, B. L. and Stainbank, R. E. (1972) *J. chem. Soc., Dalton*, 223.
36. de Vries, B. (1962) *J. Catal.*, **1**, 489.
37. Chalk, A. J. and Harrod, J. F. (1968) *Adv. organometal. Chem.*, **6**, 119.
38. Chisholm, M. H. (1978) *Transition Metal Chem.*, **3**, 321 (and references therein).
39. Harrod, J. F., Ciccone, S. and Halpern, J. (1961) *Canad. J. Chem.*, **39**, 1372.
40. Evans, D., Osborn, J. A., Jardine, F. H. and Wilkinson, G. (1965) *Nature*, **208**, 1203.
41. James, B. R., Rattray, A. D. and Wang, D. K. W. (1976) *J. chem. Soc., chem. Commun.*, 792.
42. Belluco, U., Giustiniani, M. and Graziani, M. (1967) *J. Amer. chem. Soc.*, **89**, 6494.
43. Calderazzo, F. (1977) *Agnew. Chem. Int. Ed. Engl.*, **16**, 299 (and references therein).
44. Tsutsui, M. and Courtney, A. (1977) *Adv. organometal. Chem.*, **16**, 241.
45. Forster, D. (1979) *Adv. organometal. Chem.*, **17**, 255.
46. Marko, L. (1973) In *Aspects of Homogeneous Catalysis*, ed. Ugo, R., Vol. 2, Dordrecht: Reidel.
47. Cooper, N. J. and Green, M. L. H. (1974) *J. chem. Soc. chem. Commun.*, 761.
48. Calderon, N., Lawrence, J. P. and Ofstead, E. A. (1979) *Adv. organometal Chem.*, **17**, 449.
49. Tolman, C. A. (1972) *Chem. Soc. Revs.*, **1**, 337.
50. Coates, G. E., Green, M. L. H. and Wade, K. (1968) *Organometallic*

Compounds, 3rd edition, Vol. 2, *The Transition Elements*, London: Chapman and Hall.
51. Shilov, A. E. and Shteinman, A. A. (1977) *Coord. Chem. Rev.*, **27**, 97.
52. Parshall, G. W. (1978) *J. mol. Catal.*, **4**, 243.
53. Waddams, A. L. (1978) *Chemicals from Petroleum*, 4th edition, London: John Murray.
54. Lowery, R. P. and Aguilo, A. (1974) *Hydrocarbon. Proc.*, **53** (II), 103.
55. Chuisoli, G. P. and Salerno, G. (1979) *Adv. organometal Chem.*, **17**, 195.
56. Knowles, W. S., Sabacky, M. J. and Vineyard, B. D. (1974) *Homogeneous Catalysis*-11, eds. Forster, D. and Roth, J. F., p. 274, Washington: America Chemical Society.
57. Knowles, W. S. and Sabacky, M. J. (1974) U.S. Patent, 3, 849, 480.
58. Iguchi, M. (1939) *J. chem. Soc. Japan*, **60**, 1287.
59. James, B. R. (1979) *Adv. organometal Chem.*, **17**, 319.
60. Young, J. F., Osborn, J. A., Jardine, F. H. and Wilkinson, G. (1965) *J. chem. Soc., chem. Commun.*, 131.
61. Coffey, R. S. (1965) Imperial Chemical Industries, *Brit. Pat.* 1, 121, 642.
62. Osborn, J. A., Jardine F. H., Young, J. F. and Wilkinson, G. (1964) *J. chem. Soc., (A)* 1711.
63. Lehman, D. H., Shriver, D. F. and Wharf, I. (1970) *J. chem. Soc. chem. Commun.*, 1486.
64. Eaton, D. R. and Suart, S. R. (1968) *J. Amer. chem. Soc.*, **90**, 4170.
65. Hussey, A. S. and Takeuchi, Y. (1969) *J. Amer. chem. Soc.*, **91**, 672.
66. Hussey, A. S., and Takeuchi, Y. (1970) *J. org. Chem.*, **35**, 643.
67. Tolman, C. A., Meakin, P. Z., Lindner, D. L. and Jesson, J. P. (1974) *J. Amer. chem. Soc.* **96**, 2762.
68. Clark, H. C., Jablonski, C., Halpern, J., Mantovani, A. and Weil, T. A. (1974) *Inorg. Chem.*, **13**, 1541.
69. Crabtree, R. H., Felkin, H., Khan, T. F. and Morris, G. E. (1979) *J. organometal Chem.*, **168**, 183.
70. Halpern, J. (1975) In *Organotransition Metal Chemistry*, ed. Ishida, Y. and Tsutsui, M., p. 109, New York: Plenum.
71. Montelatici, S., van der Ent, A., Osborn, J. A. and Wilkinson, G. (1968) *J. chem. Soc., (A)*, 1054.
72. Horner, L., Buthe, H. and Siegel, H. (1968) *Tett. Letts*, 4023.
73. Jardine, F. M., Osborn, J. A. and Wilkinson, G. (1967) *J. chem. Soc. (A)* 1574.
74. Candlin, J. P. and Oldham, A. R. (1968) *Discuss. Faraday Soc.*, **46**, 60.
75. Iguchi, M. (1942) *J. chem. Soc. Japan*, **63**, 634.
76. Iguchi, M. (1942) *J. chem. Soc. Japan*, **63**, 1752.
77. Kwiatex, J., Mador, I. L. and Seyler, J. Y. (1963) *Adv. Chem.*, **37**, 201.
78. Simandi, L. and Nagy, F. (1965) *Proceedings of a Symposium on Coordination Chemistry in Tihany, Hungary*, p. 83, Budapest: Akademiani Kiado.

REFERENCES

79. Vlcek, A. A. (1962) *Proceedings of the 7th International Conference on Coordination Chemistry, Stockholm*, p. 286.
80. Halpern, J. and Wong, L. Y. (1968) *J. Amer. chem. Soc.*, **90**, 6665.
81. Kwiatek, J. (1967) *Catal. Rev.*, **1**, 37.
82. Kwiatek, J. and Seyler, J. K. (1965) *J. organometal Chem.*, **3**, 421.
83. Kwiatek, J. (1971) In *Transition Metals in Homogeneous Catalysis*, ed. Schrauzer, G. N., p. 13, New York: Marcel Dekker.
84. Kwiatek, J. and Seyler, J. K. (1968) *Adv. Chem.*, **70**, 207.
85. Burnett, M. G., Conolly, P. J. and Kemball, C. (1968) *J. chem. Soc. (A)*, 991.
86. Kwiatek, J., Mador, I. L. and Seyler, J. Y. (1962) *J. Amer. chem. Soc.*, **84**, 304.
87. Simandi, L. and Nagy, F. (1965) *Acta. Chim. Akad. Sci. Hung.*, **46**, 137.
88. Halpern, J., Harrod, J. F. and James, B. R. *J. Amer. chem. Soc.*, **88**, 5150 (1966).
89. Knifton, J. F. (1975) *J. org. Chem.*, **40**, 519.
90. Knifton, J. F. (1976) *J. org. Chem.*, **41**, 1200.
91. Knifton, J. F. (1975) *Tett. Letts*, 2163.
92. Tsuji, J. and Suzuki, H. (1977) *Chem. Letts*, 1085.
93. Vriends, R. C. J., van Koten, G. and Vrieze, K. (1978) *Inorg. Chim Acta*, **26**, L29.
94. James, B. R., Markham, L. D. and Wang, D. K. (1974) *J. chem. Soc. chem. Commun*: 439.
95. McQuillin, F. J. (1976) *Homogeneous Hydrogenation in Organic Chemistry*, Dordrecht: Reidel.
96. Hallman, P. S., McGarvey, B. R. and Wilkinson, G. (1968) *J. chem. Soc. (A)*, 3143.
97. O'Connor, G. and Wilkinson, G. (1968) *J. chem. Soc. (A)*, 2665.
98. Odom, H. C. and Pinder, A. R. (1972) *J. chem. Soc. Perkin I*, 2193.
99. Nishimura, S. and Tsuneda, K. (1969) *Bull. chem. Soc. Japan*, **42**, 852.
100. Frankel, E. N., Emken, E. A., Peters, H. M., Davison, V. L. and Butterfield, R. O. (1964) *J. org. Chem.*, **29**, 3292.
101. Miyake, A. and Kondo, H. (1968) *Angew. Chem. Int. Ed. Engl.* **7**, 880.
102. Frankel, E. N., Emken, E. A., Itatani, H. and Bailar, J. C. (1968) *J. org. Chem.*, **32**, 1447.
103. Wrighton, M. S., Ginley, D. S., Schroeder, M. A. and Morse, D. L. (1975) *Pure appl. Chem.*, **41**, 671.
104. Schroeder, M. A. and Wrighton, M. S. (1976) *J. Amer. chem. Soc.*, **98**, 551.
105. Nasielski, J., Kirsch, P. and Steinert, L. W. (1971) *J. organometal. Chem.*, **27**, C13.
106. Wrighton, M. S. and Schroeder, M. A. (1973) *J. Amer. chem. Soc.*, **95**, 5764.
107. Schroeder, M. A. and Wrighton, M. S. (1974) *J. organometal Chem.*, **74**, C29.

108. Platbrood, G. and Steinert, L. W. (1974) *J. organometal. Chem.*, **70**, 393.
109. Tayim, H. A. and Bailar, J. C. (1967) *J. Amer. chem. Soc.*, **89**, 4330.
110. Adams, R. W., Batley, G. E. and Bailar, J. C. (1968) *J. Amer. chem. Soc.*, **90**, 6051.
111. Bressan, G. and Broggi, R. (1968) *Chim Ind. (Milan)*, **50**, 1194.
112. Friedman, S., Metlin, S., Svedi, A. and Wender, I. (1959) *J. org. Chem.*, **24**, 1287.
113. Feder, H. M. and Halpern, J. (1975) *J. Amer. chem. Soc.*, **97**, 7186.
114. Stuhl, L. S., Dubois, M. R., Hirsekorn, F. J., Blecke, J. R., Stevens, A. E. and Muetterties, E. L. (1978) *J. Amer. chem. Soc.*, **100**, 2405.
115. Bennett, M. A., Huang, T.-N. and Turney, T. W. (1979) *J. chem. Soc., chem. Commun.*, 312.
116. Wender, I., Levine, R. and Orchin, M. (1950) *J. Amer. chem. Soc.*, **72**, 4375.
117. Goetz, R. W. and Orchin, M. (1962) *J. org. Chem.*, **27**, 3698.
118. Tucci, E. R. (1968) *Ind. Eng. Chem. Prod. Res. Dev.*, **7**, 32.
119. Tucci, E. R. (1968) *Ind. Eng. Chem. Prod. Res. Dev.*, **7**, 125.
120. Schrock, R. R. and Osborn, J. A. (1970) *J. chem. Soc., chem. Commun.*, 567.
121. Love, C. J. and McQuillin, F. J. (1973) *J. chem. Soc. Perkins I*, 2509.
122. Jardine, I. and McQuillin, F. J. (1970) *J. chem. Soc., chem. Commun.*, 626.
123. Morrison, J. D., Masler, W. F. and Newberg, M. K. (1976) *Adv. Catal.*, **25**, 81.
124. Pearce, R. (1978) In *Catalysis*, Vol. 2, p. 176, London: Chemical Society.
125. Brittain, R. T., Jack, D. and Ritchie, A. C. (1970) *Adv. Drug Res.*, **5**, 197.
126. Patil, P. N., Miller, D. D. and Trendelenburg, U. (1975) *Pharmocol. Rev.*, **26**, 323.
127. Elliot, M. and Janes, N. F. (1978) *Chem. Soc. Revs*, **7**, 473.
128. Fischer, C. and Mosher, H. S. (1977) *Tett. Letts*, 2487.
129. Knowles, W. S. and Sabacky, M. J. (1968) *J. chem. Soc., Chem. Commum.*, 1445.
130. Horner, L., Siegel, H. and Buthe, H. (1968) *Angew. Chem. Int. Ed. Engl.*, **7**, 942.
131. Knowles, W. S. and Sabacky, M. J. (1971) *German Patent* 2, 123, 063.
132. Knowles, W. S., Sabacky, M. J. and Vineyard, B. D. (1972) *German Patent* 2, 210, 938.
133. Knowles, W. S., Sabacky, M. J. and Vineyard, B. D. (1972) *J. chem. Soc., chem. Commun.*, 10.
134. Knowles, W. S., Sabacky, M. J. and Vineyard, B. D. (1973) *Ann. N.Y. Acad. Sci.*, **214**, 119.
135. Morrison, J. D., Burnett, R. E., Aguiar, A. M., Morrow, C. J. and Phillips, C. (1971) *J. Amer. chem. Soc.*, **93**, 1301.
136. Kagan, H. B. and Dang, T. P. (1972) *J. Amer. chem. Soc.*, **94**, 6429.

REFERENCES

137. Benes, J. and Hetflejs, J. (1976) *Coll. Czech. chem. Commun.*, **41**, 2264.
138. Tanaka, M. and Ogata, I. (1975) *J. chem. Soc., chem. Commun.*, 735.
139. Dang, T. P., Poulin, J. C. and Kagan, H. B. (1975) *J. organometal Chem.*, **91**, 105.
140. Hayashi, T., Tanaka, M. and Ogata, I. (1977) *Tett. Letts*, 295.
141. Fryzuk, M. D. and Bosnich, B. (1977) *J. Amer. chem. Soc.*, **99**, 6262.
142. Achiwa, K. (1976) *J. Amer. chem. Soc.*, **98**, 8265.
143. Hayashi, T., Katsumura, A., Konishi, M. and Kumada, M. (1979) *Tett. Letts*, 425.
144. Briegen, G. and Nestrick, T. J. (1974) *Chem. Revs*, **74**, 567.
145. Kolomnikov, I. S., Kukolev, V. P. and Volpin, M. E. (1974) *Russ. Chem. Rev. (Engl. Transl.)* **43**, 339.
146. Nishiguchi, T. and Fukuzumi, K. (1974) *J. Amer. chem. Soc.*, **96**,1893.
147. Nishiguchi, T., Tachi, K. and Fukuzumi, K. (1975) *J. org. Chem.*, **40**, 237.
148. Masters, C., Kiffen, A. A. and Visser, J. P. (1976) *J. Amer. chem. Soc.*, **98**, 1357.
149. Sasson, Y. and Blum, J. (1975) *J. org. Chem.*, **40**, 1887.
150. Descotes, G. and Sinou, D. (1976) *Tett. Letts*, 4083.
151. *The Petroleum Handbook* (1966) London: Shell International Petroleum Company Ltd.
152. Evans, D., Osborn, J. A. and Wilkinson, G. (1968) *J. chem. Soc. (A)*, 3133.
153. Piacenti, F., Pino, P., Lazzaroni, R. and Bianchi, M. (1966) *J. chem. Soc. (C)*, 488.
154. Tolman, C. A. (1972) *J. Amer. chem. Soc.*, **94**, 2994.
155. Bingham, D., Webster, D. W. and Wells, P. B. (1972) *J. chem. Soc., Dalton*, 1928.
156. Werner, H. and Feser, R. (1979) *Angew. Chem. Int. Ed. Engl.*, **18**, 157.
157. Casey, C. P. and Cyr, C. R. (1973) *J. Amer. chem. Soc.*, **95**, 2248.
188. Schroeder, M. A. and Wrighton, M. S. (1976) *J. Amer. chem. Soc.*, **98**, 551.
159. Harrod, J. F. and Chalk, A. J. (1966) *J. Amer. chem. Soc.*, **88**, 3491.
160. Maitlis, P. M. (1971) *The Organic Chemistry of Palladium*, Vol. 2, p. 128, London: Academic Press.
161. Green, M. and Hughes, R. P. (1976) *J. chem. Soc. Dalton*, 1907.
162. Davies, N. R. (1967) *Rev. pure appl. Chem.*, **17**, 83.
163. Fischer, E. O. and Held, W. (1976) *J. organometal Chem.*, **112**, C59.
164. Cardin, D. J., Cetinkaya, B., Doyle, M. J. and Lappert, M. F. (1973) *Chem. Soc. Revs*, **2**, 132.
165. Paquette, L. A. (1975) *Synthesis*, 347.
166. Bishop, K. C. (1976) *Chem. Revs*, **76**, 461.
167. Hammond, G. S., Turro, N. J. and Fischer, A. (1961) *J. Amer. chem. Soc.*, **83**, 4674.

168. Hoffmann, R. and Woodward, R. B. (1968) *Accounts chem. Res.* **1**, 17.
169. Hogeveen, H. and Volger, H. C. (1967) *J. Amer. chem. Soc.*, **89**, 2486.
170. Kaiser, K. L., Childs, R. F. and Maitlis, P. M. (1971) *J. Amer. chem. Soc.*, **93**, 1270.
171. Mango, F. D. (1969) *Adv. Catal.*, **20**, 291.
172. Cassar, L. and Halpern, J. (1970) *J. chem. Soc., chem. Commun.*, 1082.
173. Cassar, L., Eaton, P. E. and Halpern, J. (1970) *J. Amer. chem. Soc.*, **92**, 3515.
174. Paquette, L. A., Boggs, R. A. and Ward, J. S. (1975) *J. Amer. chem. Soc.*, **97**, 1118.
175. Blum, J., Zlotogorski, C. and Zoran, A. (1975) *Tett. Letts*, 1117.
176. Weiss, R. and Schlierf, C. (1971) *Angew. Chem.*, **83**, 887.
177. Peelen, F. C., Rietveld, G. G. A., Landheer, I. J., de Wolf, W. H. and Bickelhaupt, F. (1975) *Tett. Letts*, 4187.
178. Landheer, I. J., de Wolf, W. H. and Bickelhaupt, F. (1975) *Tett. Letts*, 349.
179. Roelen, O. (1938) *German Patent* 849548.
180. Roelen, O. (1948) *Angew. Chem.*, **60**, 62.
181. Reppe, W. (1953) *Liebigs Ann. Chem.*, **582**, 1 (publication delayed due to World War II).
182. Bird, C. W. (1962) *Chem. Revs*, **62**, 283.
183. Kunichika, S., Sabakibara, Y. and Nakamura, T. (1968) *Bull. chem. Soc. Japan*, **41**, 390.
184. Happel, J., Umemura, S., Sabakibara, Y., Blanck, H. and Kunichika S. (1975) *Ind. Eng. Chem. Prod. Res. Dev.*, **14**, 44.
185. Falbe, J. (1970) *Carbon Monoxide in Organic Synthesis*, New York: Springer-Verlag.
186. Kaliya, O. L., Temkin, O. N., Mekrykova, N. G. and Flid, R. M. (1971) *Doklady Akad. Nauk, S.S.S.R.*, **199**, 1321.
187. Cassar, L., Chiusoli, G. P. and Guerrieri, F. (1973) *Synthesis*, 509.
188. Thompson, D. T. and Whyman, R. (1971) In *Transition Metals in Homogeneous Catalysis*, ed. Schrauzer, G. N., p. 147, New York: Marcel Dekker.
189. Hayden, P. (1969) *British Patent*, 1, 148, 043.
190. Heck, R. F. (1974) *Organotransition Metal Chemistry, A Mechanistic Approach*, p. 201, New York: Academic Press.
191. Paulik, F. E. and Roth, J. F. (1968) *J. chem. Soc., chem. Commun.*, 1578.
192. Forster, D. (1969) *Inorg. Chem.*, **8**, 2556.
193. James, B. R. and Rempel, G. L. (1967) *J. chem. Soc., chem. Commun.*, 158.
194. Roth, J. F., Craddock, J. H., Hershman, A. and Paulik, F. E. (1971) *Chem. Technol.*, **23**, 600.
195. Forster, D. (1976) *J. Amer. chem. Soc.*, **98**, 846.
196. Cornils, B., Payer, R. and Traenckner, K. C. (1975) *Hydrocarbon Processing*, **54**(6), *83*.
197. Heck, R. F. and Breslow, D. S. (1961) *J. Amer. chem. Soc.*, **83**, 4023.

198. Wender, I., Feldman, J., Metlin, S., Gwynn, B. H. and Orchin, M. (1955) *J. Amer. chem. Soc.*, **77**, 5760.
199. Orchin, M. and Rupilius, W. (1972) *Catal. Rev.*, **6**, 85.
200. Whyman, R. (1975) *J. organometal Chem.*, **94**, 303.
201. Heck, R. and Breslow, D. S. (1960) *Chem. Ind. (London)*, 467.
202. Natta, G. (1955) *Brennstoff Chem.*, **36**, 176.
203. Marko, L. (1962) *Proc. chem. Soc., (London)*, 67.
204. Marko, L. and Szabo, P. (1961) *Chem. Tech. (Leipzig)*, *13*, 482.
205. Slaugh, L. H. and Mullineaux, R. D. (1966) *U.S. Patent* 3, 239, 569.
206. Slaugh, L. H. and Mullineaux, R. D. (1966) *U.S. Patent* 3, 239, 570.
207. Slaugh, L. H. and Mullineaux, R. D. (1968) *J. organometall. Chem.*, **13**, 469.
208. Paulik, F. E. (1972) *Catal. Rev.*, **6**, 49.
209. Tucci, E. R. (1970) *Ind. Eng. Chem. Prod. Res. Dev.*, **9**, 516.
210. Ugo, R., Pregaglia, G. F., Andretta, A. and Ferrari, G. F. (1971) *J. organometal. Chem.*, **30**, 387.
211. Henrici-Olivé, G. and Olivé, S. (1976) *Transition Metal Chem.*, **1**, 77.
212. Hershman, A. and Craddock, J. H. (1968) *Ind. Eng. Chem. Prod. Res. Develop.*, **7**, 226.
213. Schiller, G. (1956) *German Patent*, 953, 605.
214. Slaugh, L. H. and Mullineaux, R. D. (1961) *German Patent Appl.*, 1, 186, 455.
    Slaugh, L. H. and Mullineaux, R. D. (1966) *U.S. Patent* 3, 239, 566.
215. Osborn, J. A., Young, J. F. and Wilkinson, G. (1965) *J. chem. Soc., chem. Commun.*, 17.
216. Craddock, J. H., Hershman, A., Paulik, F. E. and Roth, J. F. (1969) *Ind. Eng. Chem. Prod. Res. Dev.*, **8**, 291.
217. Morris, D. E. and Tinker, H. B. (1972) *Chem. Technol.*, **24**, 554.
218. Evans, D., Yagupsky, G. and Wilkinson, G. (1968) *J. chem. Soc. (A)*, 2660.
219. Oliver, K. L. and Booth, F. B. (1970) *Hydrocarbon Processing* **49**(4), 112.
220. Wilkinson, G. (1978) *U.S. Patent* 4, 108, 905.
221. Pruett, R. L. and Smith J. A. (1969) *J. org. Chem.*, **34**, 327.
222. Yagupsky, G., Brown, C. K. and Wilkinson, G. (1970) *J. chem. Soc. (A)*, 1392.
223. Brewster, E. A. V. (1976) *Chem. Eng.*, **83**(24), 90.
224. Falbe, J. (1975) *J. organometal. Chem.*, **94**, 213.
225. Fowler, R., Connor, H. and Bachl, R. A. (1976) *Hydrocarbon Processing* **55**(9), 248.
226. *Chem. Eng.*, **84**(26), 110 (1977).
227. Hershman, A., Robinson, K. K., Craddock, J. H. and Roth, J. F. (1969) *Ind. Eng. Chem. Prod. Res. Dev.*, **8**, 372.
228. Alekseeva, K. A., Vysotski, M. D., Imyanitov, N. S. and Rybakov, V. A. (1977) *Zh. Vses, Khim. Oa.*, **22**, 45.
229. Sanchez-Delgado, R. A., Bradley, J. S. and Wilkinson, G. (1976) *J. chem. Soc., Dalton*, 339.
230. Hsu, C. Y. and Orchin, M. J. (1975) *J. Amer. chem. Soc.*, **97**, 3553.

231. Schwager, I. and Knifton, J. F. (1976) *J. Catal.*, **45**, 256.
232. Bailar, J. C. and Itatani, H. (1967) *J. Amer. chem. Soc.*, **89**, 1592.
233. Tayim, H. A. and Bailar, J. C. (1967) *J. Amer. chem. Soc.*, **89**, 3420.
234. Kawabata, Y., Hayashi, T. and Ogata, I. (1979) *J. chem. Soc., chem. Commun.*, 462.
235. Tanaka, M., Ikeda, Y. and Ogata, I. (1975) *Chem. Letts*, 1115.
236. Davidson, P. J., Hignett, R. R. and Thompson, D. T. (1977) in *Catalysis*, Vol. 1, p. 369. London: The Chemical Society.
237. Consiglio, G. and Pino, P. (1976) *Helv. Chim. Acta*, **59**, 642.
238. Lefebvre, G. and Chauvin, Y. (1970) *Aspects Homog. Catal.*, **1**, 107.
239. Jolly, P. W. and Wilke, G. (1975) *The Organic Chemistry of Nickel*, Vol. 2, New York: Academic Press.
240. Wilke, G. (1966) *Agnew. Chem. Int. Ed. Engl.*, **5**, 151.
241. Tolman, C. A. (1970) *J. Amer. chem. Soc.*, **92**, 4217.
242. Bogdanovic, B., Henc, B., Kermann, H. G., Nussel, H. G., Walter, D. and Wilke, G. (1970) *Ind. Eng. Chem.*, **62**, 34.
243. Karmann, H. G. (1970) Ph.D. Thesis, University of Bochum.
244. Keim, W. (1971) In *Transition Metals in Homogeneous Catalysis*, ed. Schrauzer, G. N., p. 59, New York: Marcel Dekker.
245. Tsuji, J. (1979) *Adv. organometal Chem.*, **17**, 141.
246. Brenner, W., Heimbach, P., Hey, H., Müller, E. W. and Wilke, G. (1969) *Justus Liebigs Ann. Chem.*, **727**, 161.
247. Maitlis, P. M. (1971) *The Organic Chemistry of Palladium* Vol. 2, p. 40 London: Academic Press.
248. Barker, R. (1973) *Chem. Revs*, **73**, 487.
249. Heimbach, P., Jolly, P. W. and Wilke, G. (1970) *Adv. organometal Chem.*, **8**, 29.
250. Medema, D. and van Helden, R. (1971) *Rec. Trav. Chim. Pays-Bas.*, **90**, 304.
251. Medema, D. and van Helden, R. (1971) *Rec. Trav. Chim. Pays-Bas.*, **90**, 324.
252. Takahashi, S., Yamazaki, H. and Hagihan, N. (1968) *Bull. chem. Soc. Japan*, **41**, 254.
253. Hidai, M., Ishiwatari, H., Yagi, H., Tanaka, E., Onozawa, K. and Uchida, Y. (1975) *J. chem. Soc., chem. Commun.*, 170.
254. Cramer, R. (1967) *J. Amer. chem. Soc.*, **89**, 1963.
255. Cramer, R. (1968) *Accounts chem. Res.*, **1**, 186.
256. Su, A. C. L. (1979) *Adv. organometal. Chem.*, **17**, 269.
257. Bogdanovic, B., Henc, B., Lösler, A., Meister, B., Pauling, H. and Wilke, G. (1973) *Angew. Chem. Int. Ed. Engl.*, **12**, 954.
258. *Chemicals Information Handbook, 1977–78* (1979) London: Shell International Chemical Company Ltd.
259. Ziegler, K., Holzkamp, E., Breil, H. and Martin, H. (1955) *Angew. Chem.*, **67**, 541.
260. Wilke, G. (1975) In *Coordination Polymerization*, ed. Chien, J. C. W., p. 1, New York: Academic Press.
261. Natta, G. (1955) *J. Polymer Sci.*, **16**, 143.

REFERENCES

262. Natta, G., Pino, P., Corrandi, P., Danusso, F., Mantica, E., Mazzanti, G. and Moraglio, G. (1965) *J. Amer. chem. Soc.*, **77**, 1708.
263. Boor, J. (1967) *Macromol. Revs.*, **2**, 115.
264. Cault, A. D. (1977) In *Catalysis*, Vol. I, p. 234, London: Chemical Society.
265. Chien, J. C. W. (1975) *Coordination Polymerization*, New York: Academic Press.
266. Cossee, P. (1964) *J. Catal.*, **3**, 80.
267. Arlman, E. J. (1964) *J. Catal.*, **3**, 89.
268. Arlman, E. J. and Cossee, P. (1964) *J. Catal.*, **3**, 99.
269. Henrici-Olivé, G. and Olivé, S. (1967) *Angew, Chem. Int. Ed. Engl.*, **6**, 790.
270. Armstrong, D. R., Perkins, P. G. and Stewart, J. J. P. (1974) *Rev. Roumaine Chim.*, **19**, 1695.
271. Novaro, O., Chow, S. and Magnouat, P. (1976) *J. Catal.*, **41**, 91.
272. Novaro, O., Chow, S. and Magnouat, P. (1976) *J. Catal.*, **41**, 131.
273. Armstrong, D. R., Perkins, P. G. and Stewart, J. J. P. (1972) *J. chem. Soc., Dalton*, 1972.
274. Miller, S. A. (1969) *Ethylene and its Industrial Derivatives*, London: E. Benn Ltd.
275. (1975) *Hydrocarbon Processing*, **54**(11), 183.
276. (1975) *Hydrocarbon Processing*, **54**(11), 189.
277. (1975) *Hydrocarbon Processing*, **54**(11), 191.
278. (1975) *Hydrocarbon Processing*, **54**(11), 188.
279. Rasmussen, D. M. (1972) *Chem. Eng.*, **79**, 104.
280. (1975) *Hydrocarbon Processing*, **54**(11), 195.
281. (1975) *Hydrocarbon Processing*, **54**(11), 196.
282. Zambelli, A. (1975) In *Coordination Polymerization*, ed. Chien, J. C. W., p. 15, New York: Academic Press.
283. Natta, G. and Corradina, P. (1963) *Chim. Ind. (Milan)*, **45**, 299.
284. Pino, P. Oschwald, A., Ciardelli, F., Carlini, C. and Chiellini, E. (1975) In *Coordination Polymerization*, ed. Chien, J. C. W., p. 25, New York: Academic Press.
285. Natta, G., Pasquon, I. and Zambelli, A. (1962) *J. Amer. chem. Soc.*, **84**, 1488.
286. Waters, W. A. (1964) *Mechanisms of Oxidation of Organic Compounds*, London: Methuen.
287. Sheldon, R. A. and Kochi, J. K. (1976) *Adv. Catal.*, **25**, 272 (and references therein).
288. Stern, E. W. (1971) In *Transition Metals in Homogeneous Catalysis*, ed. Schranzer, G. N., p. 93 (and references therein), New York: Marcel Dekker.
289. Sheldon, R. A. and Kochi, J. K. (1976) *Adv. Catal.*, **25**, 272.
290. Tanaka, K. (1974) *Amer. chem. Soc., Petrol Chem. Div., Prepr.*, **19**(1), 103.
291. Haber, F. and Weiss, J. (1932) *Naturwiss.*, **20**, 948.
292. Towle, P. H. and Baldwin, R. H. (1964) *Hydrocarbon Processing*, **43**, 149.

293. Green, H. S. (1975) In *Basic Organic Chemistry, Part 5, Industrial Products*, ed. Tedder, J. M., Nechvatal, A. and Jubb, A. H., p. 61, London: John Wiley.
294. Reich, L. and Stivala, S. S. (1969) *Autoxidation of Hydrocarbons and Polyolefins*, p. 271 (and references therein), New York: Marcel Dekker.
295. Kamiya, Y. (1974) *J. Catal.*, **33**, 480.
296. Olah, G. A. (1972) *Chem. Brit.*, **8**, 281.
297. Tanaka, K. (1974) *Chem. Technol.*, 555.
298. Stobaugh, R. B., Calarco, V. A., Morris, R. A. and Stroud, L. W. (1973) *Hydrocarbon Processing*, **52**, 99.
299. Sheldon, R. A. and van Doorn, J. A. (1973) *J. Catal*, **31**, 427.
300. Sheldon, R. A. (1973) *Rec. Trav. Chim. Pay-Bas*, **92**, 253.
301. Phillips, F. C. (1894) *Amer. chem. J.*, **16**, 255.
302. Smidt, J., Hafner, W., Jira, R., Sedlmeier, J., Sieber, R., Rütlinger, R. and Kojer, H. (1959) *Angew. Chem.*, **71**(5) 176.
303. Smidt, J., Hafner, W., Jira, R., Sieber, R., Sedlmeier, J. and Sabel, A. (1962) *Angew. Chem.*, **74**(3), 93.
304. Henry, P. M. (1975) *Adv. organometal. Chem.*, **13**, 363 (and references therein).
305. Bäckvall, J. E., Åkermark, B. and Ljunggren, S. O. (1979) *J. Amer. chem. Soc.*, **101**, 2411.
306. Henry, P. M. (1964) *J. Amer. chem. Soc.*, **86**, 3246.
307. Bäckvall, J. E., Åkermark, B. and Ljunggren, S. O. (1977) *J. chem. Soc., chem. Commun.*, 264.
308. Stille, J. K. and Divakaruni, R. (1979) *J. organometal. Chem.*, **169**, 239.
309. Jira, R., Sedlmeier, J. and Smidt, J. (1966) *Ann. Chem.*, **693**, 99.
310. Smidt, J., Jira, R., Sedlmeier, J., Seiber, R., Ruttinger, R. and Kojer, H. (1962) *Angew. Chem. Int. Ed. Engl.*, **1**, 80.
311. Maitlis, P. M. (1971) *The Organic Chemistry of Palladium*, Vol. 2, p. 141, London: Academic Press.
312. Henry, P. M. (1975) *Adv. organometal. Chem.*, **13**, 363.
313. *Hydrocarbon Process*, **54**, (11) 100 (1975)
314. Henry, P. M. (1967) *J. org. Chem.*, **32**, 2575.
315. Schultz, R. G. and Cross, D. E. (1968) *Adv. chem.* **70**, 97.
316. François, P. (1969) *Ann. Chim. (Paris)* **4**(14), 371.
317. François, P. and Trambouze, Y. (1969) *Bull. Soc. Chim. Fr.* 51.
318. Maitlis, P. M. (1971) *The Organic Chemistry of Palladium*, Vol. 2, p. 89, London: Academic Press.
319. Lyons, J. E. (1977) In *Aspects of Homogeneous Catalysis*, ed. Ugo, R., Vol. 3, p. 1, Dordrecht: Reidel.
320. Rooney, J. J. and Stewart, A. (1977) In *Catalysis*, Vol. 1, p. 277, London: Chemical Society.
321. Peters, E. F. and Evering, B. L. (1960) *U.S. Patent*, 2,963,447.
322. Banks, R. L. and Bailey, G. C. (1964) *Ind. Eng. Chem. Prod. Res. Dev.*, **3**, 170.

323. Calderon, N., Chen, H. Y. and Scott, K. W. (1967) *Tett. Letts*, 3327.
324. Muetterties, E. L. and Busch, M. A. (1974) *J. chem. Soc., chem. Commun.* 754.
325. Grubbs, R. H., Burk, P. L. and Carr, D. D. (1975) *J. Amer. chem. Soc.*, **97**, 3265.
326. Wolovsky, R. and Nir, Z. (1975) *J. chem. Soc., chem. Commun.*, 302.
327. Bilhou, J. L. and Basset, J. M. (1977) *J. organometal. Chem.*, **132**(3), 395.
328. Mol, J. C., Moulijn, J. A. and Boelhouwer, C. (1968) *J. chem. Soc., chem. Commun.*, 633.
329. Clark, A. and Cook, C. (1969) *J. Catal.*, **15**, 420.
330. Bradshaw, C. P. C., Howmann, E. J. and Turner, L. (1967) *J. Catal.*, **7**, 269.
331. Lewandos, G. S. and Pettit, R. (1971) *J. Amer. chem. Soc.*, **93**, 7087.
332. Grubbs, R. H. and Brunck, T. K. (1972) *J. Amer. chem. Soc.*, **94**, 2538.
333. Cardin, D. J., Cetinkaya, B., Doyle, M. J. and Lappert, M. F. (1973) *Chem. Soc. Revs*, **2**, 139.
334. Calderon, N., Ofstead, E. A. and Judy, W. A. (1976) *Angew. Chem. Int. Ed. Engl.*, **15**, 401.
335. Katz, T. J. (1977) *Adv. organometal. Chem.*, **16**, 283.
336. Katz, T. J. and McGinnis, J. J. (1977) *J. Amer. chem. Soc.*, **99**, 1903.
337. Cardin, D. J., Doyle, M. J. and Lappert, M. F. (1972) *J. chem. Soc., chem. Commun.*, 927.
338. Casey, C. P. and Burkhardt, T. J. (1973) *J. Amer. chem. Soc.*, **95**, 5833.
339. Casey, C. P. and Burkhardt, T. J. (1974) *J. Amer. chem. Soc.*, **96**, 7808.
340. Casey, C. P., Tuinstra, H. E. and Saeman, M. C. (1976) *J. Amer. chem. Soc.*, **98**, 608
341. Katz, T. J. and Hersh, W. H. (1977) *Tett. Letts*, 585.
342. Katz, T. J., Lee, S. J. and Acton, N. (1976) *Tett. Letts*, 4247.
343. Mol, J. C. and Moulijn, J. (1975) *Adv. Catal.*, **24**, 131.
344. Archibald, J. I. C., Rooney, J. J. and Steward, A. (1975) *J. chem. Soc., chem. Commun.*, 547.
345. Grubbs, R. H. and Hoppin, C. R. (1977) *J. chem. Soc., chem. Commun.*, 634.
346. Muetterties, E. L. (1975) *Inorg. Chem.*, **14**, 951.
347. Ephritikhine, M., Green, M. L. H. and MacKenzie, R. E. (1976) *J. chem. Soc., chem. Commun.*, 619.
348. Olsthoorn, A. A. and Boelhouwer, C. (1976) *J. Catal.*, **44**, 207.
349. Schrock, R. R. (1976) *J. Amer. chem. Soc.*, **98**, 5399.
350. Schrock, R. R. (1977) Paper presented at the Central Region Meeting of the American Chemical Society (cited in [48]).
351. Krusic, P. J., Klabunde, U., Casey, C. P. and Block, T. F. (1976) *J. Amer. chem. Soc.*, **98**, 2015.

352. Brookhart, M. and Nelson, G. O. (1977) *J. Amer. chem. Soc.*, **99**, 6099.
353. Katz, T. J. and McGinnis, J. (1975) *J. Amer. chem. Soc.*, **97**, 1592.
354. Streck, R. (1975) *Chem. Ztg.*, **99**, 397.
355. *Hydrocarbon Processing* (1976) **46** (11), 232.
356. Rossi, R. (1975) *Chimica e Industria*, **57**, 242.
357. Haas, F., Nützel, K., Pampus, G. and Theisen, D. (1970) *Rubber Chem. Tech.*, **43**, 1116.
358. Dall'Asta, G. (1974) *Rubber Rev.*, **47**, 511.
359. Graulich, W., Swodenk, W. and Theisen, D. (1972) *Hydrocarbon Processing*, **51**(11), 71.
360. Ofstead, E. A. (1969) *Abstracts of the 4th International Rubber Symposium*, **2**, 42.
361. Crain, D. L. (1969) U.S. Patent 3,463,828.
362. Harrison, R. C. (1971) U.S. Patent 3,567,789.
363. Kubicek, D. H. (1973) U.S. Patent 3,728,407.
364. Hallum, A. (1973) U.S. Patent 2,723,495.
365. Solomon, P. W. (1973) U.S. Patent 3,763,240.
366. Gray, R. A. and Brady, D. G. (1973) U.S. Patent 3,755,227.
367. van Dam, P. B., Mittelmeijer, M. C. and Boelhouwer, C. (1972) *J. chem. Soc., chem. Commun.*, 1221.
368. Mol, J. C. and Woerlee, E. F. G. (1979) *J. chem. Soc., chem. Commun.*, 330.
369. Freitas, E. R. and Gum, C. R. (1979) *Chem. Eng. Progs.* **75**(1), 73.
370. Andrew, S. P. S. (1978) *Education in Chem.*, **15**(4), 114.
371. Kovaly, K. A. (1979) *Chem. Eng.*, **345**, 417.
372. Richards, R. L. (1979) *Education in Chem.*, **16**(2), 66.
373. Chatt, J. and Leigh, G. J. (1972) *Chem. Soc. Revs*, **1**, 121.
374. Chatt, J., Dilworth, J. R. and Richards, R. L. (1978) *Chem. Revs*, **78**(6), 589.
375. Hoffman, R., Chen, M. M-L. and Thorn, D. L. (1977) *Inorg. Chem.*, **16**, 503.
376. Chatt, J., Pearman, A. J. and Richards, R. L. (1977) *J. chem. Soc., Dalton*, 1853.
377. Chatt, J., Crabtree, R. H., Jeffery, E. A. and Richards, R. L. (1973) *J. chem. Soc., Dalton*, 1167.
378. Chatt, J. and Richards, R. L. (1977) *J. Less Common Metals*, **54**, 477.
379. Manriquez, J. M. and Bercaw, J. E. (1974) *J. Amer. chem. Soc.*, **96**, 6229.
380. Manriquez, J. M., Sanner, R. D., Marsh, R. E. and Bercaw, J. E. (1976) *J. Amer. chem. Soc.*, **98**, 8351.
381. Manriquez, J. M., Sanner, R. D., Marsh, R. E. and Bercaw, J. E. (1976) *J. Amer. chem. Soc.*, **98**, 3042.
382. Borodko, Yu. G., Broitman, M. O., Kachapina, K. L., Shilov, A. E. and Ukin, L. Yu (1971) *J. chem. Soc., chem. Commun.*, 1185.
383. Schrauzer, G. N. (1975) *Angew. Chem. Int. Ed. Engl.*, **14**, 514.
384. Schrauzer, G. N., Robinson, P. R., Moorehead, E. L. and Vickrey, T. M. (1976) *J. Amer. chem. Soc.*, **98**, 2815.

REFERENCES

385. Shilov, A. E., Denisov, N. T., Efimov, O. N., Shuvalov, N. R., Shuvalova, N. I. and Shilova, A. K. (1971) *Nature (London)*, **231**, 460.
386. Denisov, N. T., Shilov, A. E., Shuvalova, N. I. and Panova, T. P. (1975) *React. Kinet. Catal. Letts*, **2**, 237.
387. *Oil in Perspective* (1978) London: Shell International Petroleum Co. Ltd.
388. *The Coal Option*, (1978) London: Shell International Petroleum Company Ltd.
389. Storch, H. H., Golumbic, N. and Anderson, R. B. (1951) *The Fischer Tropsch and Related Synthesis*, New York: Wiley.
390. Büssemeier, B., Frohning, C. D. and Cornlis, B. (1976) *Hydrocarbon Processing*, **55**(11), 105.
391. Casey, C. P. and Neumann, S. M. (1976) *J. Amer. chem. Soc.*, **98**, 5395.
392. Tam, W., Wong, W.-K. and Gladysz, J. A. (1979) *J. Amer. chem. Soc.*, **101**, 1589.
393. Sweet, J. R. and Graham, W. A. G. (1979) *J. organometal. Chem.*, **173**, C9.
394. Connor, J. A. (1977) *Topics in Current Chem.*, **71**, 71.
395. Casey, C. P., Andrews, M. A. and Rinz, J. E. (1979) *J. Amer. chem. Soc.*, **100**, 741.
396. Labinger, J. A., Wong, K. S. and Scheidt, W. R. (1978) *J. Amer. chem. Soc.*, **100**, 3254.
397. Manriquez, J. M., McAlister, D. R., Sanner, R. D. and Bercaw, J. E. (1978) *J. Amer. chem. Soc.*, **100**, 2716.
398. Masters, C., van der Woude, C. and van Doorn, J. A. (1979) *J. Amer. chem. Soc.*, **101**, 1633.
399. Fischer, F. and Tropsch, H. (1923) *Brennstoff Chemie*, **4**, 276.
400. Casey, C. P. and Neumann, S. M. (1977) *J. Amer. chem. Soc.*, **99**, 1651.
401. van Doorn, J. A., Masters, C. and Volger, H. C. (1976) *J. organometal Chem.*, **105**, 245.
402. Fischer, E. O., and Plabst, D. (1974) *Chem. Ber.*, **107**, 3326.
403. Casey, C. P. and Anderson, R. L. (1975) *J. chem., Soc., chem. Commun.*, 895.
404. Schrock, R. R. and Sharp, P. R. (1978) *J. Amer. chem. Soc.*, **100**, 2389.
405. Herrmann, W. A., Plank, J., Ziegler, M. L. and Weidenhammer, K. (1979) *J. Amer. chem. Soc.*, **101**, 3133.
406. Gresham, W. F. (1951) *British Patent*, 655, 237.
407. Gresham, W. F. (1953) *US Patent* 2,636,046.
408. Pruett, R. L. and Walker, W. E. (1974) *US Patent*, 3,833,634.
409. Pruett, R. L. and Walker, W. E. (1976) *US Patent*, 3,957,857.
410. Kaplan, L. (1976) *US Patent*, 3,944,588.
411. Thomas, M. G., Beier, B. F. and Muetterties, E. L. (1976) *J. Amer. chem. Soc.*, **98**, 1296.
412. Doyle, M. J., Kouwenhoven, A. P., Schaap, C. A. and van Oort, B. (1979) *J. organometal. Chem.*, **174**, C55.

413. Chatt, J. and Davidson, J. M. (1965) *J. chem. Soc.*, 843.
414. Cotton, F. A., Frenz, B. A. and Hunter, D. L. (1974) *J. chem. Soc., chem. Commun.*, 755.
415. Webster, D. E. (1977) *Adv. organometal. Chem.*, **15**, 147.
416. Parshall, G. W. (1977) In *Catalysis*, Vol. 1, p. 335, London: Chemical Society.
417. Cheney, A. J., Mann, B. E., Shaw, B. L. and Slade, R. M. J. (1971) *J. chem. Soc. (A)*, 3833.
418. Shaw, B. L. (1975) *J. Amer. chem. Soc.*, **97**, 3856.
419. Kiffen, A. A., Masters, C. and Raynand, L. (1975) *J. chem. Soc., Dalton*, 853.
420. Garnett, J. L. and Kenyon, R. S. (1971) *J. chem. Soc., chem. Commun.*, 1227.
421. Kramer, P. A. and Masters, C. (1975) *J. chem. Soc., Dalton*, 849.
422. English, A. D. and Herskovitz, T. (1977) *J. Amer. chem. Soc.*, **99**, 1648.
423. Tolman, C. A., Ittel, S. D., English, A. D. and Jesson, J. P. (1978) *J. Amer. chem. Soc.*, **100**, 4080.
424. Ittel, S. D., Tolman, C. A., English, A. D. and Jesson, J. P. (1978) *J. Amer. chem. Soc.*, **100**, 7577.
425. Cotton, F. A., LaCour, T. and Stanislowski, A. G. (1974) *J. Amer. chem. Soc.*, **96**, 754.
426. Cotton, F. A. and Stanislowski, A. G. (1974) *J. Amer. chem. Soc.*, **96**, 5074.
427. Williams, J. M., Brown, R. K., Schultz, A. J., Stucky, G. D. and Ittel, S. D. (1978) *J. Amer. chem. Soc.*, **100**, 7407.
428. Dobson, R. C., Hayes, D. M. and Hoffman, R. (1971) *J. Amer. chem. Soc.*, **93**, 6188.
429. Bodor, N., Dewar, M. J. S. and Wasson, J. S. (1972) *J. Amer. chem. Soc.*, **94**, 9095.
430. Kochi, J. K. (1978) *Organometallic Mechanisms and Catalysis*, New York: Academic Press.
431. Gol'dshleger, N. F., Tyabin, N. B., Shilov, A. E. and Shteinman, A. A. (1969) *Zh. Fiz. Khim.*, **43**, 2174.
432. van Dongen, J. P. C. M., Masters, C. and Visser, J. P. (1975) *J. organometal. Chem.*, **94**, C29.
433. Visser, J. P., Jager, W. W. and Masters, C. (1975) *Rec. Trav. Chim. Pays-Bas*, **94**, 70.
434. Puddephatt, R. J. and Thompson, P. J. (1979) *J. organometal Chem.*, **166**, 251.
435. Muetterties, E. L. (1975) *Bull. Soc. Chim. Belg.*, **84**(10), 959 (and references therein).
436. Visser, J. P. and Masters, C. (1975) *Seventh International Conference on Organometallic Chemistry, Venice*, Abstract paper Nr. 245.
437. Schwartz, J. and Hart, D. W. (1974) *J. Amer. chem. Soc.*, **96**, 8115.
438. Smith, A. K. and Basset, J. M. (1977) *J. molec. Catal.*, **2**, 229.
439. Muetterties, E. L., Rhodin, T. N., Band, E., Brucker, C. F. and Pretzer, W. R. (1979) *Chem. Revs*, **79**(2), 91.

440. Hartley, F. R. and Vezey, P. N. (1977) *Adv. organometal Chem.*, **15**, 189.
441. Yermakov, Y. I. (1976) *Catal. Rev.*, **13**, 77.
442. Lehmkuhl, H. (1973) In *Organic Electrochemistry*, p. 621, New York: Dekker.
443. Gilet, M., Mortreux, A., Nicole, J. and Petit, F. (1979) *J. chem. Soc., chem. Commun.*, 521.
444. Sprintschnik, G., Spintschnik, N. W., Kirsch, P. P. and Whitten, D. G. (1976) *J. Amer. chem. Soc.*, **98**, 2337.
445. Okura, I. and Kim-Thuan, N. (1979) *J. molec. Catal.*, **5**, 311.
446. Brown, G. M., Brunschwig, B. S., Creutz, C., Endicott, J. F. and Sutin, N. (1979) *J. Amer. chem. Soc.*, **101**, 1298.
447. Manriquez, J. M., Fagan, P. J. and Marks, T. J. (1978) *J. Amer. chem. Soc.*, **100**, 3939.
448. Manriquez, J. M., Fagan, P. J., Marks, T. J., Day, C. S. and Day, V. W. (1978) *J. Amer. chem. Soc.*, **100**, 7112.

Index

Acetic acid, 38, 97 179
Acetaldehyde, 38, 186
Acrylamide, 51
Acrylic acid, 51, 91
Activation
  by addition, 24, 228
  by co-ordination, 23, 70, 183, 228
Adipic acid, 178, 181
Alane, 233
Alcohol carbonylation, 97
ALDOX process, 103
AlH_3, 233
Alkane activation, 181, 239
Alkene
  carbonylation, 94
  epoxidation, 182
  hydrogenation, 40
  isomerization, 70
  metathesis, 196
  oligomerization, 136, 141
  oxidation, 186
  polymerization, 159
Alkyne
  carbonylation, 90
  hydrogenation, 59
Allyl complexes, 49, 70, 76
  π-allyl, 50, 76, 79, 137, 146, 150, 205
  σ-allyl, 49
Aluminium-alkyls, 136, 159, 199
  $AlEt_3$, 57, 160
  $AlEt_2Cl$, 159, 171
  $AlEtCl_2$, 162, 197

Aluminium-alkyls (*continued*)
  $AlEt_2OEt$, 145
  $Al_2Et_3Cl_3$, 136, 157
  $AlMe_3$, 136
Ammonia, 220
Amoco oxidation process, 176
Anti Markovnikov addition, 73, 75, 92, 105
Asymmetric
  hydroformylation, 133
  hydrogenation, 39, 60
  oligomerization, 157
  synthesis, 60
Asymmetric tertiary phosphine ligands, 61, 63, 64
  ACMP, 63
  BPPFOH, 66
  DIOP, 64, 134
  NMDPP, 64
Atropic acid, 61
Autoxidation, 175

Bi-heterometallic systems, 253
Bonding, 5
  metal–alkene, 6
  metal–carbonyl, 7
  metal–nitrogen, 221
  synergy, 7
  tertiary phosphine, 9
Buta-1,3-diene, 48, 144, 150, 153, 216
Butanal, 102, 126
Butanol, 103, 112

INDEX

Caprolactam, 178
Carbene-metal species, 80, 199, 202, 207, 234, 238, 248
Carbonylation, 30, 89
  alcohol, 97
  alkene, 94
  alkyne, 90
  iridium catalysts, 100
  methanol, 97
  nickel catalysts, 90
  palladium catalysts, 92, 96
  rhodium catalysts, 97
Catalysis cycle, 33, 36, 41, 48, 52, 53, 72, 79, 91, 93, 96, 98, 100, 105, 111, 115, 122, 123, 129, 131, 138, 147, 152, 154, 180, 185, 193, 210, 225, 248, 252
Chromium complexes
  $Cr(CO)_6$, 152, 234
  $Cr(CO)_3(MeCN)_3$, 57
Cinnamic acid, 50
Citronellol, 152
Cluster catalysis, 237, 253
Cobalt catalysed
  carbonylation, 97
  ethylene glycol synthesis, 236
  hydroformylation, 104, 114
  hydrogenation, 46, 50, 58, 110
  isomerization, 74
  oligomerization, 156
  oxidation, 175, 179
Cobalt complexes
  $[Co(CN)_5]^{3-}$, 26, 46
  $Co_2(CO)_8$, 26, 58, 73, 104
  $[CoH(CN)_5]^{3-}$, 46, 47, 50
  $CoH(CO)_4$, 26, 58, 74, 104, 109, 113, 210
  $CoH(CO)_2L$, 114
  $CoH(CO)_3(PBu_3)$, 59, 119, 210
Cocyclization, 96
Codimerization, 153
Cone angle, 15, 45, 116, 131, 132, 143, 149
Copper compounds, 193, 194
  $CuCl_2$, 186, 191, 192
Cubane, 86
Cyclodimerization, 148

Cyclohexanol, 179
Cyclohexanone, 179
1,5-Cyclo-octadiene, 148
Cyclo-octadodecatriene, 145
Cyclo-oligomerization, 145
Cyclopropanation, 202
Cyclotrimerization, 146

Decobaltation, 113
Dewer benzene, 87
1-Dihydroxyphenylalanine, 39
Dimerization, 135, 138, 150, 236
Dimethylmaleate, 92
Dinitrogen, 221
  bonding, 221
  hydrogenation, 222, 225
Dioxane, 67, 247
1,2-divinylcyclobutane, 148

16/18 electron rule, 34, 41, 46, 101, 115, 123
Electronic activation, 245
Electronic parameter, ν, 13, 45, 75, 116, 131, 140, 143, 149
Elimination, 31
  α-elimination, 33, 202, 204, 248
  β-elimination, 33, 71, 73, 95, 139, 142, 155, 161, 190, 233, 248
Epoxidation, 182
Ethanal, 38, 186
Ethylene glycol, 237
Ethylene oxide, 182
2-ethylhexanol, 103, 127
2-ethylpentanal, 125

Fischer–Tropsch synthesis, 227, 234
Formyl–metal complexes, 229, 230
Free radical intermediates, 58, 166, 174
Fumaric acid, 51

H/D exchange, 73, 242
Haber–Bosch process, 220, 226
Hepta-1,5-diene, 57
Heterolytic
  addition, 27, 51, 85, 87, 233

Heterolytic (continued)
  epoxidation, 182
  oxidation, 182, 186
Hexa-1,4-diene, 153, 156
Hexa-1,5-diene, 37
Hexanal, 133
Homolytic
  addition, 26, 46
  oxidation, 174, 246
Hydrazine, 223, 224
Hydrocarboxylation, 89
Hydrogen shift
  1,2, 77, 106, 190, 205
  1,3, 77
Hydroformylation, 19, 30, 89, 217, 250
  asymmetric, 133
  CO partial pressure, 106, 110, 112, 118
  H_2 partial pressure, 107
  ligand modified cobalt catalysts, 114
  normal-iso ratio, 103, 106, 114, 116, 121, 125, 127, 129
  rhodium catalysts, 120
  ruthenium catalysts, 129
  unmodified cobalt catalysts, 104
Hydrogenation, 18, 40, 217, 251
  asymmetric, 39, 60
  cobalt catalysed, 46, 110
  rhodium catalysed, 40
  ruthenium catalysed, 51
  selective, 46, 252
  transfer, 67
Hydrogenation of
  activated alkenes, 51, 57
  aldehydes, 53, 58
  alkenes, 40, 55
  alkynes, 55
  aromatic compounds, 57
  carbon monoxide, 231
  conjugated dienes, 57
  dinitrogen, 222, 225
  heterocyclic compounds, 57
  ketones, 58
  nitro compounds, 52, 59
  steroids, 56

Hydrogenolysis, 109
Hydroperoxides, 174, 180, 183
Hydroxypalladation, 187, 189

Iridium-catalysed carbonylation, 100
Iridium complexes
  $IrCH_3I_2(CO)_3$, 100
  $[Ir(dmpe)_2]^+$, 243
  $[IrI_2(CO)_2]^-$, 100
Iron catalysed
  carbonylation, 90
  isomerization, 77
Iron complexes
  $Fe(CO)_4$, 78
  $Fe(CO)_5$, 57
  $Fe_3(CO)_{12}$, 77
  $Fe(dmpe)_2$, 245
Isomerization, 70
  alkene, 70
  skeletal, 81
Isoprene, 151

Ligand
  cone angle, 15
  electron donor acceptor properties, 13
  exchange, 251
  migration, 29
  non-participative, 11
  participative, 11
  trans-effect, 12
Ligand effects, 11, 45, 116, 125, 139, 148
Ligand migration, 29, 209
  alkene, 91, 155
  alkyl, 85, 98, 107, 123, 139, 161, 169
  carbonyl, 30
  carboxyl, 92
  hydride, 43, 71, 94, 105, 111, 155

Maleic acid, 51
Markovnikov addition, 73, 75, 91, 105, 144
Metallocyclobutane, 201, 208

INDEX 275

Metathesis, 196
  acyclic alkenes, 211
  cyclic alkenes, 213
Methanol, 97
Methoxycarboxylation, 92
3-methylbut-1-ene, 78, 212
Methyl crotonate, 92
2-Methylheptadecane, 213
Methyl methacrylate, 90
Mobil oxidation process, 176
Molecule activation, 22
  by addition, 24, 228
  homolytic, 26, 46
  heterolytic, 27, 51, 85, 87, 223
  oxidative, 24, 40, 54, 78, 85, 91,
    99, 109, 111, 121, 129, 137,
    240, 245
  by co-ordination, 23, 70, 183, 228
Molybdenum catalysed
  epoxidation, 184
  metathesis, 197
Molybdenum complexes
  $MoCl_5$, 202
  $MoCl_2(NO)_2[P(C_6H_5)_3]_3$, 199
  $Mo(CO)_6$, 234

Nickel catalysed
  carbonylation, 90
  isomerization, 75
  oligomerization, 136
Nickel complexes
  $Ni(acac)_2$, 145
  $Ni(\pi-C_3H_5)_2$, 145
  $Ni(\pi-C_3H_5)X_2$, 157
  $Ni(CO)_4$, 90
  $Ni(cyclo-octadiene)_2$, 145
  $NiH[P(OEt)_3]_3$, 75
  $Ni[P(OEt)_3]_4$, 73
Nitrogen fixation, 219, 254
Nitrogenase, 219, 225, 226
Norbornadiene, 82, 85, 157
Norbornene, 157
Normal-iso ratio, 103, 106, 114,
  116, 121, 125, 127, 129

2,7-octadiene, 150
Octanal, 130
1,3,7-octatriene, 150

Oligomerization, 135
  alkene, 136, 217
  asymmetric, 157
  butadiene, 144
  cobalt catalysed, 156
  codimerization, 153
  cyclo-oligomerization, 145
  ethene, 138
  nickel catalysed, 136
  palladium catalysed, 150
  propene, 141
Oxidation, 172
  alkyl aromatics, 175
  bromide promoters, 177, 180
  cobalt catalysed, 175
  cyclohexane, 179
  epoxidation, 182
  ethene, 186
  heterolytic, 182
  homolytic, 174
  manganese catalysed, 175
  molybdenum catalysed, 184
  palladium catalysed, 186
  propene, 182
  tungsten catalysed, 184
Oxidation state, 18, 25
Oxidative addition, 24, 40, 54, 78,
  85, 91, 99, 109, 111, 121, 129,
  137, 240, 245
Oxo process, 103

Palladium catalysed
  carbonylation, 92
  cocyclization, 96
  isomerization, 79, 82, 86
  oligomerization, 150
  oxidation, 186
Palladium complexes
  $[Pd(\pi-C_5H_5)OAc]_2$, 150
  $PdCl_2$, 186
  $PdCl_2(PPh_3)_2$, 96
Phenylanaline, 66
β-phenylpropionic acid, 50
Pheromones, 213
Phillips Triolefin process, 211
Platinum catalysed
  H/D exchange, 243, 247
  hydroformylation, 132

Platinum catalysed (*continued*)
  hydrogenation, 57
  isomerization, 82
Platinum/tin complexes, 57, 132, 134
  Pt(H)(SnCl$_3$)(PPh$_3$)$_2$, 57, 130
  Pt(H)(SnCl$_3$)(CO)(PPh$_3$), 130, 132
Platinum complexes
  [PtCl$_4$]$^{2-}$, 243, 247
  PtCl$_2$(PBu$^t_2$Pr)$_2$, 241
  PtMe$_2$(COD), 251
Polyalkenamers, 213
Polyester fibre, 176
Polymerization, 159, 213
  of carbon monoxide, 227
  of cycloalkenes, 213
  of propene, 160, 168
  of ethene, 160, 166
  titanium catalysts, 160
  vanadium catalysts, 171
  Ziegler–Natta catalysts, 159
Polyethylene, 38, 159, 166
  gas phase process, 167
  slurry process, 166
Polypropylene, 38, 159, 168
  atactic, 168
  syndiotactic, 168
  isotactic, 168
Processes,
  ALDOX, 103
  Amoco oxidation, 176
  Fischer–Tropsch, 227
  Haber–Bosch, 220, 226
  Mobil oxidation, 176
  Monsanto acetic acid, 103
  Oxo, 103
  Phillips Triolefin, 211
  Polyethylene (gas phase), 167
  Polyethylene (slurry), 166
  Ruhrchemie hydroformylation, 112, 113
  Shell higher olefin, 216
  Shell hydroformylation, 119
  Wacker, 38, 187, 191, 225
  Witten, 176
Propanal, 126

Propionic acid, 97
Propylene oxide, 182, 195

Quadricyclene, 81, 83, 85

Reductive elimination, 24, 43, 54, 99, 111, 121
Regioselectivity, 141, 242
Reppe reactions, 89
Rhodium catalysed
  carbonylation, 97
  codimerization, 153
  ethylene glycol synthesis, 237
  hydroformylation, 120
  hydrogenation, 40, 55, 59, 61, 67
  isomerization, 73, 75, 82, 86
  metathesis, 200
Rhodium complexes
  Rh(C$_2$H$_4$)$_2$(acac), 154
  RhCl$_3$, 153
  Rh$_2$Cl$_2$(CO)$_4$, 85, 121, 251
  [RhCl(C$_2$H$_4$)$_2$]$_2$, 42, 64
  RhCl(PPh$_3$)$_3$, 41, 55, 67, 121, 199
  Rh$_4$(CO)$_{12}$, 120
  [Rh$_{12}$(CO)$_{34-36}$]$^{2-}$, 237
  RhH(CO)(PPh$_3$)$_3$, 1, 55, 72, 121
  [RhI$_2$(CO)$_2$]$^-$, 32, 98
Ruhrchemie hydroformylation
  process, 112, 113
Ruthenium catalysed
  hydroformylation, 129
  hydrogenation, 51, 55, 59
Ruthenium complexes
  [RuCl$_6$]$^{3-}$, 27
  RuCl(H)(PPh$_3$)$_3$, 52
  RuCl$_2$(PPh$_3$)$_{3,4}$, 28, 52, 55, 59, 68
  Ru(CO)$_3$(PPh$_3$)$_2$, 129
  Ru$_3$(CO)$_{12}$, 233, 238
  Ru(dmpe)$_2$, 240
  [Ru(NH$_3$)$_5$N$_2$]$^{2+}$, 221

Sex attractants, 212
Silver catalysed isomerization, 87
Shell higher olefin process, 216
Shell hydroformylation process, 119
Skeletal isomerization, 81

INDEX

Stereoregulation, 171
Stereospecificity, 215
Steric activation, 240
Steric parameter, θ, 15, 45, 116, 131, 140, 149
Symmetry
　conservation, 84
　constraint, 83
Synthesis gas, 103, 227, 237

Telomerization, 151
Terpenes, 151
Thermodynamic
　control, 153
　criteria, 3
　feasibility, 3, 36, 244, 247
Tin chloride, 57, 130, 132
Titanium catalysed
　oxidation, 183
　polymerization, 160
Titanium complexes
　$Ti(\eta^5-C_5H_5)_2Cl_2$, 162
　$TiCl_3$, 160, 171
　$TiCl_4$, 160
Trans-effect, 12, 42, 132

Transfer hydrogenation, 67, 247
Tri-t-butylprismane, 82, 88
Trimerization, 148
Tungsten catalysed
　metathesis, 197
　oxidation, 183
　polymerization, 213
Tungsten complexes
　WCl_6, 197, 202
　$W(CO)_6$, 234
　$W(N_2)_2(PMe_2Ph)_4$, 222

Vanadium tetrachloride, 171
4-Vinylcyclohexene, 148, 216

Wacker process, 38, 187, 191, 225
Witten process, 176

p-xylene, 175

Ziegler catalysts, 38, 57, 136, 160
Ziegler–Natta, 38, 159, 168, 171
Zirconium complexes
　$Zr(\eta^5-C_5Me_5)_2H_2(CO)$, 231
　$Zr(\eta^5-C_5H_5)_2HCl$, 251